Organic Agriculture: A Global Overview

Organic Agriculture: A Global Overview

Edited by **Jordan Berg**

SYRAWOOD
PUBLISHING HOUSE
New York

Published by Syrawood Publishing House,
750 Third Avenue, 9th Floor,
New York, NY 10017, USA
www.syrawoodpublishinghouse.com

Organic Agriculture: A Global Overview
Edited by Jordan Berg

International Standard Book Number: 978-1-68286-259-9 (Hardback)

Printed in China.

Contents

Preface

This book has been an outcome of determined endeavour from a group of educationists in the field. The primary objective was to involve a broad spectrum of professionals from diverse cultural background involved in the field for developing new researches. The book not only targets students but also scholars pursuing higher research for further enhancement of the theoretical and practical applications of the subject.

Organic agriculture is slowly evolving as a popular method of agriculture. This book studies, analyses and upholds the pillars of organic farming and its utmost significance in modern times. This book is a valuable compilation of topics, ranging from the basic to the most complex advancements in the field of organic farming, such as crop diversity, soil management, pest control, plant pathology and breeding, etc. It includes contributions of experts which will provide innovative insights into this field. This book also contains case studies and overviews of prominent organic agricultural practices from all over the world which makes it ideal for students, scholars and organic farming practitioners alike.

It was an honour to edit such a profound book and also a challenging task to compile and examine all the relevant data for accuracy and originality. I wish to acknowledge the efforts of the contributors for submitting such brilliant and diverse chapters in the field and for endlessly working for the completion of the book. Last, but not the least; I thank my family for being a constant source of support in all my research endeavours.

Editor

1

The Influence of Shade and Organic Fertilizer Treatments on the Physiology and Establishment of *Theobroma cacao* Clones

K. Acheampong[1*], P. Hadley[2], A. J. Daymond[2] and P. Adu-Yeboah[1]

[1]Physiology/Biochemistry Division, Cocoa Research Institute of Ghana, P.O.Box 8, Akim Tafo, Ghana, West Africa.

[2]Biodiversity, Crops and Agro-ecosystems Research Division, School of Agriculture, Policy and Development, The University of Reading, Whiteknights, Reading, RG66AR, UK.

Authors' contributions

This work was carried out in collaboration between all authors. Author KA designed the study, wrote the protocol and wrote the first draft of the manuscript. Authors AJD, PH reviewed the experimental design and all drafts of the manuscript. Authors KA and PAY managed the analyses of the study. Authors KA and AJD identified the plants. Authors KA and PH performed the statistical analysis. All authors read and approved the final manuscript.

Editor(s):
(1) Marco Aurelio Cristancho, National Center for Coffee Research, CENICAFÉ, Colombia.
(2) Anonymous.
Reviewers:
(1) Anonymous, Tunisia.
(2) Charles Otieno Obiero, Egerton University, Kenya.

ABSTRACT

Aims: This experiment aimed to determine whether the soil application of organic fertilizers can help the establishment of cacao and whether shade alters its response to fertilizers.
Study Design: The 1.6 ha experiment was conducted over a period of one crop year (between April 2007 and March 2008) at the Cocoa Research Institute of Ghana. It involved four cacao genotypes (T 79/501, PA 150, P 30 [POS] and SCA 6), three shade levels ('light', 'medium' and 'heavy') and two fertilizer treatments ('no fertilizer', and 140 kg/ha of cacao pod husk ash (CPHA) plus poultry manure at 1,800 kg/ha). The experiment was designed as a split-plot with the cacao genotypes as the main plot factor and shade x fertilizer combinations as the sub-plots.
Methodology: *Gliricidia sepium* and plantains (*Musa sapientum*) were planted in different

**Corresponding author: E-mail: oyacheampong@yahoo.com;*

arrangements to create the three temporary shade regimes for the cacao. Data were collected on temperature and relative humidity of the shade environments, initial soil nutrients, soil moisture, leaf N, P and K^+ contents, survival, photosynthesis and growth of test plants.
Results: The genotypes P 30 [POS] and SCA 6 showed lower stomatal conductance under non-limiting conditions. In the rainy seasons, plants under light shade had the highest CO_2 assimilation rates. However, in the dry season, plants under increased shade recorded greater photosynthetic rates (P = .03). A significant shade x fertilizer interaction (P = .001) on photosynthesis in the dry season showed that heavier shade increases the benefits that young cacao gets from fertilizer application in that season. Conversely, shade should be reduced during the wet seasons to minimize light limitation to assimilation.
Conclusion: Under ideal weather conditions young cacao exhibits genetic variability on stomatal conductance. Also, to optimize plant response to fertilizer application shade must be adjusted taking the prevailing weather condition into account.

Keywords: Cacao; shade; organic fertilizer; photosynthesis; establishment.

1. INTRODUCTION

Cacao is cultivated through the humid tropics with over 70% of cocoa being produced in West Africa [1]. In order to meet the increasing demand for cocoa [2], it will increasingly be necessary for farmers to selectively replant their cocoa farms thereby replacing aging stock with more productive varieties.

Re-planting of cacao farms in West Africa can be challenging particularly where there is scanty overhead shade [3,4]. The difficulty has often been ascribed to diminished soil fertility partly as a result of the transfer, each year, of large amounts of nutrients as economic yield from cacao ecosystems [5-7]. Therefore, there is a need to replenish the 'mined' soil nutrients in order to sustain cocoa production. Available data on mature cacao suggest that it is practicable to address the soil fertility problem through the use of fertilizers [8-11]. However, because of health and environmental concerns there has been a steady growth in the demand for organically produced cocoa in recent times [12,13]. Therefore, it is worthwhile to investigate whether organic fertilizers aid in re-establishing cacao on soils with a previous history of cacao cultivation. Since the use of overhead shade has been shown to aid seedling survival [14] the interaction between fertilizer addition and overhead shade also was studied.

Thus, the aims of this paper were to determine whether soil application of organic fertilizers can help the field establishment of young cacao and whether shade alters plant response to fertilizers under the prevailing conditions.

2. MATERIALS AND METHODS

2.1 Experimental Site

The experiment was carried out at the Cocoa Research Institute of Ghana, West Africa (latitude 6° 13' N, Longitude 0° 22' W), and sited on a 1.6 hectare plot with an ideal soil type classification but a poor soil phase for cacao establishment. Initial soil fertility of the experimental site was determined and compared with standard soil ratings for cacao [15].

2.2 Plant Material

Four cacao genotypes T 79/501, PA 150, SCA 6 (representing, respectively, a vigorous, an intermediate, and a low vigour genotype) and P 30 [POS] (a traditionally cultivated cacao genotype in Ghana, used in this work as control), were propagated by side grafting using a common seedling rootstock (T 60/887 x Amelonado). The scion was inserted below the point of attachment of the cotyledons of the rootstock to prevent chupons growing from below the graft union.

2.3 Experimental Design

The experiment was designed as a split-plot with the cacao genotype as the main plot factor and shade x fertilizer combinations as sub-plots. Each of its four blocks contained 24 sub-plots measuring 9 m x 12 m and having a guard row (shared by adjacent sub-plots) and 6 experimental cacao plants planted at 3 m x 3 m. The experiment was re-randomized in each block. The grafted cacao plants were transplanted in June 2007.

Three temporary shade regimes were created by planting *Gliricidia sepium* and plantains (*Musa sapientum*) in different arrangements:

1. The arrangement of the 'light' shade regime consisted of 1.5 m tall *Gliricidia sepium* stakes and plantains planted at 3 m x 3 m each.
2. The 'medium' shade regime consisted of the 'light' shade treatment plus three stands of 0.5 m tall *Gliricidia sepium* stakes planted at 0.6 m away from the young cacao in a triangle to enclose the cacao.
3. The 'heavy' shade treatment arrangement was the same in all respects as the 'light' shade except that, in relation to the position of the cacao, the plantain suckers were planted towards the west and at 0.6 m (instead of 1.5 m) away from the cacao plants.

2.4 Shade Measurement

Radiation intercepted by the shade was measured twice in each of the three main seasons of West Africa ("Major rainy season"- March to July, "Minor rainy season"- August to October, "Dry season"- November to February) by means of a 'SunScan' SystemSS1 (Manufacturers: Delta-T, UK). Measurements were taken around midday with the light sensor held below the canopy of the shade plant and just above the cacao plants.

2.5 Applied Fertilizer

140 kg per hectare of cacao pod husk ash (equivalent to 125 kg muriate of potash per hectare) plus poultry manure at 1,800 kg/ha was applied per year. Prior to its application the chemical properties of the poultry manure were determined at the Soil Science Laboratory of the Cocoa Research Institute of Ghana (results summarised in Table 1).

2.6 Soil Moisture Determination

Soil moisture content in each of the experimental plots was measured twice in the dry season on January 20 2008 and February 21 2008 using a 40 cm soil moisture probe (Delta-T, U.K; model PR2/4).

2.7 Photosynthesis and Leaf Chlorophyll Fluorescence Measurements

Starting from September 2007 an infrared gas analyser (LC pro+, ADC Bioscience, U. K.) was used to measure stomatal conductance and photosynthesis on single attached leaves twice in each of the three seasons. To minimize differences caused by diurnal fluctuations in photosynthesis, measurements were taken on selected plants between 9:00 am and 11:00 am on three out of the six core plants of each treatment per replication.

2.8 Temperature and Relative Humidity Data Collection

Three data loggers (Tinytag, Gemini data loggers, U.K.) put in a miniature Stevenson screen which was placed below the cacao canopy were used to record the temperature and relative humidity at 15-minute intervals in samples of the three shade environments.

2.9 Plant Growth Measurements

Growth (stem diameter at 20 cm from ground level and height) measurements of experimental plants were carried out once every two months from September 2007. Values were transformed into stem volume indices using the formula developed by Oliet [16]:

$$Stem\ volume = (⅓)\pi(¼)(stem\ diameter)2xheight \quad (1)$$

2.10 Determination of Leaf Chlorophyll, N, P and K^+ Concentrations

In March 2008, data on leaf chlorophyll were taken. Also N, P and K^+ concentrations were determined at the Soil Science Division of the Cocoa Research Institute of Ghana on leaves that were approximately of uniform size and physiological age. The data were taken once on a leaf on each of the three plants which were used for photosynthetic data collection in each treatment.

Table 1. Chemical properties of applied poultry manure

Total N	Organic C	Organic matter	Total P	C/N ratio	Moisture content
2.45%	35.10%	59.67%	1.19%	14.3	12.30%

2.10.1 Leaf chlorophyll content determination

Leaf chlorophyll concentration was measured on intact leaves by means of an automatic-calibrated and temperature-compensated leaf chlorophyll meter (CL-01 chlorophyll content meter, Hansatech, UK) with a capability range of 0 – 2000 chlorophyll units.

2.10.2 Determination of leaf N

Samples of leaves were dried overnight at 80°C. The samples from different treatments were ground separately and then, using copper as a catalyst, 0.5 g of the samples were digested for two hours with 12 ml of nitrogen-free concentrated H_2SO_4. The digestate was then distilled by the Kjeldahl method for 3 minutes according to the method of Stuart [17] using methyl red as an indicator. A back titration was done to determine the end point. % Nitrogen (N) was calculated as:

$$(\%\,N) = [Titre\ valuexN(=0.02)\ x\ 1.401)/0.5]\% \quad (2)$$

2.10.3 Determination of leaf total phosphorus (P)

Samples of leaves (1 g each) were placed in digestion tubes. 15 ml of nitric acid (HNO_3) was added to each sample and then digested for 30 minutes at 150°C in a digester. 10 ml of perchloric acid ($HCLO_4$) was then added to each digest which was re-digested at 200°C for one hour. Total phosphorus (P) in each sample solution was then determined via the ascorbic acid method using spectrophotometry [18].

2.10.4 Determination of leaf K

15 ml of nitric acid (HNO_3) was added to each of 1 g of ground, oven-dried leaf samples and digested for 30 minutes at 150°C in a digester in a fume cupboard. 10 ml of 1 molar perchloric acid was then added to each of the digested samples and re-digested for another one hour at 200°C. The digests were allowed to cool in a fume cupboard and were then filtered through Whatman No. 42 filter papers into 250 ml volumetric flasks. The filtrates were topped up with distilled water to the 250 ml mark and mixed well. The sample filtrates in the 250 ml volumetric flasks were used to determine the leaf potassium content on an atomic absorption spectrometer [19].

3. RESULTS

3.1 Attained Shade Levels

The 'light' 'medium', and 'heavy' shade regimes attained between 50% and 52.5% shade, between 52% and 57% shade and between 58% and 61% shade, respectively. The lower shade levels for the shade regimes were obtained in the dry season as the shade plants experienced more rapid leaf senescence and leaf shedding.

3.2 Soil Fertility of Experimental Site

Table 2 shows the initial soil nutrient levels of the experimental site. Compared with Smyth's suggested standard [15], soils at the experimental site were low on % N, % O.M. and % C and the C/N ratio was below standard.

3.3 Soil Moisture

Soil moisture content was not statistically different between fertilized and non-fertilized plots. However, soil moisture increased (*P* < .001) further down the soil profile (Fig.1).

3.4 Microclimatic Data

Table 3 shows the air temperature and relative humidity under the different shade treatments across the seasons. Under increasing shade slightly higher relative humidity and lower temperature values were recorded particularly in the dry season.

Table 2. Initial soil fertility of experimental site compared with Smith's suggested standard [15]

Soil fertility parameters	Soil depth(cm)	pH	% N	% O.M.	Avail. P (μgg^{-1})	% C	C/N
Experimental site	0-15	5.7	0.13	1.67	20.55	0.97	7.46
Experimental site	15-30	5.5	0.08	1.02	16.69	0.59	7.35
Medium rating (standard range)	-	5.1-5.5	0.2-0.5	2.0-4.2	6.5-13.0	1.16-2.44	13.0- 20.0
Low rating (standard range)	-	-	0.1-0.2	1.0-2.0	3.0- 6.5	0.58-0.16	> 20.0

The dry season had the highest daytime temperature and the lowest relative humidity values whereas the minor rainy season had the lowest daytime temperature and the highest relative humidity values. Therefore, daytime water vapour pressure deficit was lowest in the minor rainy season and highest in the dry season.

3.5 Gas Exchange

3.5.1 Stomatal conductance

The mean stomatal conductance for both the major and minor rainy seasons was 7.5 times higher ($P < .001$) than that of the dry season (Fig. 2). In the major rainy season cacao plants under light shade had the highest stomatal conductance ($P = .006$). Furthermore, fertilized plants had higher stomatal conductance than non-fertilized ones (0.534 mol m^{-2} s^{-1} compared with 0.326 mol $m^{-2}s^{-1}$, $P < .001$) and P 30 [POS]) had the highest rates while PA 150 had the lowest rates ($P = .02$) in the dry season. None of the interactions was significant.

Unlike the major rainy season, a significant genotype x shade interaction ($P = .005$) was observed during the minor rainy season. Whereas higher stomatal conductance was recorded in PA 150 and P 30 [POS] under increasing shade, no clear trends were observed for T 79/501 and SCA 6.

In that season fertilized cacao had 14.5% higher stomatal conductance ($P = .05$).

In the dry season, a significant ($P = .004$) cacao genotype x shade interaction was observed in which T 79/501 had its lowest rates under the medium shade whereas the other genotypes had their lowest stomatal conductance under heavy shade. Also fertilized cacao plants had 25.4% higher stomatal conductance than non- fertilized plants ($P = .007$) in the dry season.

3.5.2 Photosynthesis

Mean photosynthetic rates were higher in the major rainy season than in the minor rainy and dry seasons by 19.26% and 71.86%, respectively (Fig. 3). Also reduced shade was associated with higher ($P =.03$) photosynthetic rates in the rainy seasons. There were no statistical differences between the cacao genotypes in any of the seasons.

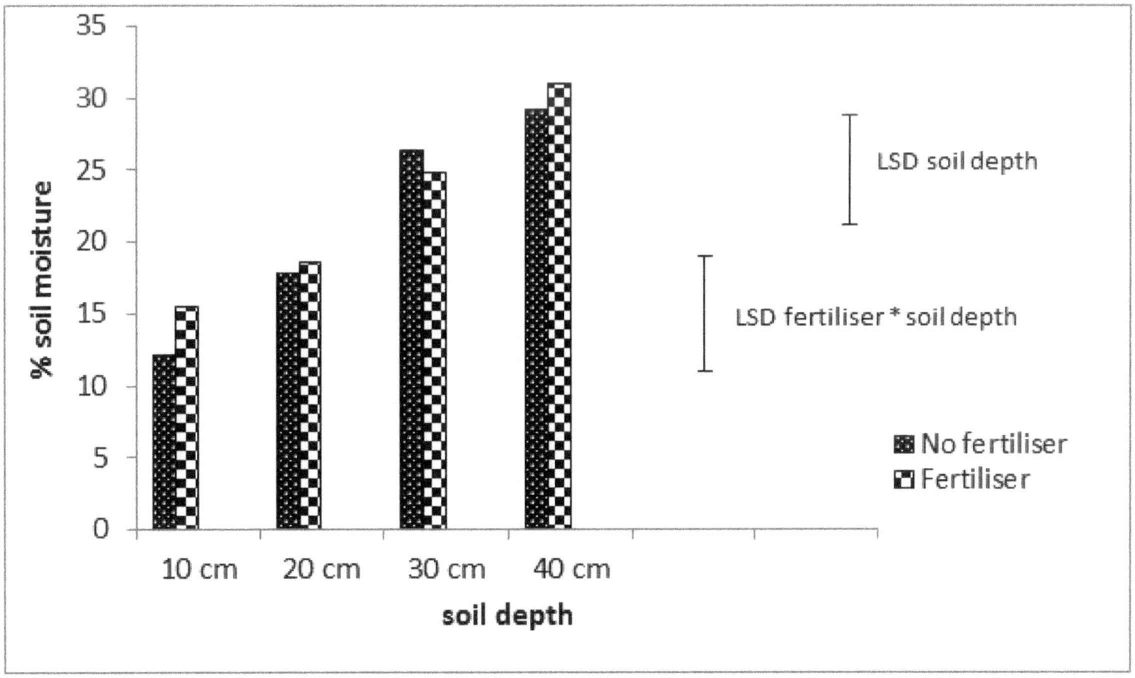

Fig. 1. Soil moisture measured under different fertilizer treatments and soil depths in the dry season

Table 3. Mean relative humidity (R.H.), daytime temperature (T °C) and vapour pressure deficit (VPD) under the shade regimes during the seasons. Each value represents the mean of data recorded every 15 minutes from 3 data loggers

	Major rainy season			Minor rainy season			Dry season		
Shade type	R.H. (%)	T °C	VPD kPa	R.H. (%)	T °C	VPD kPa	R.H. (%)	T °C	VPD kPa
Light	64.25	30.50	1.56	79.25	28.65	0.82	52.25	34.70	2.64
Medium	74.50	29.25	1.04	82.50	27.00	0.63	54.50	33.50	2.36
Heavy	78.50	30.00	0.91	83.25	26.80	0.59	56.00	33.00	2.22

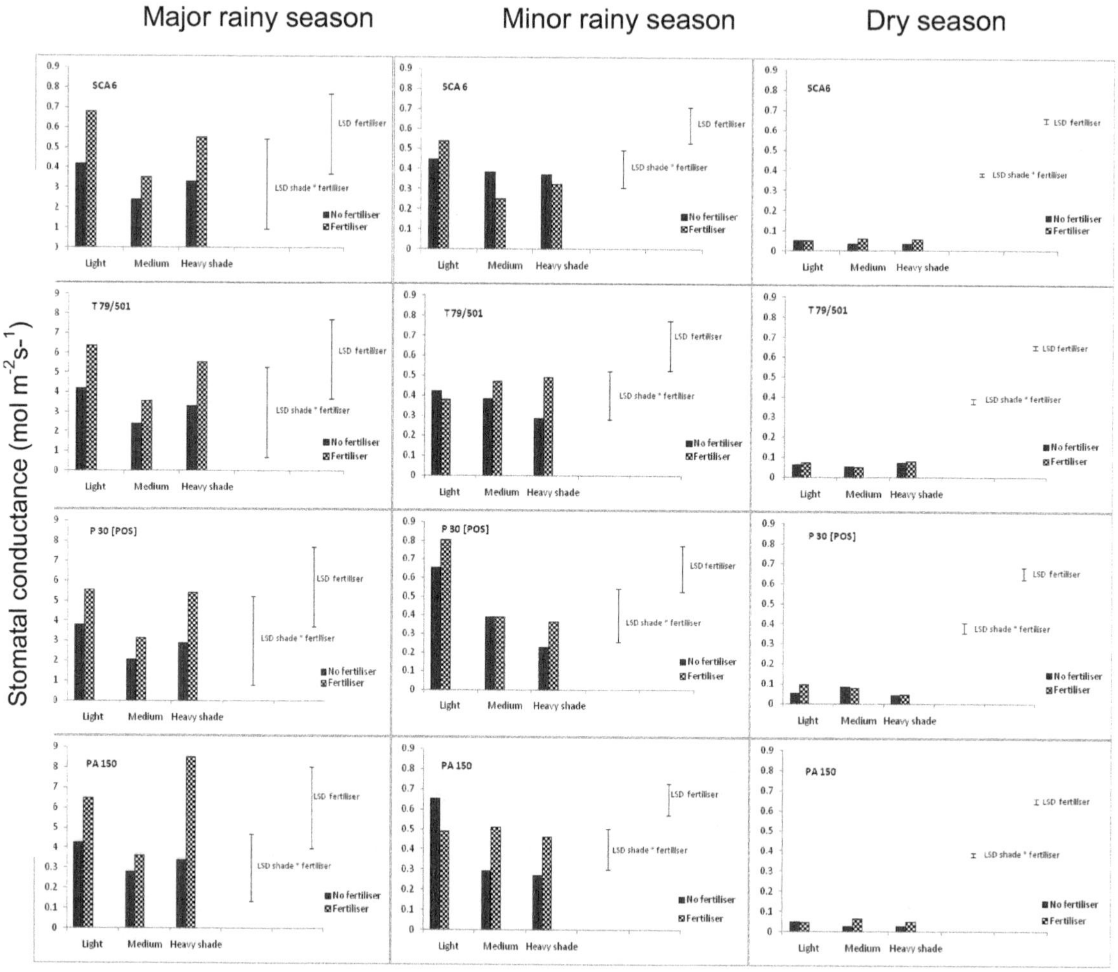

Fig. 2. Stomatal conductance of young cacao under the shade and fertilizer treatments and in different seasons. Each bar represents the mean value for 9 plants

A significant (*P* = .001) shade x fertilizer interaction was observed in the dry season with higher increases occurring in the fertilised plants under the medium (40%) and heavy (47%) shade than under the light shade (7%).

3.6 Water Use Efficiency

The mean water use efficiency of the cacao plants during the dry season was 3.2 µmol (CO_2) / mmol (H_2O) being 12% and over 100% greater than in the minor and major rainy seasons, respectively.

In the major and minor rainy seasons no significant differences were observed between the cacao genotypes or shade or fertilizer treatments on water use efficiency (Fig. 4).

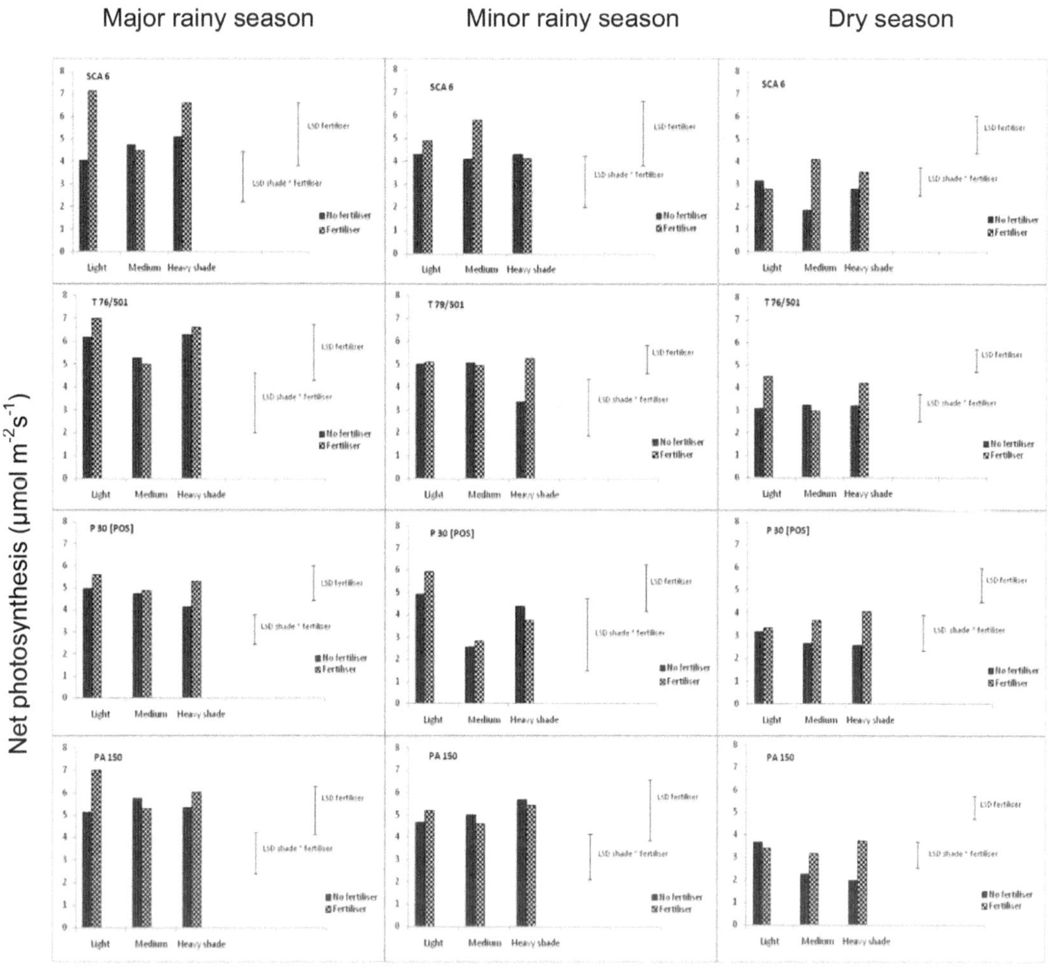

Fig. 3. Photosynthesis of young cacao under different shade and fertilizer treatments and in different seasons. Each bar represents the mean value for 9 plants

In the dry season higher (*P* = .002) water use efficiency was observed in plants under the light shade treatment than the other shade environments.

3.7 Leaf Traits

Only slight increases in leaf chlorophyll concentration were recorded under fertilizer application in the case of SCA 6 and T 79/501 and no clear trends were observed for P 30 [POS] and PA150. Similarly, no clear trends were observed in leaf chlorophyll content under the different shade regimes.

Differences (*P* = .01) were noted between cacao genotypes on leaf nitrogen concentration (Table 4a). Their values ranged from 1.78% in SCA 6 to 2.02% in PA 150. Whereas leaf nitrogen concentration increased (*P* = .001) with fertilizer application, only slight differences were recorded for the shade treatments and none of the interactions was significant.

Total leaf phosphorus content did not differ substantially between the cacao genotypes or the shade treatments (Table 4b). Plants that grew under light shade had a 6.85% higher concentration than those that grew under heavy shade. Also fertilization was associated with a marginal increase (*P* = .047) in total leaf phosphorus. No significant interactions were observed.

While no statistical differences were noticed between the genotypes on leaf potassium ion concentration increasing shade was associated with higher leaf K^+ (*P* = .01). The genotype x shade interaction term was

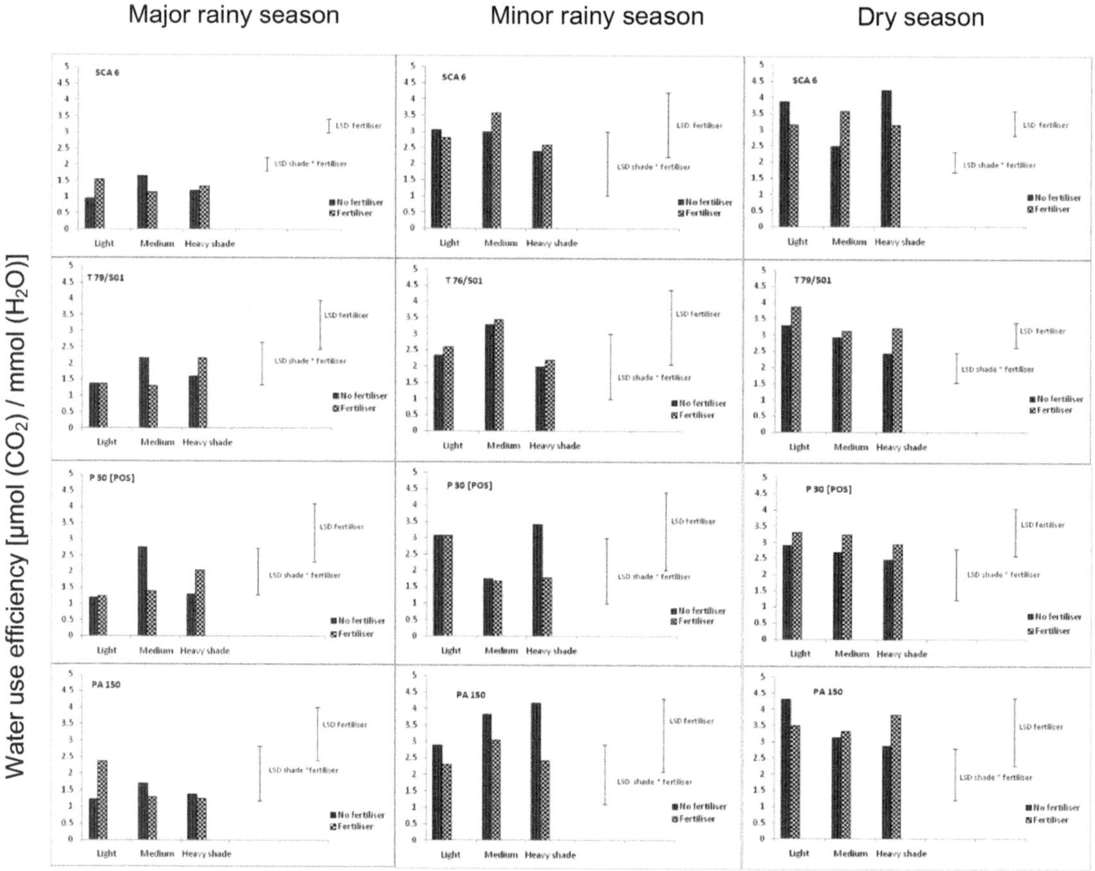

Fig. 4. Water use efficiency of young cacao under different shade and fertilizer treatments and in different seasons. Each bar represents the mean value for 9 plants

significant (P = .02) with lower leaf K^+ concentration occurring in SCA 6, T 79/501 and P 30 [POS] whereas PA 150 had higher leaf K^+ concentration under increasing shade. Also fertilized plants had a 13.7% higher leaf K^+ concentration (Table 4c) compared with the control (P = .001).

3.8 Plant Survival

Plant survival over the experimental period was different (P = .03) among the genotypes, the rates ranging from 79.9% for SCA 6 to 96.5% for PA 150 (Fig. 5). No substantial differences were recorded between the shade treatments but fertilized plants had a higher (P = .02) survival rate than the control. No significant interactions were observed.

3.9 Plant Growth

There were significant differences (P = .02) in stem volumes between the different cacao genotypes. The genotype PA 150 which had the largest stem volume (mean = 287.6 cm^3) was 2.6%, 23.5% and 53.6% bigger than T 79/501, P 30 [POS] and SCA 6, respectively. No substantial differences were recorded between the shade or fertilizer treatments (Fig. 6).

4. DISCUSSION

4.1 Impact of Shade and Nutrition on Photosynthesis

The influence of overhead shade in this study on photosynthesis differed between seasons. The observation that in the rainy seasons on average, plants under light shade had the highest stomatal conductance and CO_2 assimilation rates was probably due to light limitation under shade during the prevailing cloudy condition.

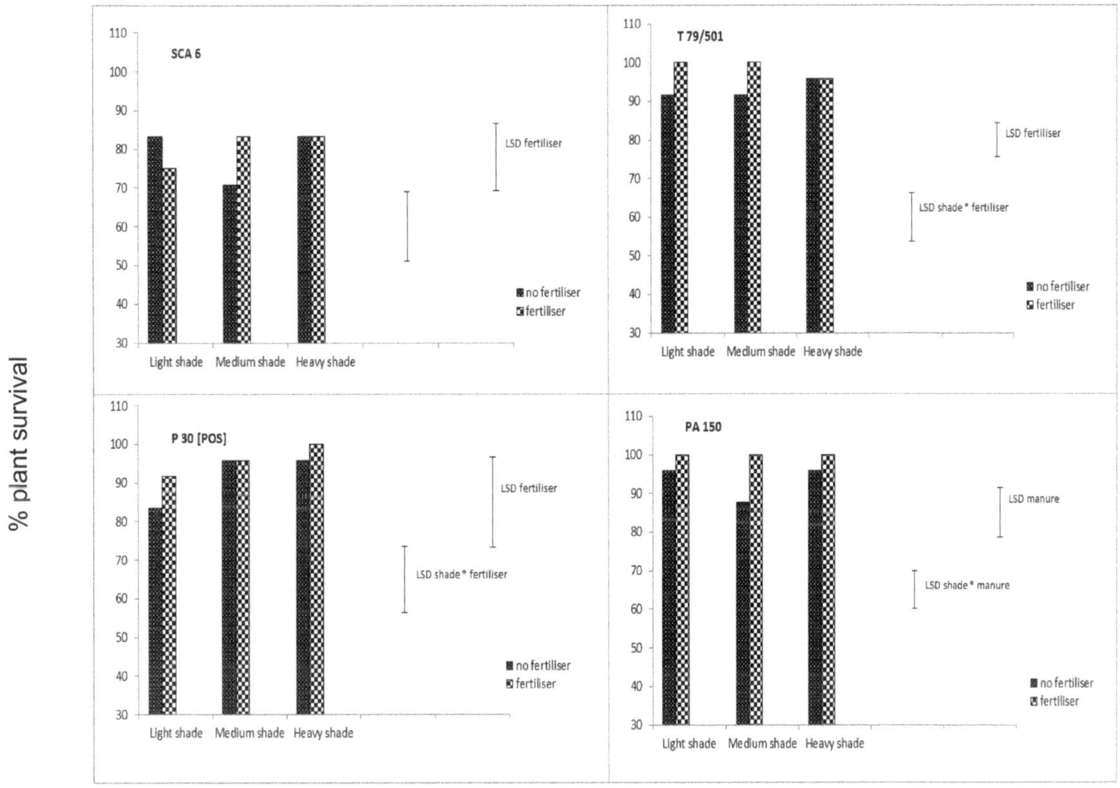

Fig. 5. % surviving cacao plants under varied shade and fertilizer treatments. Each bar represents the mean percentage value for 48 plants

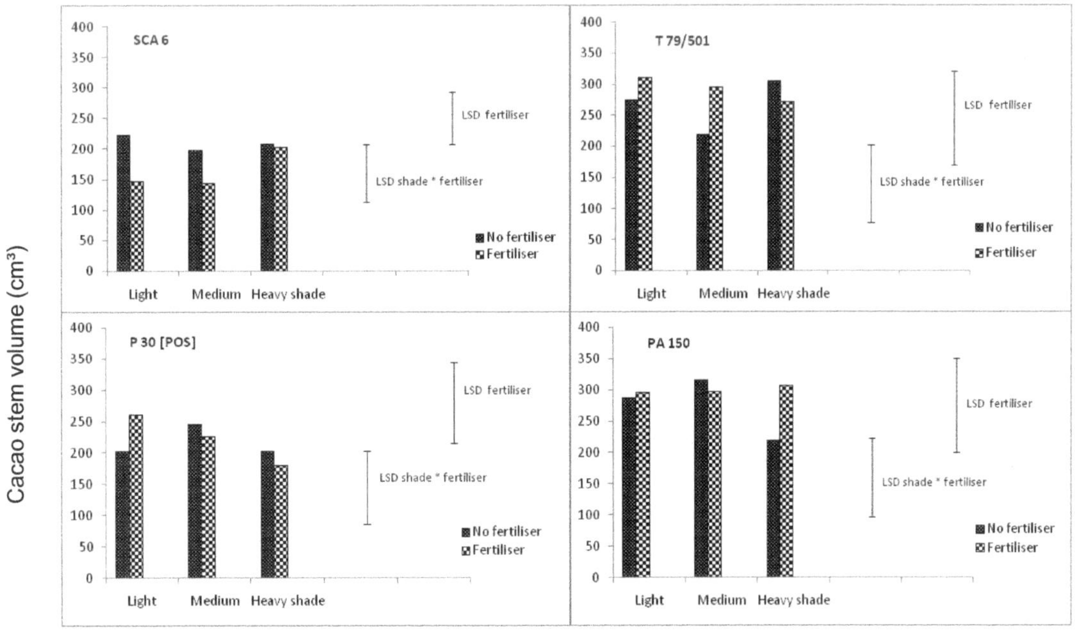

Fig. 6. Stem volume of young cacao under different shade and fertilizer treatments. Each bar represents the mean value of 48 plants

Table 4. Leaf traits of cacao under different shade and fertilizer treatments

Table 4a. % leaf nitrogen of cacao under varied shade and fertilizer treatments

Genotype	Light shade		Medium shade		Heavy shade		
	Unfertilized	Fertilized	Unfertilized	Fertilized	Unfertilized	Fertilized	Mean
SCA 6	1.713	1.854	1.899	1.937	1.568	1.718	1.781
T 79/501	1.795	1.834	1.818	2.078	1.715	1.818	1.843
P 30 [POS]	1.815	1.903	1.769	1.927	1.881	1.986	1.880
PA 150	1.829	2.164	1.982	2.177	1.945	2.014	2.019
Mean	1.788	1.939	1.867	2.030	1.777	1.884	1.881

Shade treatments: (P =.06); Cacao genotypes: (P = .01); Fertilizer treatments: (P = .001)

Table 4b. Leaf total phosphorus (μgg^{-1} of leaf tissue) of cacao under different shade and fertilizer treatments

Genotype	Light shade		Medium shade		Heavy shade		
	Unfertilized	Fertilized	Unfertilized	Fertilized	Unfertilized	Fertilized	Mean
SCA 6	1686	1836	1916	2135	1899	2073	1924
T 79/501	2103	2744	2267	2679	3030	2621	2574
P 30[POS]	1997	2066	2162	2442	2086	2248	2166
PA 150	2209	2272	2223	2201	1864	2253	2170
Mean	1998	2229	2142	2364	2219	2298	2208

Shade treatments: (P =.31); Cacao genotypes: (P = .19); Fertilizer treatments: (P = .047)

Table 4c. Leaf K^+ (meq 100 g^{-1}) of cacao under varied shade and fertilizer treatments

Genotype	Light shade		Medium shade		Heavy shade		
	Unfertilized	Fertilized	Unfertilized	Fertilized	Unfertilized	Fertilized	Mean
SCA 6	24.79	29.63	33.81	35.70	35.27	36.40	32.6
T 79/501	22.84	37.66	31.05	38.69	41.20	35.83	34.55
P 30 [POS]	33.96	35.20	31.49	39.54	37.90	45.19	37.21
PA 150	31.83	42.19	30.81	36.09	30.19	33.92	34.17
Mean	28.36	36.17	31.79	37.51	36.14	37.84	34.63

Cacao genotypes: (P = .20); Shade treatments: (P =.01); Genotype x shade (P = .02); Fertilizer treatments: (P = .001)

In contrast, in the dry season, increased shade, under which reduced water vapour pressure deficit was recorded, ameliorated the negative effect of atmospheric stress on stomatal conductance and hence photosynthetic rates were greater. A number of previous studies, including one that involved the four cacao genotypes, have shown that stomatal opening in cacao is sensitive to vapour pressure [14,20,21].

Differences in stomatal conductance were observed between cacao genotypes in the major rainy season (with PA 150 having the highest rates and P 30 [POS] the lowest). However, no such differences were observed in the dry season meaning that it was not possible to determine which of the cacao genotypes can better tolerate a condition of drought by means of a more effective stomatal regulation. Nevertheless, differences in the degree of plasticity among the genotypes were observed suggesting that P 30 [POS] and SCA 6 have an intrinsically lower stomatal conductance under non-limiting conditions.

The significant increase in stomatal conductance observed under fertilizer application in all the seasons seems to suggest that the amounts of nutrients supplied through fertilizer application were reasonable and not excessive considering the soil's water status. It has been noted in some plants that application of soil nutrients must be commensurate with available soil water. Also fertilization with N and K in their right amounts is associated with increases in transpiration rates when soil water is adequate but with a more effective stomatal control in a situation of soil water deficit [22-26]. For example, Turnbull [27] recorded a higher stomatal conductance rate in *Eucalyptus globules* plants after nitrogen fertilization. The results presented here are consistent with these earlier studies in that combining organic materials that were good sources of N and K was associated with improved water relations of the young cacao plants.

The observation that water use efficiency was highest in the dry season is consistent with that of Rada [28] who observed, in young cacao, that under conditions of soil water stress stomatal closure reduced water loss through transpiration much more than it reduced the rate of photosynthesis resulting in an increase in water use efficiency. This also explains the higher water use efficiency recorded under the more stressful 'light shade' environment than under the heavier shade during the dry season.

Mineral nutrients are known to increase photosynthetic electron flow either as "constituents of the light-harvesting complex", or as "ions facilitating electron flow" [29]. Thus fertilizer application has been linked to increased photosynthetic rates in several plant species including cacao [9,27,29-35]. As the initial N and K levels in the soil at the site of the current experiment were below the critical values for successful establishment (see Table 2), fertilizer application was expected to stimulate photosynthesis.

The fact that increased photosynthesis was recorded in fertilized plants in the dry season more than in the rainy seasons may be attributed to the ability of organic fertilizers to improve plant water status. The benefits associated with such interventions become more manifest under harsher conditions such as were prevalent in the dry season. With increased stress in the dry season, plants that had not been provided with fertilizer would exhibit greater stomatal closure in order to conserve their tissue water inevitably causing reductions in CO_2 uptake and consequently, lower photosynthetic rates.

4.2 Leaf Chlorophyll, N, P and K^+ Concentrations

Unlike leaf phosphorus and potassium contents which were not statistically different among the cacao genotypes, the significant difference in leaf nitrogen content among them indicates the presence of considerable genotypic variability as regards nitrogen uptake and/or translocation into leaf tissue. Changing the spatial arrangement of plantains or introducing additional *Gliricidia* shade plants was not found to be associated with a significant change in leaf N or P suggesting low inter-specific competition for the nutrients. A similar result was observed by Isaac [36].

The observed increase in leaf N and K in the plants following organic fertilizer application is consistent with an earlier finding [8]. The marginal increase in leaf P after fertilizer application suggests that the amount of P supplied was below the threshold quantity that was needed for a substantial increase in leaf P. Initial soil analysis showed that the soil already contained more than the critical level of P for cacao establishment [37] (see Table 2).

4.3 Plant Survival

The different survival rates for the cacao genotypes was correlated with their different growth rates. With the annual dry spell in West Africa, varieties with greater vigour establish better as they often develop deeper root systems that have greater access to soil water before the dry season begins [38,39].

Although organic fertilization was not strongly associated with plant growth within the period of study it was clearly linked with higher plant survival. The recorded enhanced plant survival could have been caused not only by improved plant nutrition but also by indirect benefits such as reduced soil temperature and increased root and stem water content during the dry season. Furthermore, the higher N and K^+ content in the plants' tissue (as evidenced by the higher leaf concentrations) likely helped plant osmoregulation [28,40,41] which contributes to plant survival under dry conditions.

Although fertilizer application failed to contribute significantly to faster cacao seedling growth, the recorded increases in assimilation rate and plant survival dispel any suggestion of nutrient antagonism [9,42,43]. Formation pruning was carried out in the experiment to remove basal chupons and unwanted developing branches and that may have confounded the results on growth to some extent. Also the time period within which the growth data were taken was, probably, too short to record any substantial difference in growth between the fertilizer treatments.

This experiment has shown a positive link between the application of organic fertilizers and the field establishment (survival) of cacao in a site that previously carried moribund cacao. Although differences in shade were not linked with plant survival, the significant shade x fertilizer interaction on photosynthetic activity in the dry season shows that during the dry season shade increases the benefits that young cacao plants get from fertilizer application. On the other hand, shade should be pruned back during the wet seasons to reduce light limitation to assimilation and growth.

5. CONCLUSION

It is concluded from this experiment that although soil application of organic fertilizers can improve cacao plant nutrition and help establishment, overhead shade must be adjusted in accordance with the prevailing weather conditions to optimize plant response to fertilizers.

ACKNOWLEDGEMENTS

The authors are grateful to Messrs Hakeem Rashied and Godwin Addo of the Cocoa Research Institute of Ghana (CRIG) for providing technical support. The authors are grateful to the Cocoa Research Association, The CFC and the Ghana COCOBOD for part-funding this work. This paper is published with the kind permission of the Executive Director of CRIG.

COMPETING INTERESTS

Authors have declared that no competing interests exist.

REFERENCES

1. ICCO. Annual forecasts of production and consumption and estimates of production levels to achieve equilibrium in the worldcocoa market, in annual report, ICCO: Berlin; 2008.
2. WCF. Introduction to cocoa market, in cocoa market update; 2012.
3. Ayanlaja SA. Rehabilitation of cocoa (Theobroma cacao) in Nigeria. Major problems and possible solutions. Causes of difficulty of seedling establishment. Plant and Soil, 1983. 73: p. 403-409.
4. Anim-Kwapong GJ, Osei-Bonsu K. Potential of natural and improved fallow using indigenous trees to facilitate cacao replanting in Ghana. Agroforest Syst. 2009;76:533-542.
5. Appiah MR, et al. The consequences of cocoa production on soil fertility in Ghana: A review. Ghana Jnl. agric. Sci. 1997;30:183-190.
6. Hartemink AE. Nutrient stocks, nutrient cycling, and soil changes in cocoa ecosystems: A review. Advances in Agronomy. 2005;86:227-253.
7. Omotoso TI. Amounts of nutrients removed from the soil in harvested Amelonado and F3 Amazon Cacao during a year. Turrialba, 1975;25:425-428.
8. Ahenkorah Y, Halm BJ, Amonoo RS. Cacao pod husk as source of potash fertilizer. Turrialba. 1981;31(4):287-291.
9. Ahenkorah Y, et al. Twenty years' results from a shade and fertilizer trial on amazon cocoa (*Theobroma cacao*) in Ghana. Experimental Agriculture, 1987;23(1):31.
10. Cunningham RK, Arnold PW. The shade, fertilizer requirements of cacao (*Theobroma cacao* L). Ghana J. Science, 1962;13(Food and Agric.):213-221.
11. Opoku IY, Ofori Frimpong K, Appiah MR. The use of fertilizer and fungicide to sustain cocoa production in severe black pod areas. Tropical Science. 2004;44(2):95-99.
12. ICCO. A study on the market for organic cocoa, London. 2006;1-16.
13. Pay E. Increasing incomes and food security of small farmers in West and Central Africa through exports of organic and fair-trade tropical products, FAO. 2009;1-14.
14. Acheampong K, Hadley P, Daymond AJ. Photosynthetic activity and early growth of four cacao genotypes as influenced by different shade regimes under West African dry and wet season conditions. Experimental Agriculture. 2013;49(1):31-42.
15. Smyth AJ. The selection of soils for cacao, in Soils Bulletin No. 51966. Food and Agriculture Organization of The United Nations: Rome. 52-53.
16. Oliet J, et al. Mineral nutrition and growth of containerized *Pinus halepensis* seedlings under controlled-released fertilizer. Scientia Horticulturae: International Society for Horticultural Science. Elsevier; 2004.
17. Stuart NW. Adaptation of the Micro-Kjeldahl method for the determination of nitrogen in plant tissues. Plant Physiology. 1936;11(1):173-179.
18. Twine JR, Williams CH. The determination of phosphorus in Kjeldahl digests of plant material by automatic analysis. Communications in Soil Science and Plant Analysis, 1971;2(1532-2416)(6):485-489.
19. de Almeida M, et al. Interfacing multisyringe flow injection analysis to flame atomic emission spectrometry: An

intelligent system for automatic sample dilution and determination of potassium. Journal of Analytical Atomic Spectrometry. 2009;24(3):340-346.

20. Raja Harun RM, Hardwick K. The effect of different temperatures and water vapour pressure deficits on photosynthesis and transpiration of cacao leaves, in 10th International Cocoa Research conference: Santo Domingo. 1988;211-214.
21. Sena Gomes AR, Kozlowski TT, Reich PB. Some physiological responses of *Theobroma cacao* var. catongo seedlings to air humidity. New Phytol. 1987;107:591-602.
22. Boruah HPD, et al. Non-uniform, patchy stomatal closure of a plant is a strong determinant of plant growth under stressful situation. Current Science. 2008;94(10):1310-1314.
23. Bosshart RP, von Uexkull HR. Some occasionally overlooked criteria for assessing fertilizer requirements of high yielding cacao in Seminar on palm kernel utilization and recent advances in cacao cultivation. Sawan, Sabah Malaysia; 1987.
24. Cramer MD, Hoffmann V, Verboom GA. Nutrient availability moderates transpiration in *Ehrharta calycina*. New Phytologist. 2008;179(4):1048-1057.
25. Otoo E, Ishii R, Kumura A. Interaction of nitrogen supply and soil water stress on photosynthesis and transpiration in rice. Japanese Journal of Crop Science. 1989;58(3):424-429.
26. de Melo AS, et al. Alteration of the physiologic characteristics in banana under fertirrigation conditions. Ciencia Rural. 2009;39(3):733-741.
27. Turnbull TL, et al. Within-canopy nitrogen and photosynthetic gradients are unaffected by soil fertility in field-grown *Eucalyptus globulus*. Tree Physiology. 2007;27:1607-1617.
28. Rada F, et al. water relations and gas exchange in *Theobroma cacao* var. Guasare under periods of water deficits. Rev. Fac. Agric. (LUZ). 2005;22:105-112.
29. Cakmak I, Engels C, eds. Role of mineral nutrition in photosynthesis and yield formation. Mineral Nutrition of crops. Fundamental mechanisms and iimplications., ed. Z. Rengel. Published by Food Products Press, and impring of the Haworth Press Inc.: Binghmton; 1999,
30. Lockard RG, Asomaning EJA. Composition of complete nutrient solutions, in Rep. W. Afr. Cocoa Res. Inst. West African Cocoa Research Institute: Tafo, Ghana. 1963;56-57.
31. Efthimiadou A, et al. Effects of cultural system (organic and conventional) on growth, photosynthesis and yield components of sweet corn (*Zea mays* L.) under semi-arid environment. Notulae Botanicae Horti Agrobotanici Cluj-Napoca. 2009;37(2):104-111.
32. Ahmed N, et al. Impact of zinc fertilization on gas exchange characteristics and water use efficiency of cotton crop under arid environment. Pakistan Journal of Botany. 2009;41(5):2189-2197.
33. Rana NK, et al. A CsGS is regulated at transcriptional level during developmental stages and nitrogen utilization in *Camellia sinensis* (L.) O. Kuntze. Mol Biol Rep. 2009;37(2):703-10.
34. Bown HE, et al. Chlorophyll fluorescence response of *Pinus radiata* clones to nitrogen and phosphorus supply. Ciencia E Investigacion Agraria, 2009;36(3):451-464.
35. Liu HG, et al. Interactive effects of molybdenum and phosphorus fertilizers on photosynthetic characteristics of seedlings and grain yield of *Brassica napus*. Plant and Soil. 2010;326(1-2):345-353.
36. Isaac ME, et al. Early growth and nutritional response to resource competition in cocoa-shade intercropped systems. Plant and Soil, 2007;298:243-254.
37. Akirinde EA, Ayegboyin KO. Performance of *Theobroma cacao* seedlings irrigated with water from different sources. Biotechnology. 2006;5(3):330-336.
38. Oppong FK, et al. The effect of time of planting at stake on cocoa seedling survival. Ghana Jnl. Agric. Sci, 1999;32(1):79-86.
39. Oppong FK, Opoku-Ameyaw K. Comparison of methods of field planting on cocoa seedling survival and early growth in a marginal cocoa-growing area of Ghana. Ghana Journal of Agricultural Science, 2008;40(2):199-205.
40. de Almeida, et al. Some water relations aspects of *Theobroma cacao* clones. in 13th International Cocoa Research Conference; 2002.

41. Orlova YV, et al. Contributions of inorganic ions, soluble carbohydrates, and multiatomic alcohols to water homeostasis in *Artemisia lerchiana* and A-pauciflora. Russian Journal of Plant Physiology. 2009;56(2):200-210.
42. Hardy F. Cacao ecology and physiology: Nutrient relations of cacao Cacao Manual (English Edition), ed. F. Hardy, Turrialba, Costa Rica: Inter-American Institute of agricultural Sciences; 1960.
43. Bonaparte EENA. Yield gradients in cacao (*Theobroma cacao* L.) shade and fertilizer experiments Acta Hort. 1975;49:251-257.

2

Effect of Shelterbelt and Land Management on Soil Carbon Sequestration in Shelterbelt-pasture System at Charles Sturt University, Orange Campus New South Wales Australia

A. Bangura[1*], Y. L. Oo[2], C. S. Kamara[1], A. Raman[2], D. S. Hodgkins[2] and H. I. Nicol[2]

[1]*Agricultural Research Institute, Rokupr Agricultural Research Centre, Rokupr, PMB73, Freetown, Sierra Leone.*
[2]*School of Agricultural and Wine Sciences, Charles Sturt University, Orange Campus, New South Wales, Australia.*

Authors' contributions

This work was carried out in collaboration between all authors. Author AB designed the study, wrote the protocol. Authors AB and CSK wrote the first draft of the manuscript. Authors CSK and AR reviewed the experimental design and all drafts of the manuscript. Authors AB, YLO and DSH managed the analyses of the study. Authors HIN, AB and YLO performed the statistical analysis. All authors read and approved the final manuscript.

Editor(s):
(1) Peter A. Roussos, Lab. Pomology, Agricultural University of Athens, Greece.
(2) Hab. Mariusz Cycoń, Department and Institute of Microbiology and Virology, School of Pharmacy, Division of Laboratoty Medicine, Medical University of Silesia, Poland.
Reviewers:
(1) Anonymous, Cameroon.
(2) Aweng A/L Eh Rak, Faculty of Earth Science, Universiti Malaysia Kelantan, Malaysia.
(3) Nagendra Bastakoti, Wageningen University, Organic Farming Systems Group, P.O. Box 563, 6700 AN Wageningen, The Netherlands.
(4) Anonymous, Serbia.
(5) Anonymous, Brazil.

ABSTRACT

Measurements of soil organic carbon (SOC) levels were made on soils from 0.00–0.10m and 0.10-0.20m soil depth that were collected from three 12 - years old shelterbelts integrated with pastures in new South wales, Australia to determine whether there was any effect of shelterbelts on SOC

Corresponding author: E-mail: abdulaibags@yahoo.com;

levels in the adjacent pasture. The samples were collected in the spring (September 2011) and autumn (March 2012) at increasing distance from the midpoint of shelterbelts. To determine the SOC level in the sample, two permanganate oxidisable methods were used: the Total Organic Carbon (TOC) and labile carbon analysis method. Regression analysis of our result indicated no significant difference (P>0.005) in SOC along the sampling points in an increasing distance from the midpoint of the shelterbelts in both seasons. However, there was a significant difference (P<0.005) in SOC level between the two depths. During autumn at 0.00-0.10m depth the TOC was higher at Weston 1 (1.386%) than College 4 and Leeds Parade site with 1.146% and 1.11% respectively. For the same depth, Weston 1 had 0.061% labile C and 0.054% labile C for both College 4 and Leeds Parade sites. This may be attributed to the topographic difference among the sites.SOC was higher during autumn sampling than spring sampling in all three sites and at both depths due to microbial activity, higher vegetation and warmer climate in autumn.

Keywords: Oxidisable permanganate; integrated pastures; carbon sequestration.

1. INTRODUCTION

The importance of organic carbon with respect to the physical, chemical and biological aspect of soil quality is well established. In addition to organic carbon being a crucial sustainability factor in temperate, as well as tropical land use. Soil organic carbon (SOC) sequestration is also an efficient pathway for mitigating agricultural CO_2 emission and even in reducing anthropogenic emission of CO_2 [1]. Labile organic carbon is a critical component of SOC because of its intimate association with soil microbial respiration and role in the decomposition of resistant soil organic matter (SOM) [2].

All agricultural soils have been altered from their natural state by human interventions which are designed to maximizing production functions and which, to some degree, result in a loss of other ecosystem functions [3]. After the natural vegetation has been cleared to establish agricultural fields, all the major soil properties that describe its health are changed, largely negatively.

The decline in soil organic matter content starts immediately after clearing and the initiation of cultivation, after a period of continuous cultivation they reach a new dynamic equilibrium [4].

There is evidence that conservation practices, well-managed pasturelands, planted forests, and agroforestry systems can considerably reduce carbon loss, maintaining SOM levels or even increasing them [5]. Since SOM has no definite chemical composition, SOC has been measured to indicate soil health [6]. Soil organic carbon stock and pools are usually highly variable on both spatial and temporal scales [7]. The presence of SOC at any location is influenced by complex interactions among plant growth, climate, soil type or parent material, topography and site management.

Attributes such as increase water–holding capacity, higher infiltration rates and higher nitrogen availability arising from increased SOC can enable farming system to be more resilient to climate change [8].

It has been reported that the average SOC in the 0.00–0.5m layer within the shelterbelt was significantly greater than in cultivated fields, and this increase was as a result of organic inputs by tree litter and deposition of wind–blown sediments accumulating beneath the shelterbelt [9].

Storing carbon in soil (soil carbon sequestration) is a key process in relation to global warming and climate change, which is believed to be caused primarily by elevated concentration of atmospheric CO_2 with other greenhouse gases. Soil carbon sequestration restores degraded soils, enhances biomass production, purifies surface and ground waters, and reduces the rate of enrichment of atmospheric CO_2 by offsetting fossil fuel emissions [10]. Every tonne of carbon sequestered in the soil is a tonne of carbon reduced from the atmosphere, so good management of soil carbon is an important tool in the reduction of greenhouse gases in the atmosphere. Soil carbon sequestration is vital in agricultural systems and it is gaining global attention because of the growing concern to reduce the rapidly increasing atmospheric CO_2. [11] Estimated the potential carbon sequestration in pasturelands and rangelands in USA and concluded that half of the sequestration was derived from changes in pasture management, i.e. fertility management, application of manure, improved pasture species, and grazing management [11].

In Australia, pastures are important to extensive livestock industries and are vital to crop–rotation systems. Despite growing interest in the soil carbon sequestration in recent times and the importance of pastures in the mitigation of climate change, little information is available on the soil carbon–sequestration potential of pastures in New South Wales agriculture [12].

The present study examines shelterbelt and pasture management and the extent to which they can be used to sequester soil organic carbon. The objective of this study was to compare the quantity of labile carbon and total organic carbon in soils managed under long term pasture and adjacent soil under shelterbelt plantings of native species at different depths.

2. MATERIALS AND METHODS

2.1 Study Site and Experimental Design

The study was done at the Orange campus of Charles Sturt University (33°15'S; 149°07'E; 875 m above sea level (asl)) in the Central Tablelands of New South Wales. The Campus farm extends across two soil landscapes identified as North Orange and Macquarie. Soil is shallow, well-structured and well drained with clay–loam top soil texture. It ranges from red–brown silty clay loam dermosols along the mid–lower slopes to red chromosols along the upper slopes and ridges [13]. From the records of the weather station of the University Campus, the climate is characterized by cold—wet winters (2—10°C) and mild summers (12—25°C). Usually, rainfall occurs uniformly through the year (700—950 mm). Mean-monthly maximum temperatures in 2011 ranged from 15.1 to 18.4°C in 2011 indicating that temperatures were warmer than the comparable 30-year average of 15.6°C; average rainfall was 603.4 mm in 2011, nearly 90 mm more than the preceding 30-year [14].

Three shelterbelts (*Leeds Parade*, *College 4*, and *Weston 1*; hereafter referred to as Sites 1, 2, and 3, respectively), each about 100 m long and 15—25 m wide, established in 1998—2000 and situated within previously established pasture land, were used (Fig. 1). The mean above-sea elevations are 885m asl for Sites 1 and 2, and 907 m for Site 3. These sites were chosen for their similarity in being narrow, elongated shelterbelts with a northerly—southerly orientation, receiving the same easterly downwind, in which the sampling transects were constructed.

The principal purpose of establishing these shelterbelts with a mixture of seedlings of Australian native trees was to capitalize on the advantages the shelterbelts provide when occurring in a pasture ecosystem of perennial-pasture taxa. Populations of *Poaceae* and Fabaceaewere the principal species in the pasture. Population of *Boraginaceae*occurred in fewer numbers than the previously listed taxa of *Poaceae* and *Fabaceae*. Shelterbelt taxa include; *Myrtaceae*, *Mimosaceae*, *Casuarinaceae*, *Myrtaceae* and large native shrubs. Major tree and grass species found at the study site are listed in Table 1 to show the flora at the study sites.

Table 1. Types of trees and grass species at the study sites

Study site	Tree species	Grass species
Leeds Parade	*Eucalyptus blakelyi, Eucalyptus macrorrhyncha, Eucalyptus viminalis, Acaciaimplexa, Acaciavestita, Casuarina cunninghamiana, Callistemon sieberi.*	*Phalaris aquatic, Trifolium subterraneum, Vulpia bromoides,*
College 4	*Eucalyptus blakelyi, Eucalyptus melliodora, Eucalyptus pauciflora, Acacia dealbata, Acacia vestita, Callistemon sieberi*	*Lolium rigidum, Holcus lanatus, Trifolium repens, Echium plantagineum*
Weston 1	*Eucalyptus blakelyi, Eucalyptus macrorrhyncha, Acacia dealbata, Acacia implexa, Casuarina cunninghamiana, Leptospermum myrtifolium*	*Phalaris aquatic, Lolium rigidum, Lolium perenne, Hordeum glaucum*

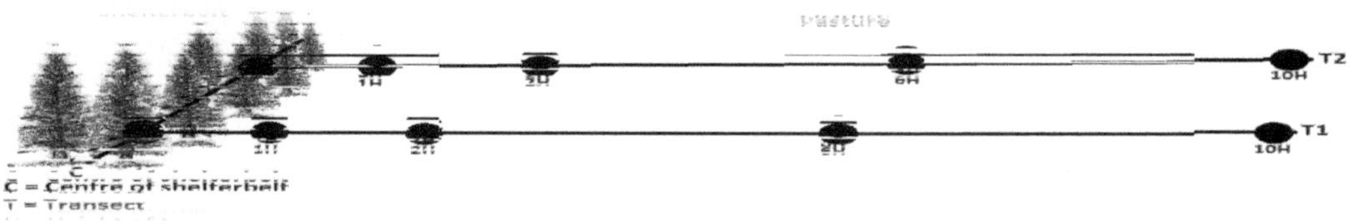

Fig. 1. Layout for soil sampling in each shelterbelt-pasture system (Not to scale)

Due to minor differences at the times of planting, the mean-tree heights of plants at Sites 1, 2, and 3 varied from 4.6 to 6.4 m. Therefore, average tree heights at each site were used as a pertinent measure to determine the distances for various sampling points to extract soil samples at two different depths. Tree heights were determined by obtaining average clinometric readings for randomly selected 15 trees, following the procedure described by [15].

Two parallel-running transects (T_1, T_2) situated at 90° to the midline of the shelterbelt, separated by a distance of 30 m, were constructed at Sites 1, 2, and 3. Five sampling points, named Zero, 1H, 2H, 6H and 10H, starting from the midline of each shelterbelt were used (Fig. 1). The mean tree height at Site 1 was 4.6 m; that at Site 2 was 6.4 m, and that at Site 3 was 5.6 m; the sampling points — the distance variables — 1H, 2H, 6H and 10H, were calculated based on mean tree heights. Two of the five sampling points (zero and 1H) fell within the shelterbelt vegetated area; the remaining sampling points fell within the adjacent pasture (Fig. 1). A maximum of the equivalent of 10 tree heights was chosen as the most-distant sampling point, following [16].

2.2 Soil Sampling and Soil Organic Carbon and Total Organic Carbon Analysis

Sixty samples were collected(0.00–0.10 m and 0.10–0.20 m depth in each sampling point, 5 sampling points in each transect, two transects in each sites) in September, 2011 for spring sampling another sixty samples in March, 2012 for autumn sampling from three shelterbelt-pasture sampling sites. The vegetation and litter fall were removed from the sampling points and the soil samples were taken at 0.00-0.10m and 0.10-0.20m depth using a soil auger of 0.10m diameter. The soil samples were put in paper bags and labelled. The soil samples were crushed and rock and plant materials were removed before drying at 40°C for 24 hr. in a desiccator. Additional soil sampling was done at every sampling point with soil cylinder to obtain intact soil core for bulk density measurements.

To measure different fractions of SOC: labile carbon and TOC in shelterbelt-pasture system at Charles Sturts University paddock in Orange Campus, the $KMnO_4$ oxidation methods of [6] and [17] were followed respectively. These methods have been found to effectively measure different fractions of soil organic carbon [6].

Following method of [6], 5g ofdried sieved soil was placed in screw–capped centrifuge tubes and 20.0 mL of 0.02 M $KMnO_4$ solution was added into each centrifuge tube. The tubes were shaken on a mechanical shaker for 2 min at 200 rpm. The tubes were then centrifuged (Avanti® J–E, Beckman Coulter Inc., USA) at 704.34 G for 5 min to separate soil particles from the solution. The clear supernatant (0.20 mL) was transferred to conical flasks and diluted with 10.0 mL deionized water. Absorbance was read at 550 nm using a UV–VIS spectrophotometer (Lambda 35 UV/VIS Spectrometer, Perkin Elmer Inc., USA) and absorbance readings were compared with a standard curve and the result carbon (mg/g) were reported as air–dried basis percentage carbon.

To measure the TOC, the method of [17] was followed. Two g of the sieved soil mentioned above was placed in a tube and 10 mL of 0.333 M $KMnO_4$ solution was added. The tubes were then shaken on a mechanical orbital shaker for 24 h at 12 rpm to be able to oxidize all TOC with $KMnO_4$. The content was then transferred to centrifuge tube and the polycarbonate tubes were washed with 10 mL of deionized water to minimize loss of materials. The tubes were centrifuged at 312 G for 5 min to separate the soil particles from the solution. After centrifuging, the clear supernatant of 0.2 mL was diluted with 25 mL deionized water and absorbance was read in the same UV–VIS spectrophotometer at 565nm. The absorbance readings were compared with standard curves and the quantity of carbon were converted to air dry basis percentage carbon.

2.3 Calculation of Carbon Percentage

The concentration of carbon in soil sample was calculated from the concentration $KMnO_4$ lost in the reaction base on the assumption that each 1.0 mM of MnO_4^- consumed 0.75 mM of carbon (or 9.008 mg of C) by reducing Mn^{7+}derived from $KMnO_4$ to Mn^{2+} [18] Therefore the concentration of labile carbon in soil sample for [6] method was calculated by the following formula:

$$C_{labile}\ (mg/g) = \frac{(M_0 - M_1) \times 20 \times 9}{5}, \text{ then } C_{labile}\ (\%) = \frac{(M_0 - M_1) \times 20 \times 9}{5 \times 10} \quad \text{(Equation 1)}$$

Where:

M_0 = initial concentration of $KMnO_4$ (0.02)
M_1 = concentration of $KMnO_4$ after oxidation with carbon in the soil sample.
5 = dry weight of soil sample 20 and 9, are constants derived from the equation

Then the concentration of TOC in [6] method was calculated by the following equation:

$$TOC\ (mg/g) = \frac{(M_0 - M_1) \times 10 \times 9}{2}, \text{ then } TOC\ (\%) = \frac{(M_0 - M_1) \times 10 \times 9}{2 \times 10} \quad \text{(equation 2)}$$

M_0 = initial concentration of KMnO4 (0.333)
M_1 =concentration of $KMnO_4$ after oxidation with carbon in the soil sample.
2 = dry weight of soil

The mass of carbon (Mg ha^{-1}) was calculated following [19].

3. RESULTS AND DISCUSSION

Regression analysis for percentage of labile C and TOC were carried out using GenStat for Windows [20] for both spring and autumn samplings. For spring sampling, the TOC% showed no significant difference ($p>0.005$) between shelterbelt and pasture sampling points along transacts at 0.00-0.10m depth for all three sites (Leeds Parade, College 4 and Weston 1). Similarly, there was no significant difference ($p>0.005$) of TOC% along the transect for all sampling sites at the 0.10-0.20m depth. The TOC % ranged from 0.8% to 1.5% at 0.00-0.10m depth and from 0.2% to 1.3% at 0.10-0.20m depth with the mean of 1.1%, 1.1% and 1.4% at 0.00-0.10m depth and of 0.9%, 0.4% and 1.2% at 0.10-0.20m depth for Leeds Parade, Collage 4 [21]. The tree density at Leeds Parade shelterbelt was 20 plants per 250 m^{-2} and and Weston 1 respectively (Table 2). However, there was a slight increase of TOC% at 0.10-0.20m depth from shelterbelt to pasture (see Fig. 2).

The percentage of labile carbon showed similar trend as in TOC in spring sampling. There was no significant difference ($p>0.005$) of labile carbon percentage along the transect from the shelterbelt to pasture at 0.00-0.10m depth which varies from 0.05% to 0.06%. Moreover, there was no significant difference ($p>0.005$) of percentage labile carbon at 0.10-0.20m depth along transect from shelterbelt to pasture paddock which ranged from 0.03% to 0.06% (see Table 2). However, there were significant differences between both TOC% and labile C%, and sites ($p<0.001$), depth ($p<0.001$) and site-depth interaction ($p<0.001$) were also significant.

The seasonal difference between the mean value of TOC% and labile C%, indicates that the TOC% in autumn was higher than that of spring (see Fig. 2 and Fig. 3) and particularly Collage 4 site showed significantly lower TOC% in spring (mean 0.43%) than that of autumn (mean 0.97%) (Table 2).

From our result and analysis provided, it is clear that the TOC and labile C% did not change significantly along transects and hance the shelterbelt did not have significant effect on the mass of carbon in the adjacent pasture paddock. According to [21], the age and type of shelterbelt tree significantly affect the change in soil carbon down to 0.3m depth. The changes in soil C was found to be declining (-3.25%) under 7 years old trees, while 4.37% increased under 26 years old trees at 0.00-0.10cm depth. For 0.10-0.20cm depth, the change in carbon was 0.24% increased with 7 years old trees and 11.39% increased with 30 years old trees from around 100 observations [21]. Therefore, it is obvious that the trees in the shelterbelt under investigation were still young (10 years) to have effect on the mass of carbon sequestration in the shelter belts and pasture paddocks.

The effect of tree species is also an important factor to consider. For trees under 10years old, there was only a little change (0.23%) in soil carbon under eucalyptus compared to hard wood (4.67%) such as poplar (*Populous angustifolia*) and mahogany (*Swietenia mahagoni*) and decreased (2.39%) under pine (*Pinus radiate*)

dominated by eucalyptus (50%) with larger canopy coverage, tree density at Collage 4 was 9 plants per 250 m^{-2} with pure eucalyptus (100%), and Weston 1 with 14 plants per 250 m^{-2} and dominated by acacia (70%) with understory shrubs. In addition to the shelterbelt species, the topography of the sites was also different: The mean above-sea elevations are 885 m (asl) for Sites 1 and 2, and 907 m for Site 3. Furthermore, the shelterbelts at Leeds Parade and Collage 4 are located down the hill, the transect therefore move towards the top of the hill unlike that of Weston 1 wherein the shelterbelt is located on top of the hill, and the transects moves down the hill.

From Table 3, it is clear that TOC (Mgha^{-1}) is higher at the 0.00-0.10 m depth than 0.10-0.20 m depth. Furthermore, at the 0.00-0.20 m depth, more TOC is recorded for the autumn sampling than the spring sampling for both Leeds parade and College 4 however, Weston 1 recorded a higher TOC in spring than autumn.

The shelterbelt species and topography might contribute to the significant difference of TOC% and labile C% at the three sites of the shelterbelt and pasture paddocks. The finding of [22] is consistent with our result due to the fact that the values of TOC in autumn were higher than that of spring when measured in a pasture. The result of lower carbon in 0.10–0.20m depth is consistent with the finding reported by [23] in New Zealand that 43.3 Mg ha^{-1} at 0.00–0.10m depth and 31.7 Mg ha^{-1} at 0.10–0.20m depth under pasture. From Table 3, the mean TOC at 0.00-0.10m depth for autumn sampling at college 4 transect 2 is 17.7 Mg ha^{-1} and the mean TOC at 0.10-0.20m depth is 14.9 Mg ha^{-1}.

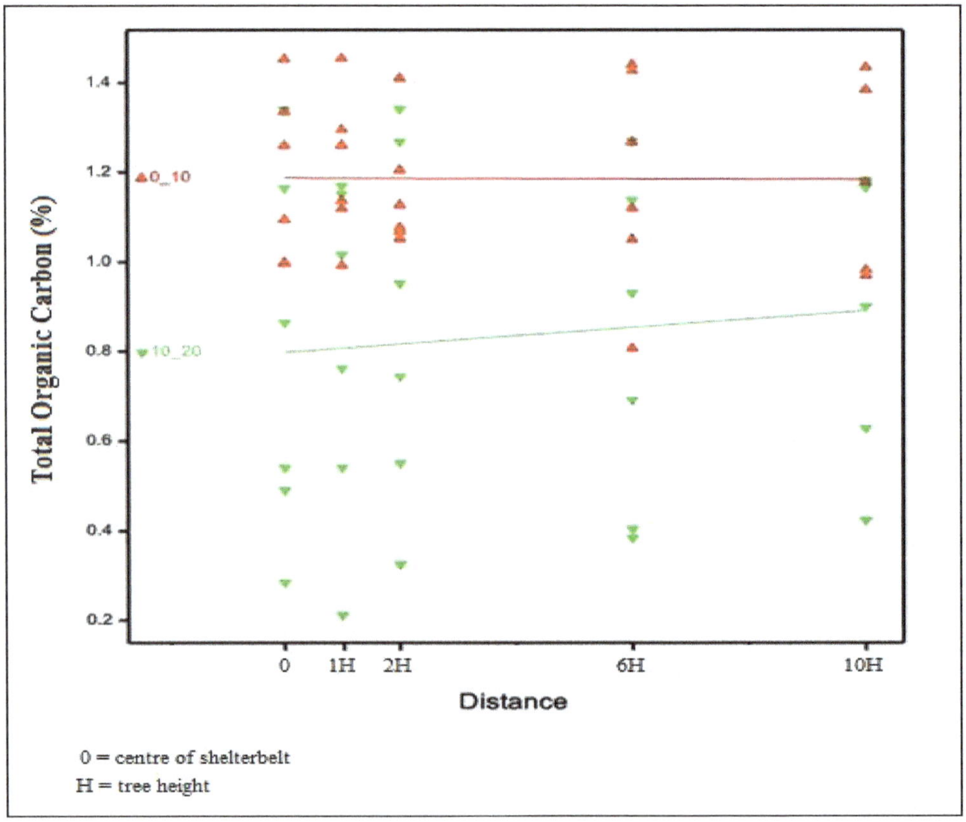

Fig. 2. GenStat output of fitted and observed relationship for TOC. (Spring sampling 2011)

Table 2. The mean and standard deviation of TOC% and labile C% for spring and autumn sampling

Site	TOC %				labile C %			
	Spring		Autumn		Spring		Autumn	
	0.00-0.10 m	0.10-0.20 m	0.00-0.10 m	0.10-0.20 m	0.00-0.10 m	0.10-0.20 m	0.00-0.10 m	0.10-0.20 m
Leeds Parade	1.083±0.062*	0.853±0.192	1.111±0.057	0.908±0.048	0.052±0.005	0.036±0.005	0.054±0.005	0.046±0.012
Collage 4	1.091±0.154	0.430±0.133	1.146±0.071	0.968±0.119	0.051±0.005	0.038±0.006	0.054±0.005	0.043±0.005
Weston 1	1.384±0.079	1.219±0.078	1.386±0.056	1.209±0.103	0.049±0.005	0.051±0.006	0.061±0.013	0.054±0.005

Denote ± standard deviation

Table 3. Mean and Standard deviation of TOC ($Mgha^{-1}$) for spring and autumn sampling period

Sampling period	Site	Transect	TOC ($Mgha^{-1}$)		
			0.00–0.10 m	0.10–0.20 m	0.00–0.20 m
Spring 2011	Leeds	1	16.6±1.31*	13.92±2.94	30.52±3.84
	Parade	2	15.36±1.94	11.84±2.87	27.2±4.18
	College 4	1	20.6±1.64	6.32±3.13	26.92±4.08
		2	19.8±1.34	7.37±0.91	27.17±2.06
	Weston 1	1	20.7±2.25	17.6±1.81	38.3±3.91
		2	21.51±1.97	19.56±1.54	41.07±2.51
Autumn 2012	Leeds	1	16.01±2.13	12.63±0.99	28.64±3.09
	Parade	2	17.7±0.79	14.9±0.89	32.6±1.68
	College 4	1	16.85±1.22	14.49±2.19	31.34±3.23
		2	18.0±0.39	14.94±1.03	32.94±1.16
	Weston 1	1	18.52±3.18	16.04±3.34	34.56±6.47
		2	19.12±2.21	16.87±2.48	35.99±4.62

Denote ± standard deviation

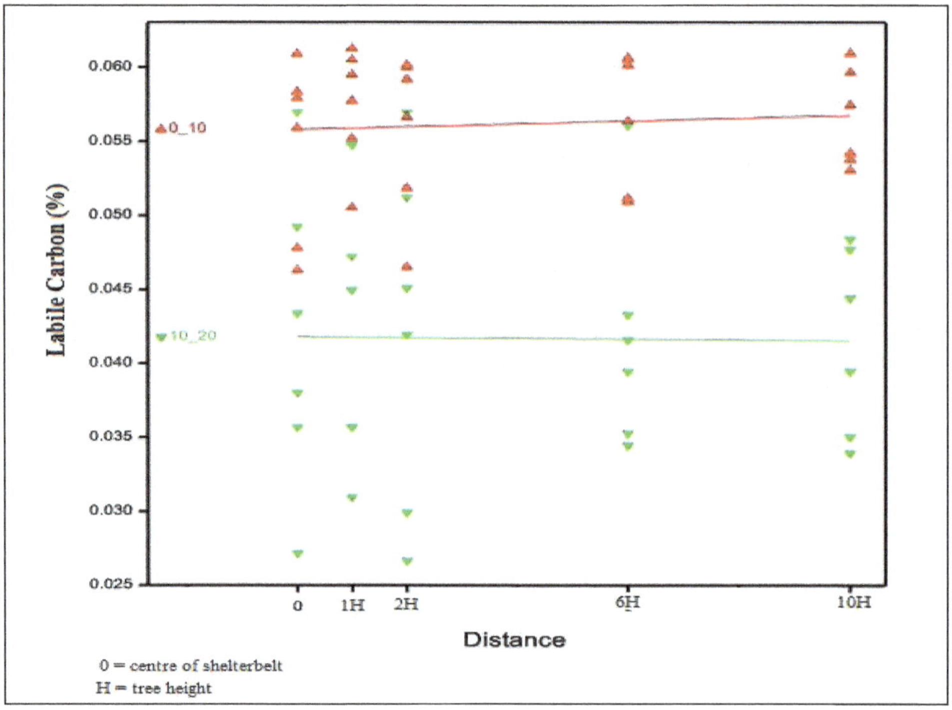

Fig. 3. GenStat output of fitted and observed relationship for labile carbon. (Spring 2011 sampling)

4. CONCLUSION

The TOC and labile-carbon has not changed significantly among sampling points along the transect as we move from the shelterbelt into the pasture paddock. However, both fractions of carbon decreased with depth. This is as result of the fact that SOC levels are dynamic and generally decline with depth because most sources of organic matter from which it is derived are either on the surface or near the surface of the soil and because plant roots are less dense as they grow deeper into the soil. Moreover, there was significant difference between sampling sites- Weston 1 site showed higher level of carbon than Leeds Parade and Collage 4. This may be attributed to topographic difference among the sites. The transect at Weston 1 moves down the slope while those for the other two sites move up the slope. In addition the proportion of carbon was higher in autumn sampling than spring sampling, which reflect the higher vegetation, microbial activity and warmer climate in autumn.

ACKNOWLEDGEMENT

We thank Steve Mannix (Farm manager, Charles Sturt University Orange campus, New South Wales, Australia) for providing information (age and specie) of shelter-belt and pasture and the Trustees of the Rural-Management Research Institute (Charles Sturt University, Orange, NSW) for financial support to the first author during his studies.

COMPETING INTERESTS

Authors have declared that no competing interests exist.

REFERENCES

1. Lal R. Black and buried carbon impacts on soil quality and ecosystem services. Soil & Tillage Research. 2008;99:1–3.
2. Feng W, Schaefer D, Zou X, Zhang M. Shifting sources of soil labile organic carbon after termination of plant carbon inputs in a subtropical moist forest of southwest China. Ecological Research. 2011;26(2):437–444.
3. Saunders DW. International activities in assessing and monitoring soil degradation (Special Issue on Soil Quality). Journal of Alternative Agriculture. 1992;7:17–24.
4. Oldeman LR. The global extent of soil degradation. In: Greenland DJ, Szabolcs I.

(Eds.), Soil Resilience and Sustainable Land Use. CAB International, Wallingford, Oxon, UK. 1994;99–118.

5. Freitas PL, Blancaneaux P, Gavinelli E, Larre´–Larrouy M, Feller C. Nı'vel and nature of the stock of organic Oxisols under different land use systems and management. Pesqagropec Bras. 2000;35:157–170.
6. Weil RR, Islam KR, Stine MA, Gruver JB, Samson-Liebig SE. Estimating active carbon for soil quality assessment: A simplified method for laboratory and field use. American Journal of Alternative Agriculture. 2003;18: 3–17.
7. Allen DE, Pringle MJ, Page KL, Dalal RC. A review of sampling designs for the measurement of soil orange carbon in Australian grazing lands. The Rangeland Journal. 2010;32:227–246.
8. New South Wales Industry and Investment; 2010. Department of Primary Industry (DPI) online material accessed 5th February 2012. Available: http://www. Dpi.nsw.gov.au.
9. Sauer TJ, Cambardella CA, Brandle JR. Soil carbon and tree litter dynamics in a red cedar-scotch pine shelterbelt. Agroforestry Systems. 2007;71:163–174. DOI:10.1007/s10457-007-9072-7. (accessed on 29.8.2011).
10. Lal R. Soil carbon sequestration impacts on global climate change and food security. Science. 2004;304(5677):1623-1627.
11. Follett RF, Kimble JM, Lal R. The potential of US grazing lands to sequester soil carbon. The Potential of US Grazing Lands to Sequester Carbon and Mitigate the Greenhouse Effect. 1998;401–430.
12. ChanKy, Oates A, Li GD, Conyers MK, Prangnell RJ, Poile G, Liu DL, Barchia IM. Soil carbon stocks under different pastures and pasture management in the higher rainfall areas of south-eastern Australia. Australian Journal of Soil Research. 2010;4:7–15.
13. KovaćM, Lawrie JA. Soil landscapes of the Bathurst 1:250000 sheet. Soil Conservation Service of NSW, Sydney. 1990;323-324.
14. Bureau of Meteorology New South Wales, Annual climate summary 2011. Accessed 15th March; 2012. Available: http://www.bom.gov.au/climate/annual.
15. Department of Environment, Climate Change and Water; 2010. Accessed April 2012. Available: http://www.ecosteps.com.au.
16. Cleugh H. Trees for shelter – windbreaks for Australian farms. Rural Industries Research and Development Corporation. 2003;06:7-9.
17. Blair GJ, Lefroy RDB, Lise L. Soil carbon fractions based on their degree of oxidation, and the development of a carbon management index for agricultural systems. Australian Journal of Agricultural Research. 1995;46:1459–1466.
18. Rayment GE, Lyons DJ. Soil Chemical Methods–Australia. Commonwealth Scientific and Industry Research Organisation. Collingwood, Australia; 2011.
19. Ellert BH, Bettany JR. Calculation of organic matter and nutrients stored in soils under contrasting management regimes. Canadian Journal of Soil Science. 1995;75:529–538.
20. Payne R, Murray D, Harding S, Baird D, Soutou D, Lane P. GenStat for Windows—introduction. VSN International. Hemel Hempstead, Oxford. 2003;12.
21. Paul KI, Polglase PJ, Nyakuengama JG, Khanna PK. Change in soil carbon following afforestation. Forest Ecology and Management. 2002;168(1–3):241–257.
22. Lodge GM, King KL. Soil microbial biomass, labile and total carbon levels of grazed sown and native pastures in northern New South Wales. Australian Journal of Agricultural Research. 2006;57(8):837–845.
23. Parfitt RL, Percival HJ, Dahlgren RA, Hill LF. Soil and solution chemistry under pasture and radiata pine in New Zealand. Plant and Soil. 1997;191(2):279–290.

3

Effect of Bio-organic Amendments on the Infestation of Major Pests & Foliar Disease, Leaf Productivity in Mulberry (*Morus alba* L)

Barna Chakraborty[1*], Asit Kumar Chanda[2] and Susanta Kumar Chakraborty[1]

[1]Department of Zoology, Vidyasagar University, Midnapore-721102, India.
[2]Central Sericultural Research and Training Institute, Berhampur, West Bengal, India.

Authors' contributions

This work was carried out in collaboration between all authors. Author BC designed the study, wrote the protocol and wrote the first draft of the manuscript. Authors AKC and SKC reviewed the experimental design and all drafts of the manuscript. Authors AKC and SKC managed the analyses of the study. Author AKC identified the plants. Author BC performed the statistical analysis. All authors read and approved the final manuscript.

Editor(s):
(1) Masayuki Fujita, Department of Plant Sciences, Kagawa University, Japan.
(2) Daniele De Wrachien, State University of Milan, Italy.
Reviewers:
(1) Raul Leonel Grijalva Contreras, Instituto Nacional de Investigaciones Forestales Agrícolas y Pecuarias, Mexico.
(2) A. Tikader, Central Sericultural Germplasm and Resources Centre, India.
(3) Anonymous, Indonesia.

ABSTRACT

Massive utilization of inorganic fertilizers for a prolonged period leads to alteration of the soil chemical properties, the availability and uptake of nutrients by the plants. The field experiments aimed at investigating the effect of bio-organic amendments on infestation of major pests, foliar disease and leaf productivity of mulberry (*Morus alba* L) have been conducted during March 2004-February 2007 in lateritic soil condition (having around pH of 5.6) with limited irrigation support in some areas of Midnapur (West) Districts, West Bengal, India. The research was set up as randomized block design, consisted of five treatments and provided with three replications. (T_1 = 20 MT/ha FYM + 336 kg N+180 kg P_2O_5+ 112 kg K_2O /ha /yr, T_2 = 10 MT/hc Vermicompost + *Azotobacter* +AMF + 50% N + 33% P_2O_5 and K_2O /ha /yr of T_1, T_3 = 7 MT/hc of Poultry-manure + *Azotobacter* +AMF + 50% N + 33% P_2O_5 and K_2O /ha /yr of T_1, T_4 = 5 MT/hc Vermicompost + *Azotobacter* +AMF + 50% N + 33% P_2O_5 and K_2O /ha /yr of T_1, T_5 = 3.5 MT/hc Poultry-manure +

**Corresponding author: E-mail: barnack2020@gmail.com;*

Azotobacter +AMF + 50% N + 33% P_2O_5 and K_2O /ha /yr of T_1). The experimental results revealed that the effect of vermicompost in combination with bio-fertilizers followed by reduced doses of inorganic fertilizers imparted significant effect on leaf productivity and incidence of pests and diseases of mulberry plants. However, maximum leaf yield was recorded with the application of poultry manure along with bio-fertilizers and reduced dose of inorganic fertilizers (T_3) in the third year after the establishment of plants (16.98 MT/ha/yr). Study of surveillance on the incidence of disease and insect pests clearly indicated that no other diseases than foliar disease was recorded under different manurial and fertilizer treatments. Further, population of the insect pests like white fly and thrips were found to be significantly reduced below the economic threshold level in the experimental plots treated with different organic manures along with bio fertilizers over control.

Keywords: Mulberry; organic manures; bio-fertilizers; lateritic soil; pests; disease.

1. INTRODUCTION

Mulberry being deep rooted perennial crop tends to grow and produce leaves throughout the years in the tropical region and represents the only prime food sources for mulberry silkworms (*Bombyx mori.* L). Mulberry leaf is a major economic component in sericulture, since the quality and quantity of leaves produced per unit area have a direct bearing on cocoon harvest [1]. Production of leaves in the tropical countries is largely dependent on the application of inorganic fertilizers. Pests and diseases play significant roles in determining the leaf yield of mulberry per unit area [2]. Fertilization with inorganic nitrogen can often encourage succulent and excessive vegetative growth of the plants that may enhance the multiplication rate of pests leading to more damage to plants. The plants having good growth coupled with biomass impart profound influence on insect population [3]. In order to reduce the chemical pesticides load, adoption of some agronomical practices like inter cultural operation, proper plant spacing and pruning are supposed to reduce the infestation of diseases without adversely affecting the ecosystem. Proponents of organic agriculture have asserted that plants grown with biological sources of nutrients such as manures and composted organic wastes are less susceptible to the attack of the insect pests than other conventionally grown plants [4]. Recently, a great attention has been paid towards the bio organic farming to avoid the heavy use of agrochemicals which resulted in numerous environmental troubles [5]. The coincident application of organic manures and bio-fertilizers is frequently recommended for improving soil properties and obtaining clean agricultural products [6]. Vermicompost is an organic manure obtained in the form of casting of ingested biomass by earthworms after undergoing physical, chemical and microbial transformations. . The water soluble components of vermicompost such as humic acid, growth regulators, vitamins, micronutrients and beneficial microorganism increases the availability of plant nutrients, resulting increased growth, higher yield in crop [7] especially significant growth in tomato [8]. Besides, macro and micronutrients, it also contains humic acids, plant growth promoting substances like auxins, gibberellins, and cytokinins [9]. In addition to vermicompost other organic manures and bio-fertilizers also play a vital role as organic nutrient sources for sustainable soil health and crop growth [10). Kundaswamy et al. [11] reported an increase nitrogen level of mulberry was found due to Arbuscular Mycorrhizal fungal (AMF) inoculation. Beneficial effect of application of Azotobactor in mulberry leaf production was established and demonstrated by Das et al. [12-14]. Therefore, the focus of this investigation was to evaluate the response of bio-organic nutrition through the application of different organic manures and bio- fertilizers in the lateritic soil of Midnapore (West). West Bengal, India and to identify the best suited eco friendly nutrient management practice for sustainable leaf yield of mulberry and suppression of diseases and pest attack.

2. MATERIALS AND METHODS

The plantation was raised in lateritic soil of Vidyasagar University campus (22° 25′ N latitude and 87° 17′ longitude) of Midinapur (West), West Bengal, India with eight months old mulberry saplings of S-1635 variety, which is a triploid variety with better rooting capacity having high leaf yielding ability. The pH of the soil was 5.6 having 0.41% organic carbon as determined following Jackson [15]. The plantation was made with 49 plants in each plot ((5.4 x 5.4 m) with a spacing of 60cm. x 60cm between the plants and was laid out in randomized complete block design having five treatments and three replications. In the present study, different organic manures and bio-fertilizers were applied after bottom pruning at 15 cm height and they were as follows.

A. Organic manures: Farm yard manure (FYM), Vermicompost and Poultry manures.
B. Biofertilizers: *Azotobacter* (AZB) *chroococcum* and Arbuscular mycorrhizal fungi (AMF)

After pruning and cultural operation, five different plots were earmarked for giving five different treatment combinations (two variable doses of NPK along with farm yard manure, vermicpompost, poultry manure, *Azotobacter* and Arbuscular mycorrhizal fungi). The uniqueness of this standardisation process involves the reduction of recommended dose into half for nitrogen and 1/3rd for phosphorus although recommended dose for potassium remained same and was supplemented with different organic manures and bio-fertilizers. A total of 735 plants were imposed with five different treatments viz., T_1=20 MT/ha FYM+336 kg N+180 kg P2O5+112 kg K2O /ha /yr, T_2=10 MT/ha Vermicompost+*Azotobacter*+AMF+50% N+33% P2O5 and K2O /ha /yr of T_1, T_3=7 MT/ha Poultry-manure + *Azotobacter*+AMF+50% N+33% P2O5 and K2O /ha /yr of T_1, T_4=5 MT/ha Vermicompost + *Azotobacter*+AMF+50% N+33% P2O5 and K2O /ha /yr of T_1, T_5=3.5 MT/ha Poultry-manure + *Azotobacter*+AMF+50% N+33% P2O5 and K2O /ha /yr of T_1. In Treatment-1, plots were maintained with, recommended fertilizer NPK (336 kg/ha/yr nitrogen,180 kg/ha/yr phosphorus and 112 kgha/yr potassium) and 20 MT/ha/yr farm Yard which treated as control, for Treatment-2, plots were provided with 10 Mt/ha/yr vermicompost along with two bio-fertilizers namely *Azotobacter* (20 kg/ha/yr) and *Arbuscular mycorrhizal* fungi (100 kg/ha/4yr) and reduced doses of nitrogen(168kg/ha/yr), phosphorous (60 kg/ha/yr) and normal dose of potassium. In Treatment -3, plots were provided with 7 MT/ha/yr poultry litter along with two bio-fertilizers namely *Azotobacter* and *Arbuscular mycorrhizal* fungi and reduced dose of nitrogen, phosphorous and normal dose of potassium were applied like T_2 plots. In Treatment- 4, 5 Mt/ha/yr vermicompost along with two bio-fertilizers namely *Azotobacter* and *Arbuscular mycorrhizal* fungi with reduced doses of nitrogen, phosphorous and normal dose of potassium were applied likeT_2 and T_3 plots.In Treatment -5, soils of the plots were added with 3.5 MT of poultry litter along with two bio-fertilizers namely *Azotobacter* and *Arbuscular mycorrhizal* fungi(AMF) and reduced dose of nitrogen, phosphorous and normal dose of potassium as recommended for irrigated package like T_2,T_3 and T_4 plots. All the above mentioned organic manures required for organic farming were readily available from the surrounding rural sectors which were almost free of cost. The Azotobactor and AMF bio-fertilizer were obtained from the cultural collection of Central Sericultural Research and Training Institute (CSR&TI) of Berhampur, West Bengal. Though all the bio-fertilizers selected for the study were low cost and indigenous but they differed in their microbial population. As per recommendation of Sudhakar et al. [16] *Azotobactor chroococcum* cells blended with peat soil/charcoal/FYM in powdered form as carrier material containing 10^{8-9} cells per gram was applied at the rate of 20kg/ha/yr which were optimum for the application in mulberry garden. These were was applied at the rate of 20kg/ha/yr in four equal splits to the soil after each crop harvested following the recommended norms of CSR&TI, Berhampur. The Mycorrhizal product was used for the mycorrhizal inoculants, which is a low cost and commercially available product and this containing multi-strain Arbuscular mycorrhizal fungi (Spore density 100 numbers/5 gm soil) and it was reported to retain at the root zone for many years [17]. AMF was applied only once at the rate of 100kg/ha/4years following the same process. The garden was pruned four times in a year. Inorganic fertilizers were applied after 20 days of the bio fertilizer application. Irrigation was provided as and when required. After 65 days of growth, the observations were recorded during the August, November, February and May, 2005-07. Leaf yield was assessed by harvesting the leaves from all the plants available in net plots except the border line effect and then converted into yield per hectare. Before conducting the experiment, the chemical characteristics of organic manures were recorded and presented in the Table-1 following the standard analytical method of Jackson [15], Walkley & Black [18], Subbiah & Asija [19].

The incidence of occurrences of the diseases and out breaks of pests per plant were recorded for two consecutive years (2005-06 to 2006-07) after 55-60 days of pruning , by randomly taking 10 plants in each replication. Foliar disease such as leaf spot caused by *Myrothaium roridum* was observed during September 2006 and

Table 1. Chemical properties of organic manures

Organic manures	pH	Organic carbon %	N%	P%	K%
Farm Yard Manure (FYM)	7.2	39.4	1.1	0.42	1.82
Vermicompost	7.2	15.4	1.65	0.92	1.2
Poultry Litter	7.0	7.7	3.1	2.84	2.70

September 2007. Observations on Percent Incidence (PI) in respect of leaf spot disease were recorded and the data during both the years were pooled and analyzed. The Percent Incidence (PI) was calculated by applying formula given by Wheeler [20].The total number of leaves (normal leaves + infected leaves) and infected leaves were counted on each shoot of mulberry; three shoots per plant were taken for the count and the mean scoring was expressed in percentage:

$$\text{Disease Incidence (\%)} = \frac{\text{Total number of infected leaves}}{\text{Total number of leaves}} \times 100$$

During this period, two types of pests populations were recorded namely White fly (*Dialeuropora decempuncta*); December-06 and September-07 and Thrips (*Pseudodendrothrips mori*); April-06 and May-07). For those pests incidence ,numbers of nymphs/adults for White fly as well as Thrips were recorded from ten plants in each plot. Leaf Yields from those plots were recorded and calculated in respect of per hectare. Data on leaf yield were recorded at the end of every crop and were subjected to analysis of variance (ANOVA) for a randomized block design with three factors (Treatment, Season & Year). Data on disease incidence and pest attack were also taken into consideration and were subjected to analysis of variance (ANOVA). The significance of different sources of variation was tested using Fisher and Snedecor's F-Test at 5% and 1% level of significance.

3. RESULTS AND DISCUSSION

Leaf Yield showed an increasing trend in different treatments for the three consecutive years (Table 2). Soil productivity of mulberry had been recorded in the tune of 16.98 MT/ha/yr after application of poultry litter along with biofertilizers (*Azotobacter*+AMF) and curtailing dose of NPK (T_3). Analysis of three years leaf yield data under different nutrient sources indicated that bio-organic amendments recorded higher leaf yields which were found to be antagonistic against diseases and pests attack. Among the treatments, a maximum leaf yield was recorded in T_3 in the third year after the establishment of plants (16.98 MT/ha/yr.), Overall 16.69% increase in leaf yield was recorded. Pooled data recorded for three years with respect of total leaf yield of the plants raised under different treatments revealed that treatments were found to be differed significantly from each other. Leaf yields were not hampered due to attack of pests and foliar disease incidence were supposed to be due to effect of organic usage.

Incidence of occurrence of leaf spot disease and pests of mulberry (variety S-1635) under different treatments has revealed that there were no significant differences among the treatments with regard to leaf spot disease and attack of pests like thrips. Regarding infestation of white fly, significant differences were observed among the treatments (CD value 2.265 at 5% level) (Table 3) It was also observed that maximum infestation of white fly was recorded in T_1 (control). The infestation in case of T_3 was 17.17% which was followed by T_2 (16.71%), T5 (16.66%) and T_4 (15.97%). But the infestation in these three (T_2, T_3 and T_4) and T_5 differed significantly from each other. Similarly in foliar disease (leaf spot), the infestation was maximum in control (9.48%) and minimum in T_3 (8.43). Maximum infestation with thrips was noticed in T_2 (14.23%) followed by T_4 (13.10%) and T_5 (12.40%) respectively (Table 3). In the treatment combination of (T_2-T_5) the treatment T_4 was found consistently effective in reducing the incidence of pests and recorded the lowest whitefly population. The treatments T_5 and T_2 were the next in order of effectiveness in reducing the population of white fly and sustainable yield in mulberry. The result is more or less in conformity with Ravi et al. [21] who also recorded significant reduction of the incidence of sucking pests namely whitefly and leaf hoppers under organic manures and bio-fertilizer treated plots. Similarly in foliar diseases, infestation was maximum in control and minimum in the plot treated with organic composts like poultry litter along with bio-fertilizers. These findings also corroborates with the study of Ghorbani et al. [22] who stated that organic composts must have direct anti-disease effects or stimulate competitor micro-organisms and/or induce plant resistance .The use of organic soil amendments can result in a better soil quality and greater plant disease suppressiveness [23]. Sharma et al. [2] also concluded that reduction in disease development might be due to the production of various growth substances or hormones by *azotobacter*, azospirillum and partial application of nitrogen which enables plants to develop resistance power in the plant against certain pathogens. In the present investigation, highest abundance of white fly in control plot (FYM+NPK) has also corroborated the research outcomes of Prestidge & Neill [24] who found that plants grown with high levels of nitrogen fertilizer resulted in larger infestation of pest. Integration of different sources of nutrients tended to release essential plant nutrients gradually throughout the growth period that might have induced development of resistance which subsequently helped in escaping whitefly infestation in tomato. Walkley & Black [25] highlighted the pest suppressing ability of vermicompost and found that vermicompost based treatments harboured lowest population of sucking pests namely jessids and thrips in the plantation of chilli.

The lowest disease incidence was recorded in the plots treated with poultry manure at the dose of 7 MT/ha. The results generated out of the present research studies are in conformity to the findings of Sullivan [26] who concluded that excessive nitrogen levels may further boost different diseases which are rarely a problem in organic production.

Table 2. Effect of bio-organic fertilizers on leaf productivity (MT/ha/yr) of mulberry (Var. S-1635)

Treatments	2004-05 (MT/ha/yr)	2005-06 (MT/ha/yr)	2006-07 (MT/ha/yr)	Average leaf Yield of three years (MT/ha/yr)	Total Leaf yield (MT/ha/yr)	% of increase in yield
T_1 (FYM +NPK)	13.14	13.45	14.49	13.70	41.08	-
T_2 (Vermicompost+ *Azotobacter*+AMF+50% N+33% P_2O_5 and K_2O)	14.71	14.42	15.56	14.90	44.69	8.77
T_3 (Poultry-manure+ *Azotobacter*+AMF+50% N+33% P_2O_5 and K_2O)	15.09	15.88	16.98	15.98	47.95	16.69
T_4 (Vermicompost+ *Azotobacter*+AMF+50% N+33% P_2O_5 and K_2O)	14.42	14.48	15.45	14.78	44.34	7.94
T_5 (Poultry-manure+ *Azotobacter*+AMF+50% N+33% P_2O_5 and K_2O)	14.02	14.09	15.19	14.43	43.30	5.39
CD at 5%	NS	NS	NS	-	2.834	-
CD at 1%	NS	NS	NS	-	4.123	-

Table 3. Effect of bio-organic fertilizers on incidence of disease and pests of Mulberry (Var. S-1635)

Treatments	Disease of Mulberry	Pests of Mulberry (number/leaf)	
	Leaf spot disease incidence (%)	White fly	Thrips
T_1 (FYM +NPK)	9.48	20.02	13.10
T_2 (Vermicompost + *Azotobacter* +AMF + 50% N + 33% P_2O_5 and K_2O)	9.43	16.71	14.23
T_3 (Poultry-manure + *Azotobacter* +AMF + 50% N + 33% P_2O_5 and K_2O)	8.43	17.14	11.57
T_4 (Vermicompost + *Azotobacter* +AMF + 50% N + 33% P_2O_5 and K_2O)	8.69	15.97	11.27
T_5 (Poultry-manure + *Azotobacter* +AMF + 50% N + 33% P_2O_5 and K_2O)	9.30	16.66	12.40
CD at 5%	NS	2.265	NS
CD at 1%	NS	NS	NS
CV %	4.69	6.953	11.49

4. CONCLUSION

The present study has revealed that Bio-organic amendments with the application of organic manures and bio-fertilizers have sound effect on the protection of natural enemies of mulberry under field condition. Bio-organic fertilizers have contributed the higher leaf yield of mulberry as compared to chemical fertilizers and thereby tend to improve the fertilizer use efficiency as well as reducing the inorganic pesticide load on soil and ground water which subsequently ensure sustainable agriculture.

ACKNOWLEDGEMENTS

The authors thankfully acknowledge the financial support received from The Department of Bio-technology, Government of India for this research work. Thanks are due to the authority of Vidyasagar University for laboratory and field facilities. The authors are thankful to the Director, CSR&TI, Behrampore, West Bengal for providing library facilities.

COMPETING INTERESTS

Authors have declared that no competing interests exist.

REFERENCES

1. Chandraju S, Nagendraswamy Girija, Chidan Kumar CS, Nagendra Swamy R. Studies on the yields of cocoons of CSR19, Kolar gold and CSR2 x CSR4 reared fed with M5 variety of mulberry leaves cultivated by spent wash irrigation. International Journal of Pharmaceutical and Biological Research. IJPBR, ISSN. 2012;3(2).
2. Sharma DD, Govindaiah, Das PK, Philip Tommy, Choudhury PC, Datta RK. Influence of bacterial bio-fertilizers under graded levels of nitrogen on the incidence of major mulberry diseases. Indian Journal of Sericulture. 1994;33:31-35.
3. Chatterjee Ranjit, Choudhuri Partha, Laskar Nripendra. Influence of nutrient management practices for minimizing whitefly (*Bemisia tabaci* Genn.) Population in tomato (*Lycopersicon esculentum* Mill. International Journal of Science, Environment ISSN 2278-3687 (O) and Technology. 2013;2(5):956–962.
4. Lotter DW. Organic Agriculture. J. of sustain. agric. 2003;21:59-128.
5. Lampkin N. Organic farming. Ipswhich, England: Farming Press;1990.
6. Gomaa AM. Response of certain vegetable crops to biofertilization. Ph.D. Thesis, Fac. Agric., Cairo Univ; 1995. 1996;35:35–38.
7. Atiyeh RM, Lee SS, Edwards CA, Arancon NQ, Metzger J. The Influence of Humic Acid Derived from Earthworm-Processed Organic Waste on Plant Growth, Biores. Tech. 2002;84:7-14.
8. Chanda GK, Goutam B, Chakraborty SK. The effect of vermicompost and other fertilizers on cultivation of Tomato plants. J Hortic For. 2011;3(2):42–45.
9. Krishnamoorthy RV, SN Vajrabhiah. Biological activity of earthworm casts, an assessment of plant growth promoter levels in casts, Proceedings of the Indian Academy of Science, Animal Science. 1986;95:341–351.
10. Mali MB, Musmade AM, Kulkarni SS, Prabu T, Dirade RM. Effect of organic manure on yield and nutrient uptake of cucumber cv. Himangi, South Indian Horticulture. 2005; 53(1/6):110-115.
11. Kundaswamy D, Radha NV, Oblisami G. Response of different mulberry varieties to the inoculation of VA-mycorrhizal fungi. . Indian. J. Seric. 1986;25(1):36-39.
12. Das PK, Ghosh A, Choudhury PC, Katiyar RS, Sengupta K. 1990. In:Gangawane, L.V. (Ed.), Response of irrigated mulberry to *Azotobacter* and Azospirillum bio-fertilizers under graded levels of nitrogen. Indian Bio-fertilizer Technology Transfer. Associated Publication Co, New Delhi.1990;71–77.
13. Das PK, Gouda H, Katiyar RS, Madav Rao AR, Mathur VB, Mazumdar MK. Studies on the effect of bacterial biofertlizers in irrigated mulberry (*Morus alba* L). Indian J. Sericult. 1994;33:170–173.
14. Das PK, Gouda H, Katiyar RS, Ghosh A, Chaudhury PC. Response of different mulberry cultivars to *Azatobacter* bio-fertilizers under irrigated conditions. Indian J Sericult. 1996;3535-38.
15. Jackson ML. Soil chemical analysis. Prentice Hall of India Pvt. Ltd., New Delhi; 1973.
16. Sudhakar P, Chattopadhyay GN, Gangwar SK, Ghosh JK. (2000). Effect of *Azotobacter* bio-fertilizer with inorganic nitrogen on leaf yield and quality of

mulberry (*Morus alba* L) Trop. Sci. 2000; 40(2):75 - 82.
17. Katiyar RS, Das PK, Choudhury PC,Ghosh A, Singh GB, Datta RK. Response of irrigated mulberry (*Morus alba* L.) to VA mycorrhiza inoculation under graded doses of Phosphorus. Plant Soil. 1995;170:331-337.
18. Walkley A, Black IA. (1934). An examination of Degtjareff method for determining soil organic matter, and proposed modification of the chromic acid titration method. Soil Science. 1934;37:29-38.
19. Subbiah BV, Asija GL. A rapid procedure for estimation of available nitrogen in soils. Current Science. 1956;25:259-260.
20. Wheeler BEJ. An Introduction to Plant Diseases. John Wiley and Sons Ltd., London, UK. 1969;254.
21. Ravi N, Dhandatani N, Sathiah N, Murugan M. Influence of organic manures and fertilizers on the incidence of sucking pest of sunflower, Annals of Plant Protection Sciences. 2006;14(1):41-44.
22. Ghorbani R, Wilcockson S, Leifert C. Alternative treatments for late blight control in organic potato: Antagonistic micro-organisms and compost extracts for activity against *Phytophthora infestans*,Potato Res. 2006;48:171–179.
23. Bulluck LR, Ristaino JB. Effect of syntheticand organic soil fertility amendments on Southern blight soil microbial communities,and yield of processing tomatoes. Phytopathology. 2002;92:181-189.
24. Prestidge RA, Mc Neill S. Auchenorrhyncha–host plant interaction Leaf hoppers and grasses, Ecological Entomology. 1983;8:331–339.
25. Panda S, Samal MK, Patnaik HP. Effect of oilcake based vermicompost on the incidence of sucking insect pests and fruit yield in chilli, Journal of Applied Zoological Researches. 2005;16(2):184-185.
26. Sullivan P. Organic rice production. 2003; Available:http://www.attra.ncat.org/attrapub/rice.html

4

Effect of Vermiwash Obtained from Different Sources (Neem, Rice Straw and Bagasse) and Standardised Hydroponics Solution on the Growth of *Colocasia esculenta* (Australian Poi) in Guyana

A. A. Ansari[1*], M. Pereira[1] and S. Jaikishun[1]

[1]Department of Biology, University of Guyana, Georgetown, Guyana.

Authors' contributions

This work was carried out in collaboration between all authors. Authors AAA and SJ designed the study, wrote the protocol and wrote the first draft of the manuscript. Author AAA and SJ reviewed the experimental design and all drafts of the manuscript. Author MP managed the analyses of the study. Author MP conducted and monitored the experiments. Author AAA performed the statistical analysis. All authors read and approved the final manuscript.

Editor(s):
(1) Edward Wilczewski, University of Technology and Life Sciences in Bydgoszcz, Poland.
(2) Daniel de la Torre Llorente, Department of Biotechnology-Plant Biology, Technical School of Agronomy Engineers, Universidad Politecnica de Madrid, Spain.
(3) Daniele De Wrachien, State University of Milan, Italy.
Reviewers:
(1) Keshav Singh, Department of Zoology, DDU Gorakhpur University, Gorakhpur-273009, UP, India.
(2) Anonymous, India.

ABSTRACT

The chief economic activity of Guyana is agriculture. In the agricultural industries, the use of fertilizers for crop cultivation has been in common practice and has affected the crop quality over the years. The use of fertilizers accounts for about 75% of crop cultivation and quantity has basically become the major concern rather than quality. The present research carried out during 2012-13, indicated that chemical fertilizers improve quantity but not quality while organic fertilizers maintain both quantity and good quality of crops. The physico-chemical analysis of verminwash showed the necessary elements for plant growth are present in sufficient quantity. T8 (neem + rice straw + cattle dung) followed by T9 (neem+ rice straw+ bagasse+ cattle dung) and H (chemical treatment) were the most effective treatments in hydroponics compared to C (distilled water) and

**Corresponding author: E-mail: abdullah.ansari@uog.edu.gy;*

T1(cattle dung) that were minimally effective. This is substantiated by 2-factor ANOVA, LSD and composite index. Plants grown using different vermiwash in comparison to hydroponically grown showed less moisture, better shoot and root growth, number of leaves, nodes and resistance to insect damage.

Keywords: Hydroponics; bagasse; vermiwash; earthworms; Colocasia esculenta.

1. INTRODUCTION

The revolution of the agricultural industry has become noticeable over the past twenty years. However seventy five percent (75%) of the agricultural industries incorporate fertilizers as a necessity for production. Fertilizers use have led to the success in the production of quantity rather than quality and also distorted the organic and inorganic components of our environment. In today's society, we are faced with the challenge of effective waste management and as such we can utilize vermin technology [1] to actively manage our organic waste disposal crises and simultaneously utilize the products of vermin technology as a nutrient source for modern agricultural technique.

Vermiculture is the culture of earthworms and vermicasting is the faecal matter released by the earthworms [2,3]. Chemical fertilizers are synthetic (chemical) sources of nutrients {nitrogen (N), phosphorous (P), and potassium (K)} that help plants grow faster and healthier. Fertilizers must be formulated so that nutrients are easily absorbed by crops [4, 5]. Vermicomposting is the biological degradation and stabilization of organic waste by earthworms and microorganisms to produce vermicompost [4]. The earthworms fragment the organic waste substrates, stimulate microbial activity greatly and increase rates of mineralization. These rapidly convert the waste into humus-like substances with finer structure than thermophilic composts but possessing a greater and more diverse microbial activity [6,7]. Vermiwash is a liquid fertilizer which is collected after it passes through a layer of soil, with earthworms in it [2,3]. Vermiwash is composed of secretions of earthworms and micronutrients from organic breakdown as nutrients can be readily absorbed. Vermiwash, if collected properly, is a clear and transparent, pale yellow coloured fluid [2,3]. The regular inputs of feed materials for earthworms can be in the form of agro waste, kitchen waste and nitrogen rich materials such as cattle dung, goat dung and pig manure [8].

Hydroponics is a useful technology in locations where the land areas are minimal or there are affected by excessive flooding. The process requires the use of hydroponics solution which is composition of plant nutrients in appropriate quantity is present. This research work was conducted to compare the growth of Australian poi with the use of hydroponics solution and vermiwash produced from different sources (neem, rice straw and bagasse).

2. MATERIALS AND METHODS

Vermitech units were set up at the University of Guyana (2011-2012) at the Faculty of Natural Sciences. Chemical analysis of vermiwash and other laboratory work was done at the Guyana Sugar Corporation Central Laboratory. Plants were grown at the University Of Guyana at the Faculty of Natural Sciences.

2.1 Vermiwash Units

The vermiwash unit was set up using nine buckets (size: 115 mm x 115 mm x 100 mm). A tap was fixed on the lower side of each bucket. About 25 cm of broken pebbles were placed at the bottom of each bucket. A 25 cm layer of coarse sand was placed after the pebbles. Water was allowed to flow through these layers to enable the settling of the basic filter unit. A 20-30 cm layer of loamy soil was placed on top of the filter bed. Earthworms were introduced into the soil for seven buckets. Cattle dung and selected organic materials in specified quantities were placed on top of the soil for the buckets with earthworms. One bucket contained cattle dung without earth worms and one with a combination of all organic materials without earthworms. Units were watered and drained at approximately every two days interval so as to keep moist. Units were monitored and additional specified quantity of organic materials was added when needed. After about six weeks, vermiwash from different treatments was harvested at needed intervals and used to grow *Colocasia esculenta,* (Australian poi). Table 1 shows the different treatments used for the experiment.

Table 1. Treatments and abbreviations

Treatment	Symbol
Distilled water	C
Cattle dung only (300 g) without earthworms	T1
Neem (25 g)+ Rice straw (25 g) + Bagasse (25 g) + cattle dung (300 g) without earthworms	T2
Rice straw (75 g) + cattle dung (300 g) + earthworms (100 in number)	T3
Bagasse (75 g) + cattle dung (300 g) + earthworms (100 in number)	T4
Neem (75 g) + cattle dung (300 g) + earthworms (100 in number)	T5
Neem (37.5 g) + Bagasse (37.5 g) + cattle dung (300 g) + earthworms (100 in number)	T6
Neem (37.5 g) + Rice straw (37.5 g) + cattle dung (300 g) + earthworms (100 in number)	T7
Rice straw (37.5 g) + Bagasse (37.5 g) + cattle dung (300 g) + earthworms (100 in number)	T8
Neem (25 g)+ Rice straw (25 g) + Bagasse (25 g)+ cattle dung (300 g)+ earthworms (100 in number)	T9
Standardize hydroponic solution	H

For each treatment collected, 500 ml was diluted with distilled water to five litres and used to cultivate poi *Colocasia esculenta* (Australian poi) seedlings at three weeks old in a tray. Each tray was drained manually at two weeks interval by carefully removing the plants and new treatment of standard measurement was replaced. Initial and final solutions from all trays were stored for analysis. Observations were recorded at weekly where growth parameters (shoot length, number of leaves and number of nodes) were recorded. Three seedlings were placed in a tray and supported by Styrofoam around the stems had sponge which allowed expansion during growth. Styrofoam was used to keep the upper part of the seedling suspended in the treatments while the roots were immersed. Three seedlings were placed in each tray and in duplicate giving rise to twenty two trays for the nine treatments along with two trays containing distilled water only and two more containing hydroponic solution. Final growth parameters (root length, fresh and dry biomass) of the plants were recorded after six weeks. The Random Block Design (R.B.D) was used for precision. Physico-chemical analysis of vermiwash (pH, Electrical conductivity, organic carbon, nitrogen, phosphorous and potassium) was carried out using standard methods [9].

2.2 Data Analysis

Data recorded were subjected to statistical tools like Student's t-test, ANOVA and LSD and results were interpreted in the light of statistical justifications.

3. RESULTS AND DISCUSSION

3.1 Shoot Length

The results obtained showed that all the plants grown on vermiwash had significant shoot growth (Table 2). The treatment with all combinations (neem, bagasse and rice straw) showed maximum growth even above the commercial hydroponic treatment H. Other vermiwash treatment with two combinations and with only one material had fairly constant shoot growth although they did not exceeded treatments T9 and H respectively. The works of earthworms was proven very effective since the treatments without earthworms (T1 and T2) had poor growth of plants shoot length compared to the shoot of plants grown on vermiwash. The control treatment showed the minimum growth since nutrients were not available in sufficient quantity for the plants. The results are supported by ANOVA indicating significant impact of different treatments on shoot growth [Fcal ($4.17 \geq 2.03$) for treatments and Fcal ($29.73 \geq 2.40$) for time duration]. LSD analysis suggests that progression of growth is not significant in Control whereas all the other treatments show significant progression in terms of shoot growth. T-test analysis suggest that the growth in Control is not significant compared to all the other treatments. Studies show that earthworms' excreta provide plant growth hormones which promote growth and works as organic pesticides [10]. The casts of earthworm is one of the most useful and active agents in introducing changes by directly increasing the fertility and crop producing power in the soil [11].

3.2 No. of Leaves

The combination treatment of vermiwash with all materials (T9) had maximum results for number of leaves (Table 3). Although treatment T9 exceeded the others, there was fairly constant growth for the other vermiwash treatment when compared to the commercial hydroponic treatment (H) which was used. The leaf quality for the treatments without earthworms was poor which again placed emphasis on the effectiveness of the vermiwash.

The results are supported by ANOVA indicating significant impact of different treatments on number of leaves [Fcal (8.89≥2.03) for treatments and Fcal (126.63≥2.40) for time duration]. LSD analysis suggests that progression in number of leaves is not significant in Control whereas all the other treatments show significant progression in terms of number of leaves. T-test analysis suggest that the increase in number of leaves in Control is not significant compared to all the other treatments. Plants grown on treatments T1, T2, H and C showed lots of leaf damage by leaf miners (*Lepidoptera*) which fed on the leaf tissues. Both earthworms and its body liquid (vermiwash) are scientifically proving as both 'growth promoters and protectors' for crop plants [12].

3.3 No. of Nodes

Similarly like the number of leaves and growth of shoot length, the number of nodes showed significant increase for the vermiwash treatment T9 (Table 4). The hydroponic treatment (H) showed a significant increase in the number of nodes also while the other vermiwash treatments (T6, T8) showed consistent number of nodes. Because of the readily available nutrients present in the vermiwash, node increase was successful throughout the course of the experiment while the plants grown without the influence of earthworms showed poor results.

The results are supported by ANOVA indicating significant impact of different treatments on number of nodes [Fcal (9.56≥2.03) for treatments and Fcal (106.81≥2.40) for time duration]. LSD analysis suggests that progression in number of nodes is not significant in Control whereas all the other treatments show significant progression in terms of number of nodes. T-test analysis suggest that increase in number of nodes in Control is not significant compared to all the other treatments. Earthworms restore and improve vermiwash fertility and significantly boost crop productivity [12]. Throughout the experiment for the parameters that are being influenced by treatments of vermiwash of both combination and single treatment it can be highlighted that vermiwash are also enriched in certain metabolites and vitamins that belong to the B group or provitamin D which also help to enhance plant growth [13,14].

3.4 Root Length

Plants grown on treatment T6 had the highest mean percentage increase while plants grown on treatments T8 followed by plants grown on T9, H and C had a significant increase in root growth also from initial to final where the control treatment C was not minimal for this parameter (Table 5). All treatments used in the experiment had a positive impact on root growth throughout the course of the experiment. There was significant root growth for all treatments except C. Even the plants that had poor leaves shoot and node growth showed significant root growth and this point out a major flow of nutrients going towards root growth. The increase in the growth of treated plants is because of the uptake of macro and micronutrients available [15]. T-test analysis suggest that increase in root length in Control is not significant compared to all the other treatments.

3.5 Chlorophyll Content

The reason which accounts for the variation in chlorophyll content random set up of plants using diffused sunlight and not direct sunlight. Light is critical for chlorophyll synthesis and because of the arrangement of the plants chlorophyll varied. However, the plants grown with vermiwash (T4) showed maximum chlorophyll content since nutrients available in vermiwash, contributed to healthier plants (Table 6). Studies revealed that 10% vermiwash has potential to be used in hydroponic culture and also exhibits superiority in growth as well as in content of photosynthetic pigments [16].

Table 2. Shoot length over six the week period (mean±SD) (cm)

Treatment	Weeks							
	1	2	3	4	5	6	Increase	t- test
T1	7.58±0.46^{a}	7.98±0.55^{b}	8.52±0.37^{c}	8.88±0.19^{d}	9.3±0.19^{e}	9.70±0.20^{f}	2.12	**S**
T2	7.67±0.41^{a}	8.06±0.32^{b}	8.28±0.28^{b}	8.74±0.21^{c}	9.06±0.11^{d}	9.46±0.11^{e}	1.79	**S**
T3	7.38±0.39^{a}	7.76±0.48^{b}	8.1±0.59^{c}	8.94±0.36^{d}	9.74±0.31^{e}	10.34±0.27^{f}	2.96	**S**
T4	8.05±0.25^{a}	8.47±0.14^{b}	8.71±0.12^{c}	9.20±0.17^{d}	10.0±0.20^{e}	10.88±0.30^{f}	2.83	**S**
T5	7.67±0.35^{a}	7.97±0.48^{a}	8.21±0.66^{a}	9.17±0.22^{b}	10.05±0.13^{c}	10.8±0.36^{d}	3.13	**S**
T6	7.42±0.52^{a}	7.8±0.59^{b}	8.25±0.72^{c}	9.14±0.36^{d}	10.1±0.29^{e}	10.98±0.19^{f}	3.56	**S**
T7	7.50±0.33^{a}	8.03±0.29^{b}	8.53±0.37^{c}	9.28±0.29^{d}	10.5±0.69^{e}	11.75±0.58^{e}	4.25	**S**
T8	7.33±0.53^{a}	7.6±0.62^{a}	8.12±0.84^{b}	9.55±0.54^{c}	9.97±0.29^{c}	11.07±0.46^{c}	3.74	**S**
T9	7.75±0.26^{a}	8.33±0.29^{b}	9.01±0.24^{c}	10.13±0.48^{d}	12.65±0.53^{e}	15.87±0.95^{f}	8.12	**S**
C	7.43±0.27^{a}	7.53±0.29^{a}	7.58±0.35^{a}	8.1±0.28^{b}	8.35±0.21^{b}	8.65±0.21^{b}	1.22	**NS**
H	7.53±0.38^{a}	7.85±0.54^{b}	8.26±0.84^{c}	9.45±0.24^{d}	12.82±0.99^{e}	14.75±0.48^{f}	7.22	**S**

Confidence level 95%: S {Significant), NS {Not Significant}

Table 3. No. of leaves over the six week period (Mean±SD)

Treatment	Weeks						Increase	t - test
	1	2	3	4	5	6		
T1	4.33±0.82^{a}	5.5±1.38^{b}	7.4±0.55^{c}	7.80±0.83^{c}	9.40±0.55^{d}	10.2±0.45^{e}	5.87	**S**
T2	4.33±0.82^{a}	5.3±0.52^{b}	6.33±1.03^{c}	7.00±0.70^{c}	8.80±0.44^{d}	9.20±0.45^{e}	4.87	**S**
T3	4.83±0.75^{a}	5.67±1.03^{b}	6.83±1.60^{c}	8.8±0.84^{d}	10.6±0.89^{e}	13.00±1.00^{f}	8.17	**S**
T4	5.00±0.00^{a}	5.83±0.75^{b}	7.33±1.03^{c}	8.8±1.30^{d}	11.67±1.21^{e}	13.17±1.47^{f}	8.17	**S**
T5	4.5±0.55^{a}	5.33±1.03^{b}	6.67±1.86^{c}	10.25±0.5^{d}	11.25±0.95^{e}	13.5±0.58^{f}	9.00	**S**
T6	4.67±0.82^{a}	5.83±0.98^{b}	7.33±1.75^{c}	9.00±1.87^{d}	11.4±1.52^{e}	13.6±1.67^{f}	8.93	**S**
T7	4.83±0.75^{a}	6.00±0.89^{b}	7.33±1.37^{c}	10.0±0.71^{d}	11.3±1.21^{e}	13.5±1.05^{f}	8.67	**S**
T8	4.83±0.75^{a}	5.5±1.22^{b}	6.8±1.92^{c}	9.5±0.58^{d}	11.75±0.5^{e}	13.75±0.96^{f}	8.92	**S**
T9	4.33±0.52^{a}	6.67±0.82^{b}	8.5±1.05^{c}	10.8±0.84^{d}	14.33±1.21^{e}	16.33±1.21^{f}	12.00	**S**
C	4.5±0.54^{a}	4.67±0.82^{a}	5.00±0.89^{a}	6.50±0.71^{b}	8.00±0.00^{c}	8.50±0.71^{c}	4.00	**NS**
H	5.3±0.82^{a}	6.5±1.22^{b}	7.00±1.79^{c}	10.0±0.82^{d}	12.75±0.5^{e}	14.25±1.5^{f}	8.95	**S**

Confidence level 95%: S {Significant), NS {Not Significant}

Table 4. No. of nodes over the six weeks period (Mean±SD)

Treatment	Weeks						Increase	t-test
	1	2	3	4	5	6		
T1	4.67±1.03^{a}	6.83±2.04^{b}	7.6±0.89^{b}	9.00±0.70^{c}	10.2±0.55^{d}	10.2±0.44^{e}	5.53	**S**
T2	4.67±1.03^{a}	6.33±1.03^{b}	6.5±0.84^{b}	8.8±0.45^{c}	9.00±0.00^{c}	9.6±1.34^{c}	4.93	**S**
T3	5.5±0.84^{a}	7.33±1.21^{b}	7.67±1.03^{b}	10.2±0.84^{c}	11.8±1.48^{d}	13.00±1.00^{e}	7.50	**S**
T4	5.67±0.82^{a}	7.33±1.02^{b}	7.83±1.17^{b}	10±1.41^{c}	11.67±1.50^{d}	13.16±1.47^{e}	7.49	**S**
T5	5.00±0.00^{a}	6.67±1.86^{b}	7.17±2.32^{b}	10.5±0.58^{c}	12.25±0.5^{d}	14.25±1.26^{e}	9.25	**S**
T6	5.33±0.52^{a}	7.17±1.47^{b}	7.67±1.63^{b}	10.2±0.84^{c}	11.6±1.52^{d}	14.4±1.95^{e}	9.07	**S**
T7	5.5±0.84^{a}	7.00±1.09^{b}	8.17±1.17^{c}	10.3±1.51^{d}	11.83±0.98^{e}	13.83±1.33^{f}	8.33	**S**
T8	5.17±0.75^{a}	6.50±1.87^{b}	7.8±1.64^{c}	10.75±0.96^{d}	12.5±0.58^{e}	14.75±1.5^{f}	9.58	**S**
T9	5.67±0.52^{a}	8.00±1.26^{b}	9.33±0.82^{c}	12±1.41^{d}	15.17±1.47^{e}	17.83±2.56^{f}	12.16	**S**
C	4.83±0.75^{a}	5.17±0.75^{a}	5.33±1.03^{a}	8.00±0.00^{b}	9.00±1.41^{c}	9.00±1.41^{c}	4.17	**NS**
H	5.83±0.91^{a}	7.00±1.79^{b}	7.67±2.16^{c}	11.25±0.5^{d}	14.5±1.29^{e}	16.75±0.96^{f}	10.92	**S**

Confidence level 95%: S {Significant}, NS {Not Significant}

Table 5. Initial and final root length (Mean±SD)

Treatment	Initial (cm)	Final (cm)	Increase	t-test
T1	3.10±0.65	34.06±5.81	30.96	S
T2	2.50±0.00	32.30±8.43	29.80	S
T3	2.80±0.45	55.08±21.5	52.28	S
T4	3.20±0.57	45.18±21.76	41.98	S
T5	2.70±0.27	61.85±16.49	59.15	S
T6	2.50±0.00	100.42±9.85	97.92	S
T7	2.80±0.27	69.48±19.13	66.68	S
T8	2.80±0.27	76.72±17.79	73.92	S
T9	2.90±0.22	58.40±25.58	55.50	S
C	2.90±0.65	45.75±21.28	42.75	NS
H	2.9±0.41	53.37±12.40	50.47	S

Confidence level 95%: S {Significant}, NS {Not Significant}

The results show that the plants grown on treatment H had poor quality due to very high moisture content (Table 7). Presently consumers no longer wish to utilize plants that have been saturated with fertilizers and pesticides and producers of agricultural products are being encouraged towards organic farming [17]. This theorem supports the theme for better quality crops and this experiment proves that most crops grown on vermiwash have maintained quality and quantity unlike the chemical treatment H used that provided quantity rather than a good quality crop.

Table 6. Chlorophyll content (mg/g)

Treatment	Chlorophyll content (mg/g)
T1	4.40
T2	2.70
T3	7.96
T4	9.57
T5	6.22
T6	6.89
T7	7.35
T8	8.16
T9	8.23
C	8.39
H	9.05

From the summary composite index (Table 8) of plant growth parameters, vermiwash showed excellent results because vermiwash improves fertility and boost crop productivity, Sinha et al. [12]. Based on 2- factor ANOVA conducted on the plant growth parameters of *Colocasia esculenta* (Australian Poi) grown in the experiment, there is a significant difference between the different treatments applied on the plant growth parameters investigated ($p \leq 0.05$). According to [2,10] earthworms improves physico-chemical and biological properties. The research revealed that vermiwash had favourable influence on yield parameters where high levels of bio-available nutrients in balanced form for plant can be readily obtained [10].

Table 9 below shows the nutrient status for the initial treatment given to the plants in study (*Colocasia esculenta, Australian poi*). Physico-chemical analysis of the vermiwash (T3, T4, T5, T6, T7, T8, and T9) showed an excellent content of the factors that contribute significantly to plant growth. Earthworms excreta (vermicast) is a nutritive 'organic fertilizer' rich in humus, NKP, micronutrients, beneficial soil microbes, nitrogen-fixing and phosphate solubilizing bacteria and growth hormones auxlns, gibberellins and cytokinins [12].

Table 7. Percentage moisture (Mean±SD)

Treat-ment	Fresh weight (g)	Dry weight (g)	% Moisture
T1	17.42±3.53	8.49±2.66	51.25
T2	13.55±4.10	5.16±2.94	61.94
T3	30.42±7.05	16.14±4.25	46.94
T4	19.79±5.72	9.71±4.21	50.96
T5	28.59±9.09	15.93±5.59	44.28
T6	24.73±9.69	10.952±6.53	55.72
T7	31.20±9.13	17.36±6.37	44.37
T8	24.53±2.69	9.19±1.08	62.56
T9	28.77±6.09	15.93±5.65	44.63
C	26.36±8.52	10.62±4.10	59.71
H	15.41±1.63	5.06±0.39	67.16

Table 8. Composite index based on plant growth parameters

Treatment	Shoot length	Root length	Number of nodes	Number of leaves	Chlorophyll	Moisture %	Total	Rank
T1	9	11	9	9	10	6	54	10
T2	10	6	10	10	11	3	50	9
T3	7	10	7	6	6	8	44	7
T4	8	9	8	8	1	7	41	6
T5	6	4	4	2	9	11	36	5
T6	5	1	5	3	8	5	27	3
T7	3	3	6	5	7	10	34	4
T8	4	2	3	4	5	2	20	1
T9	1	5	1	1	4	9	21	2
C	11	8	11	11	3	4	48	8
H	2	7	2	7	2	1	21	2

Table 9. Physico-chemical analysis of vermiwash obtained from different sources

Treatment	pH	EC(µs/cm)	OC(%)	N(mg/l)	P(mg/l)	K(mg/l)
T1	6.22	1.87	0.20	2.90	2.36	64.61
T2	6.65	1.79	0.03	2.64	1.16	54.54
T3	7.20	7.64	0.02	3.04	0.95	254.90
T4	7.07	7.22	0.02	2.60	2.04	226.60
T5	7.30	7.71	0.03	3.96	0.91	239.20
T6	7.73	5.31	0.02	2.42	3.65	191.50
T7	7.00	5.32	0.04	4.66	0.76	208.20
T8	5.36	7.60	0.02	6.65	1.71	138.40
T9	7.22	6.48	0.03	4.59	0.96	215.30
C	6.77	0.11	0.01	4.79	0.40	3.51
H	6.5	0.98	0.03	11.2	1.08	58.00

In most of the treatments pH values fluctuated around a neutral pH (7) while treatment T8 showed acidity. Electrical conductivity varied where the highest values was observed for vermiwash treatments. Organic Carbon, Phosphorus, Potassium and nitrogen varied for all treatments as the table clearly indicates that treatments of vermiwash had significant level of the physico-chemical properties investigated. The interesting fact about vermiwash is that is contains organic nitrogen that does not accumulate in food products unlike food grown on chemical fertilizers. Using vermiwash may significantly increase nitrogen content due to the presence of nitrogen fixing bacteria, which increase the nitrogen content of the soil [10,13]. The vermiwash becomes alkaline because of high pH and micro-and macronutrients present in the vermiwash which may become essential source of nutrients that are easily available to the plants [12]. This concept sets a strong foundation for hydroponics because of the nutrient rich compounds and trace elements readily available for plants to thrive.

4. CONCLUSION

This research revealed that vermiwash is an effective treatment which can be used in hydroponics to grow plants. During this study, it was found that growth of plants using vermiwash had consistency with the growth parameters investigated. Some of the plants grown on vermiwash exceeded the growth parameters of plants grown on the commercial hydroponic solution and this indicates that organic treatment (vermiwash) maintained both quantity and quality being eco-friendly simultaneously. Also, plants grown on vermiwash exhibited resistance to insect damage unlike plants grown on chemical treatment and treatment without earthworms. Use of vermiwash as feticide in growing plants would eliminate the use of chemical fertilizers that remains as residue in the environment. The physico-chemical properties of vermiwash had significant mineral content to support plant growth where necessary requirements (nitrogen, phosphorus and potassium) were present in significant quantities. Compared to the commercial hydroponic solution, most treatments of vermiwash showed a more significant composition of nutrients and elements available. Vermiwash was found to have outstanding properties which is an asset to organic agriculture and sets a strong foundation for hydroponics.

ACKNOWLEDGEMENTS

The authors acknowledge the support and facilities rendered by Department of Biology, University of Guyana and Central Laboratory, Guyana Sugar Corporation Inc. Guyana.

COMPETING INTERESTS

Authors have declared that no competing interests exist.

REFERENCES

1. Ansari AA. Vermitech: An innovation in organic solid waste management. Journal of Sustainable Development and Environmental Protection. 2011;1(1):107-113.
2. Ismail SA. Vermicology: The biology of earthworms. Orient Longman Ltd., Chennai, India; 1997.
3. Ismail SA. The Earthworm Book. Other India Press: Goa; 2005.

4. Edwards CA. Breakdown of animal, vegetable and industrial organic waste by earthworms. Agric. Ecosyst. Environ. 1998; 24:21-31.
5. Asha A, Tripathi AK, Soni P. Vermicomposting: A better option for organic solid waste management ecology and environment division, Forest Research Institute (FRI), P.O. New Forest, Dehradun 248 006, Uttaranchal, India; 2008.
6. Elvira C, Samperdo L, Mato S. Nogales R, Bioconversion of solid paper-pulp mill sludge earthworms. Bioresource Technology. 1996;57(2):173-177.
7. Atiyeh RM, Dominguez J, Subler S. Metzger JD. Earthworm-processed organic waste as components of horticultural potting media for growing marigold and vegetable seedlings. Compost Science and Utilization; 2000.
8. Ismail SA. Vermitech: institute of research in soil biology and biotechnology Chennai, India; 2000.
9. Homer F. Soil analysis manuel. Central Analytical and Environmental Monitoring Services, Agriculture Research Department. LBI, Guyana; 2003.
10. Ansari AA. Effect of Vermicompost and Vermiwash on the Productivity of Spinach (*Spinacia oleracea*), Onion (*Allium cepa*) and Potato (*Solanum tuberosum*). World Journal of Agricultural Sciences. 2008; 4(5), 554-557.
11. Nath G, Singh K. Singh DK. Chemical Analysis of Vermicomposts / Vermiwash of Different Combinations of Animal, Agro and Kitchen Wastes. Australian Journal of Basic and Applied Sciences. 2009;3(4): 3672-3676.
12. Sinha RK, Agarwal S, Chauhan K. Valani D. The wonders of earthworms and its vermicompost in farm production: Charles Darwin's 'friends of farmers', with potential to replace destructive chemical fertilizers. Agricultural Sciences. 2010;76-94.
13. Lalitha R, Fathima K, Ismail SA. The impact of biopesticide and microbial fertilizers on productivity and growth of *Abelmoschus esculentus*. Vasundara the Earth. 2000;1&2:4–9.
14. Ansari AA, Jaikishun S. Vermicomposting of sugarcane bagasse and rice straw and its impact on the cultivation of *Phaseolus vulgaris* L. in Guyana, South America. Journal of Agricultural Technology. 2011;7(2):225-234.
15. Hatti SS, Londonkar RL, Patil SB, Gangawane AK, Patil CS. 2010. Effect of *Eisenia fetida* vermiwash on the growth of plants. Journal of Crop Science. 2010;1:6-10.
16. Shlrene Q, Embrandiri A, Parveen F, Singh R. Effect of Vermiwash and Vermicomposting Leachate in Hydroponics Culture of Indian Borage (*Plectranthus ambionicus*) Plantlets. Proceedings of 2nd International Conference on Environment and Bio Science. IACSIT Press, Singapore. 2012;44:43-47.
17. Budhan R. Diversifying agriculture: Guyana's biodiversity for food security. Government Information Agency, Georgetown Guyana; 2004.

5

Evaluation of Tillage Practices on Weed Control and Yield of Groundnut (*Arachis hypogaea* L.) in Sudan Savanna Agro-ecology of Nigeria

A. Lado[1*], M. A. Hussaini[1] and S. G. Mohammed[1]

[1]*Department of Agronomy, Bayero University, Kano, Nigeria.*

Authors' contributions

This work was carried out in collaboration between all authors. All authors read and approved the final manuscript.

Editor(s):
(1) Masayuki Fujita. Department of Plant Sciences, Faculty of Agriculture, Kagawa University, Japan.
Reviewers:
(1) Jamilu Garba, Department of Agricultural Education, Zamfara State College of Education Maru-Zamfara State, Nigeria.
(2) Rusu Teodor, University of Agricultural Sciences, Department of Technical and Soil Sciences and Veterinary Medicine Cluj-Napoca, Romania.
(3) Tarirai Muoni, University of Zimbabwe, Zimbabwe.

ABSTRACT

Three years field studies were conducted at teaching and research farm of faculty of Agriculture, Bayero University Kano to evaluate the effect of various tillage practices on weed control and yield of groundnut. The study consisted of nine (9) treatments comprising of two (2) zero tilled, three (3) minimum and four (4) conventional tillage practices arranged in randomized complete block design with three (3) replications. Identification of weeds was made to species level and data were collected on weed count, cover score and weed dry weight. At maturity, groundnut stand and kernel yield were assessed. Data were analyzed using SAS (version 9.3). The result showed a higher weed infestation in T1, T2, T3, T4 and T6 (control) than in T9, T8 and T7 tillage practices from all the three years of study. Perennial weeds species were more frequent in T1 and T2 tilled plots than in T9 and T8 tilled plots. T9 had the least cover score while T6 (control) had the highest weed cover score in all the trials. In all the trials, T1 consistently recorded the highest weed dry matter while T9 recorded the lowest. The highest stands of 35333 ha^{-1} and 37852 ha^{-1} were observed at T9 in 2008 and 2010 trials respectively while control (T6) was among the least. The highest and lowest stands of 41556 and 17778 ha^{-1} were observed at T8 and T5 respectively in 2009 trial. Generally T9 had the highest kernel yield of 1067.7 kg ha^{-1} in 2008 and 1151 kg ha^{-1} in 2010 season. Tillage

**Corresponding author: E-mail: abdulladodoka@yahoo.com;*

practices modified weed floral composition of the study area. T9 and T8 reduced the level of weed occurrences and resulted in better weed control with higher kernel yield than T1, T2, T3, T4, and T6 tillage practices. T8 and T9 tillage practices are recommended for weed control in the study area.

Keywords: Tillage; weed control; groundnut; yield; Sudan savanna.

1. INTRODUCTION

Groundnut (*Arachis hypogaea*) is an important tropical legume and oil crop produced for both domestic and industrial purposes. It used to be one of the most important export crops in Nigeria. It is the sixth most important oil seed crop in the world. It contains 48-50% oil and 26-28% protein, and is rich on dietary fiber, mineral and vitamins [1]. Tillage and weed control are among the most important practices in groundnut production [2,3]. Weed control and tillage practices in crop production are inseparable [4]. When the land is properly tilled, weeds are effectively controlled and this leads to improved crop yield. Soil tillage, which dated back to the beginning of agriculture, is aimed at optimizing soil condition for seed germination, seedling emergence and growth; as well as weed control [3]. Weed shifts are known to occur in continuously cultivated land in response to tillage practices, cropping systems, weed control and other changes in the habitat [2].

Agriculture still relies heavily on mechanical forms of soil disturbances from plowing, harrowing and on use of herbicides for weed control both prior and after planting [2]. These tools reduce weed density and result in favorable yield response. Soil tillage was reported to have determined the absolute and relative abundance of weeds with different traits in a given cropping system [5]. Tillage practices control weeds by breaking, cutting or tearing them from the soil thus exposing the vegetative parts to desiccation and smothering them with soils. Tillage assists greatly in breaking seed dormancy and stimulating seed germination through mechanical scarification [2]. In soils with uniform seed density, cultivation generally stimulates weed seedling emergence causing a faster decline of the soil seed reserves compared to treatment without cultivation. Where weeds are permitted to re-seed, population of annual grasses increased faster in zero-tillage than in conventional tillage systems [6]. Thus, tillage practices strongly influence number and diversity of seeds in soil weed seed bank, depending predominantly on period of induced seed dormancy [2]. Conventional tillage operations primarily invert the soil burying viable weed seeds in topsoil thereby lowering their germination [2].

The production of Groundnut in the study area involves insufficient tillage which leads to high weed infestation. Yield looses in groundnut due to weed infestation can be as high as 77% [6]. Apart from competition for moisture, nutrient and sunlight, weeds may inhibit pegging in groundnut and interfere with crop harvest. Weed competition in groundnut is critical up to 45 days after sowing and weed free environment during this period may likely give higher yield [6]. Manual hoe weeding and row cultivation after emergence but before pegging are effective in controlling weeds establishment but do not provide complete weed free environment necessary for high yield. Manual hoe cultivation during pegging destroys roots and pegs. Thus, once pegging begins, soil disturbance should be avoided or kept to a minimum so as not to interfere with the developing pods. It is important therefore to determine the best methods of tilling the soils that will effectively control weeds and produce high groundnut yield. Very little was done in the study area on the effect of tillage practices on weed control and yield of groundnut. This study was therefore initiated to investigate the effect of different tillage practices on weed control and yield of groundnut.

2. MATERIALS AND METHODS

The trials were conducted at the Teaching and Research Farm of Faculty of Agriculture Bayero University, Kano in 2008, 2009 and 2010 rainy seasons. Kano (11°58N, 8°25E and 475m above sea level) is located within the Sudan savanna zone of northern Nigeria. It has a monomodal rainfall. The total annual rainfall was 990, 699.06 and 731.1mm in 2008, 2009 and 2010 seasons respectively. The monthly mean temperature range from 14.6°C in January to 39.5°C in March 2008; 15.87°C in January to 41.7°C in April, 2009; 13.96°C in January to 46.7°C in April, 2010. The textural class of the experimental soils is loamy sand. Total Nitrogen and organic carbon were 0.041 and 0.759% with cation exchange capacity of 7.33 cmol kg^{-1}. The experiment involved nine (9) tillage treatments as shown in Table 1 below:

T1 and T2 represent zero tillage where herbicide was applied as pre-plant application, followed by post-emergence application of Fluazifop. During sowing only small cut necessary to bury the seed into the soil was made. T3, T4 and T5 represent minimum tillage which involved incomplete loosening of soil layer with hoe, plow or harrow. It was followed by pre-planting and post-emergence application of herbicide. T6, T7, T8 and T9 represent conventional tillage where the soil was loosened and turned with hoe, plow, harrow and ridger. It was followed by supplementary hoe weeding. The treatments were laid out in randomized complete block design replicated three times. The gross and net plots sizes were 22.5 m^2 and 15 m^2, respectively. The gross plots were demarcated and pegged with iron rods for easy identification in subsequent years. The same site for each treatment was maintained for the whole experimental period. Pre-plant herbicide application was achieved by spraying plots with paraquat at 0.8 kg a.i.ha^{-1} one week before sowing; pre-emergence application was done with pendimethalin at 1.0 kg a.i. ha^{-1} immediately after sowing and post-emergence application with fluazifop at 0.4 kg a. i ha^{-1} at 4 weeks after sowing (WAS).

An extra early maturing groundnut variety (SAMNUT 23) was obtained from Institute for Agricultural Research, Ahmadu Bello University Samaru, Zaria. Two seeds were sown 2cm deep at inter-row and intra-row spacing of 0.75 m and 0.25 m, respectively and thinned to 1 plant per hole at 2 WAS. Fertilizer was applied at 2 WAS using a mixture of NPK (15-15-15%) and single super phosphate (18% P_2O_5) to supply 20 kg N ha^{-1}, 40 kg P_2O_5 ha^{-1} and 20 kg K_2O ha^{-1}.

Weed sample was collected from 1M^2 quadrant. Weeds within the quadrant were identified, counted and recorded. Samples within the quadrant were separated by species and the level of infestation determined and cut from the ground level using sickle and oven dried at 75°C to constant weight for determination of dry weight. Weed cover score was recorded at crop's (groundnut) physiological maturity by visual observation on scale of 1-10; where 1 represented uncovered plots while 10 represented plots completely covered by weeds. Relative weed frequency was estimated using the equation of [7]:

Relative weed frequency = (Frequency of individual weed species/Total frequency of all weed species) x 100

Stands count were recorded by counting the number of groundnut stands at physiological maturity from each net plot area and extrapolated to per hectare basis. Kernel yield was obtained by threshing and winnowing pods; and weighing the kernels obtained from the net plot. The values were extrapolated to kg per hectare. Data generated were subjected to statistical analysis of variance using SAS package (version 9.3). Where treatment means were significant, they were separated using Duncan's Multiple Range Test at 5% level of probability. Correlation analysis was carried out among the groundnut and weed attributes as suggested by [8].

Table 1. Experimental treatments

Treatment number	Treatment description
T1	Paraquat at 0.8 kg a. i.ha^{-1} followed by Fluazifop at 0.4 kg a.i. ha^{-1}
T2	Paraquat at 0.8 kg a. i. ha^{-1} followed by Pendimethalin at 1.0 kg a. i. ha^{-1} followed by Fluazifop at 0.4 kg a.i. ha^{-1}
T3	Hoe tilling followed by Paraquat at 0.8 kg a. i. ha^{-1} followed by Fluazifop at 0.4 kg a. i. ha^{-1}
T 4	Ploughing followed by Paraquat at 0.8 kg a.i. ha^{-1} followed by Fluazifop at 0.4 kg a. i ha^{-1}
T5	Harrowing followed by Paraquat at 0.8 kg a. i ha^{-1} followed by Fluazifop at 0.4 kg a. i. ha^{-1}
T 6	Animal Traction followed by supplementary hoe weeding (Control)
T 7	Manual Cultivation followed by supplementary hoe weeding
T8	Harrowing followed by Ridging followed by Pendimethalin at 1.0 kg a. i. ha^{-1} followed supplementary hoe weeding
T 9	Ploughing followed Harrowing followed by Ridging followed by Pendimethalin at 1.0 kg a.i. ha^{-1} followed by Supplementary hoe weeding

3. RESULTS

A total of 35 weed species belonging to 29 genera and 22 families were identified in the first year (Table 2). Result shows that, T8, T2 and T4 had the highest relative frequency of 12.4% each followed by T7 and the control (T6) while T9 had the least. From the second year trial, a total of 32 weed species belonging to 26 genera and 19 families were identified (Table 3). Result shows that, T2 had the highest (16.8%) relative frequency followed by T1 (15%) and the control (T6) while T9 had the least (6.4%). A total of 20 weed species belonging to 17 genera and 10 families were identified from the third year trial (Table 4). Result shows that, T1, T2, T3, T4 and T6 (control) had the highest relative frequency of 12.3% each followed by T7 (11.1%), T5 and T8 (9.8% each) while T9 had only 7.4%. Perennial weed species comprising of both grasses, sedges and broadleaf were the most predominant species associated with most of the treatments in all the years. The frequency of occurrences and level of infestations of individual species were generally higher in zero tillage (T1 and T2), T3, T4, and T6 (control) than in T8 and T9 treatments in all the years (Tables 2-4).

Table 5 shows the effect of tillage practices on number of weeds per m^2, weed cover scores and weed dry weight during the experimental period. It was observed that T9 had the lowest number of weeds than other tillage practices in all the trials. The results showed that T9 had the least cover score while T6 (control) had the highest weed cover score in all the trials. Influence of tillage practices on weed dry weight is presented in Table 6. The results revealed a significant effect ($P \leq 0.05$) of tillage practices on weed dry weight. In all the trials, T1 consistently recorded the highest weed dry matter while T9 recorded the lowest. Significant ($P \leq 0.05$) treatment effects were also observed on number of plant stands at harvest and kernel yield (Table 6). The highest stands of 35333 ha^{-1} and 37852 ha^{-1} were observed at T9 in 2008 and 2010 trials respectively while control had the least. The highest and lowest stands of 41556 and 17778 ha^{-1} were observed at T8 and T5 respectively in 2009 trial. Generally, T9 had the highest kernel yield of 1067.7 kg ha^{-1} in 2008, 1188.3 kg ha^{-1} in 2009 and 1151 kg ha^{-1} in 2010 season. Yield of T6 (control), T1, T2, T3, T4 and T5 were lower than T8 and T9 in all the trials. The correlation analysis revealed that kernel yield was significantly and positively correlated with number of plant stands ha^{-1} but negatively correlated with weed density, cover score and dry weight (Table 7).

4. DISCUSSION

Three (3) years of consistent tillage practices on the same piece of land led to modification of weed floral composition. Similar observations were reported by [2] and [9]. General decrease in number and frequency of weed occurrence were observed in all the treatments across the years. The decrease was more pronounced among conventional than minimum or zero tillage. This is because conventional tillage resulted in frequent cutting of weed seedlings and burial of weed seeds deep into the soil layer. These destroyed the established weeds and prevented viable seeds from geminating. Destruction of established weeds also helped in reduction of seeds build up in the seed bank. Perennial weeds were observed to be more persistent than the other species but their level of infestation were generally lower in conventional (T9, T8 and T7) than in minimum tillage (T4 and T5). Frequent cultivation in T9, T8 and T7 might have depleted the food reserve of perennating organs of these species of weed eventually causing starvation and death of the plant as reported by [10]. It was observed that minimum tillage stimulated the growth of perennial species. [5] reported that in shallow tilled conditions, perennial weeds respond to breakage by producing increased number of shoots and consequently resulted in increased biomass and productivity. Similarly, [11] reported high occurrences of perennial weeds in conventional than zero tilled plots where glyphosate was used. Broadleaf species were more in zero and minimum tillage treatments indicating that the post emergence application of fluazifop was ineffective against these species. The level of infestation and the total relative frequency of T9 were generally lower than other conventional treatments indicating that pre-emergence application of pendimethalin and supplementary hoe weeding was effective in controlling germinating weeds.

Table 2. Influence of tillage practices on weed species compositions and their level of infestation (%) of groundnut field in 2008 Season

	Family	Growth habit	Zero tillage			Minimum tillage			Conventional tillage		
Weed species			T_1	T_2	T_3	T_4	T_5	T_6	T_7	T_8	T_9
Acanthospermum hispidium	*Asteraceae*	ABL	-	-	13	13	-	-	-	13	-
Alysicarpus vaginalis	*Fabaceae*	PBL	-	-	-	-	-	13	-	-	-
Aneilema beninience	*Commelinaceae*	ABL	-	13	-	13	13	13	13	-	13
Cassia mimosoides	*Caesalpiniaceae*	ABL	13	38	-	13	13	13	13		-
Cassia obtusiforlia	*Caesalpiniaceae*	ABL	63	38	13	13	-	13	13	13	-
Chamaecristus absus	*Caesalpiniaceae*	ABL	-	13	13	13	38	13	13	-	-
Chonchorus oliteruis	*Tliaceae*	ABL	-	13	13	13	13	13	13	-	-
Chloris pilosa	*Poeceae*	AG	-	-	-	38	-	38	13	13	38
Cleome viscose	*Cleomaceae*	ABL	-	-	13	-	-	-	-	-	-
Commilina species	*Commelinaceae*	AN	-	-	13	13	-	-	-	-	-
Cyperus difformis	*Cyperaceae*	PS	63	38	38	63	63	13	13	13	13
Cyperus rotundus	*Cyperaceae*	PS	38	-	13	38	-	-	-	-	-
Crotelera retusa	*Mimosoidae*	ABL	-	-	13	-	13	-	-	-	-
Cynodon dactylon	*Poaceae*	PG	38	38	38	38	38	13	13	13	13
Dactylocteniumm aegytium	*Poaceae*	AG	-	13	13	-	-	-	-	-	-
Digitaria deblis	*Poaceae*	AG	38	13	13	-	-	-	-	-	13
Euphorbia hirta	*Euphorbiaceae*	ABL	-	-	-	-	13	-	-	-	-
Heteropogon contortus	*Poceae*	PG	-	38	38	-	63	38	13	63	-
Hibiscus subdariffa	*Malvaceae*	ABL	-	-	-	-	13	-	-	-	-
Hyphis lanceolata	*Lamaceae*	ABL	-	-	38	-	-	-	-	-	-
Imperata cylindrica	*Poaceae*	PG	38	38	13	13	-	38	-	-	-
Indigofera hirtusa	*Fabaceae*	ABL	-	-	-	-	13	-	13	13	-
Ipomea acquatica	*Convovulaceae*	PBL	13	13	-	13	13	13	13	13	13
Ipomea asorifolia	*Convovulaceae*	PBL	-	-	-	-	13	-	-	-	-
Ipomea vagans	*Convovulaceae*	ABL	-	-	13	13	-	-	-	-	-
Kyllinga pumilla	*Cyperaceae*	PS	38	-	-	-	-	-	13	-	-
Kyllinga squamulata	*Cyperaceae*	PS	63	13	63	13	13	38	38	13	38
Leucas martinicensis	*Labiatae*	ABL	13	38	38	38	13	38	13	13	13
Melochia cordifolia	*Sphenocaceae*	ABL	13	13	13	38	38	38	-	13	13
Mitracarpus villosus	*Portulacaceae*	ABL	-	13	13	-	-	-	-	-	-
Monechum ciliatum	*Acanthaceae*	ABL	38	38	13	13	38	38	13	38	13

	Family	Growth habit	Zero tillage		Minimum tillage			Conventional tillage			
Oldenlandia herbacia	*Rubiaceae*	ABL	13	13	-	-	-	-	-	-	13
Pennisetum pidicelatum	*Poaceae*	AG	13	-	-	-	13	-	13	38	-
Stirga gesneriodes	*Scrophurariaceae*	ABL	-	-	13	-	-	-	-	-	-
Tridax procumbens	*Astreraceae*	ABL	13	-	-	-	-	-	**	13	-
RF %			9.6	12.4	11.0	12.4	11.0	11.0	11.7	12.4	8.2

T_1= Paraquat at 0.8 kg a. i.ha-[1] followed by Fluazifop at 0.4 kg a.i. ha-[1] T_2= Paraquat at 0.8 kg a. i. ha-[1] followed by Pendimethalin at 1.0 kg a. i. ha-[1] followed by Fluazifop at 0.4 kg a.i. ha-[1] T_3 = Hoe tilling followed by Paraquat at 0.8 kg a. i. ha-[1] followed by Fluazifop at 0.4 kg a. i. ha-[1] T_4 = Ploughing followed by Paraquat at 0.8 kg a.i. ha-[1] followed by Fluazifop at 0.4 kg a. i ha-[1] T_5 = Harrowing followed by Paraquat at 0.8 kg a. i ha-[1] followed by Fluazifop at 0.4 kg a. i. ha-[1] T_6= Animal Traction followed by supplementary hoe weeding (SHW) T_7= Manual Cultivation followed by SHW, T_8 = Harrowing followed by Ridging followed by Pendimethalin at 1.0 kg a. i. ha-[1] followed by SHW T_9= Ploughing followed by Harrowing followed by Ridging followed by Pendimethalin at 1.0 kg a.i. ha-[1] followed by SHW ABL= Annual broadleaf; AG= Annual grass; PBL= Perennial broadleaf; PG= Perennial grass; PS= Perennial sedges

Table 3. Influence of tillage practices on weed infestation % of groundnut field in 2009 season

Weed species	Family	Growth habit	Zero tillage		Minimum tillage			Conventional tillage			
			T_1	T_2	T_3	T_4	T_5	T_6	T_7	T_8	T_9
Acanthospermum hispidium	*Asteraceae*	ABL	-	-	-	13	-	-	-	-	-
Alysicarpus vaginalis	*Fabaceae*	PBL	13	13	13	-	-	13	-	13	-
Aneilema beninience	*Commelinaceae*	ABL	-	13	-	13	-	13	-	-	-
Cassia mimosoides	*Caesalpiniaceae*	ABL	13	13	-	13	-	13	-	13	-
Cassia obtusiforlia	*Caesalpiniadeae*	ABL	13	13	-	13	13	-	13	-	13
Chamaecristus absus	*Caesalpiniadeae*		-	13	-	13	-	13	-	-	-
Chonchorus oliteruis	*Tiliaceae*	ABL	13	13	-	13	13	13	13	13	-
Chloris pilosa	*Poeceae*	AG	13	-	13	13	-	-	-	-	-
Cleome viscose	*Cleomaceae*	ABL	13	-	13	-	-	-	-	-	-
Commilina species	*Commelinaceae*	AN	-	-	13	-	-	-	-	-	-
Cyperus difformis	*Cyperaceae*	PS	38	13	13	13	13	13	-	13	13
Cyperus rotundus	*Cyperaceae*	PS	13	13	-	13	-	-	13	13	-
Crotelera retusa	*Mimosoidae*	ABL	13	-	-	-		-	-	-	13
Cynodon dactylon	*Poaceae*	PG	38	38	-	38	38	13	13	13	38
Digitaria deblis	*Poaceae*	AG	13	-	-	-	-	38	-	-	13
Euphorbia hirta	*Euphorbiaceae*	ABL	13	-	-	-	-	13	-	-	-
Heteropogon contortus	*Poceae*	PG	13	13	-	-	13	-	-	13	-
Imperata cylindrica	*Poaceae*	PG	13	13	38	38	13	38	-	13	-

Weed species	Family	Growth habit	Zero tillage		Minimum tillage			Conventional tillage			
			T_1	T_2	T_3	T_4	T_5	T_6	T_7	T_8	T_9
Indigofera hirtusa	*Fabaceae*	ABL	13	13	-	-	13	13	13	-	-
Ipomea acquatica	*Convovulaceae*	PBL	13	13	13	13	13	13	-	-	-
Ipomea asorifolia	*Convovulaceae*	PBL	13	-	-	-	-	13	-	-	-
Ipomea vagans	*Convovulaceae*	ABL	-	13	13	-	-	-	-	-	-
Kyllinga pumilla	*Cyperaceae*	PS	-	13	-	-	-	-	13	13	-
Kyllinga squamulata	*Cyperaceae*	PS	13	38	13	13	13	38	38	13	13
Leucas martinicensis	*Labiatae*	ABL	13	38	38	38	13	38	-	13	-
Melochia cordifolia	*Sphenocaceae*	ABL	13	13	13	13	13	-	13	-	13
Mitracarpus villosus	*Portulacaceae*	ABL	-	-	-	-	-	13	13	-	13
Monechum ciliatum	*Acanthaceae*	ABL	13	13	13	-	13	-	-	-	-
Oldenlandia herbacia	*Rubiaceae*	ABL	-	13	13	-	-	-	-	-	-
Pennisetum pidicelatum	*Poaceae*	AG	-	-	-	-	-	13	13	-	-
Tridax procumbens	*Astreraceae*	ABL	13	13	13	-	-	-	-	-	-
RF%			15	16.8	9.6	11.2	9.6	14	8.0	8.8	6.4

T_1= Paraquat at 0.8 kg a. i.ha-[1] followed by Fluazifop at 0.4 kg a.i. ha-[1] T_2= Paraquat at 0.8 kg a. i. ha-[1] followed by Pendimethalin at 1.0 kg a. i. ha-[1] followed by Fluazifop at 0.4 kg a.i. ha-[1] T_3 = Hoe tilling followed by Paraquat at 0.8 kg a. i. ha-[1] followed by Fluazifop at 0.4 kg a. i. ha-[1] T_4 = Ploughing followed by Paraquat at 0.8 kg a.i. ha-[1] followed by Fluazifop at 0.4 kg a. i ha-[1] T_5 = Harrowing followed by Paraquat at 0.8 kg a. i ha-[1] followed by Fluazifop at 0.4 kg a. i. ha-[1] T_6= Animal Traction followed by supplementary hoe weeding (SHW) T_7= Manual Cultivation followed by SHW T_8 = Harrowing followed by Ridging followed by Pendimethalin at 1.0 kg a. i. ha-[1] followed by SHW; T_9= Ploughing followed by Harrowing followed by Ridging followed by Pendimethalin at 1.0 kg a.i. ha-[1] followed by SHW ABL= Annual broadleaf; AG= Annual grass; PBL= Perennial broadleaf; PG= Perennial grass; PS= Perennial sedges

Table 4. Influence of tillage practices on weed infestation of groundnut Field in 2010 season

Weed species	Family	Growth	Zero tillage		Minimum tillage		Conventional tillage				
			T1	T2	T3	T4	T5	T6	T7	T8	T9
Alysicarpus vaginalis	*Fabaceae*	PBL	-	-	-	-	-	-	-	38	-
Cassia mimosoides	*Caesalpiniaceae*	ABL	13	13	13	13	-	-	-	-	13
Cassia obtusiforlia	*Caesalpiniaceae*	ABL	-	13	-	13	13	-	-	-	-
Chonchorus oliteruis	*Tiliaceae*	ABL	-	-	-	-	-	-	13	-	-
Cleome viscose	*Cleomaceae*	ABL	-	-	-	-	-	-	13	-	-
Cyperus difformis	*Cyperaceae*	PS	63	38	13	13	38	13	13	38	13
Cyperus rotundus	*Cyperaceae*	PS	13	13	38	38	13	13	13	-	13
Cynodon dactylon	*Poaceae*	PG	38	13	38	38	13	13	13	13	13
Euphorbia hirta	*Euphorbiaceae*	ABL	13	-	-	-	-	-	-	-	-

Weed species	Family	Growth	Zero tillage		Minimum tillage		Conventional tillage				
			T1	T2	T3	T4	T5	T6	T7	T8	T9
Heteropogon contortus	*Poceae*	PG	38	38	13	-	13	-	-	-	-
Imperata cylindrical	*Poaceae*	PG	-	13	38	13	-	13	-	13	-
Ipomea acquatica	*Convovulaceae*	PBL	-	-	-	-	-	-	-	13	-
Kyllinga pumilla	*Cyperaceae*	PS	13	-	13	-	-	-	13	-	-
Kyllinga squamulata	*Cyperaceae*	PS	38	13	38	38	-	13	13		-
Leucas martinicensis	*Labiatae*	ABL	13	38	13	13	13	13	-	-	-
Mitracarpus villosus	*Portulacaceae*	ABL	-	-	-	-	-	38	13	13	-
Monechum ciliatum	*Acanthaceae*	ABL	13	13	13	13	38	13	-	13	13
Stirga gesnerioides	*Scrophurariaceae*	ABL	-	-	-	13	-	13	13	13	13
Tridax procumbens	*Astreraceae*	ABL	-	-	13	-	-	13	-	-	-
Physalis angulata	*Scrophurariaceae*	ABL	-	-	-	-	13	13	-	-	-
Relative Frequency %			12.3	12.3	12.3	12.3	9.8	12.3	11.1	9.8	7.4

T_1= Paraquat at 0.8 kg a. i.ha-^{1} followed by Fluazifop at 0.4 kg a.i. ha-^{1}; T_2= Paraquat at 0.8 kg a. i. ha-^{1} followed by Pendimethalin at 1.0 kg a. i. ha-^{1} followed by Fluazifop at 0.4 kg a.i. ha-^{1} T_3 = Hoe tilling followed by Paraquat at 0.8 kg a. i. ha-^{1} followed by Fluazifop at 0.4 kg a. i. ha-^{1} T_4 = Ploughing followed by Paraquat at 0.8 kg a.i. ha-^{1} followed by Fluazifop at 0.4 kg a. i ha-^{1} T_5 = Harrowing followed by Paraquat at 0.8 kg a. i ha-^{1} followed by Fluazifop at 0.4 kg a. i. ha-^{1} T_6= Animal Traction followed by supplementary hoe weeding (SHW) T_7= Manual Cultivation followed by SHW T_8 = Harrowing followed by Ridging followed by Pendimethalin at 1.0 kg a. i. ha-^{1} followed by SHW T_9= Ploughing followed by Harrowing followed by Ridging followed by Pendimethalin at 1.0 kg a.i. ha-^{1} followed by SHW ABL= Annual broadleaf; AG= Annual grass; PBL= Perennial broadleaf; PG= Perennial grass; PS= Perennial sedges

Table 5. Effect of tillage practices on weed attributes in 2008, 2009 and 2010 rainy seasons in groundnut field

Treatments	Weed density M^{-2}			*Weed cover scores			Weed dry weight Kg ha^{-1}		
	2008	2009	2010	2008	2009	2010	2008	2009	2010
T_1	108a	74.6a	70.2a	7.5ab	7.3ab	7.7a	6752a	3443a	4665a
T_2	93ab	67a	52.8a	6.66cd	6.5bc	7.1a	2493bc	1920b	2253bcd
T_3	97a	83a	68.1a	6.3bcd	6.6bc	6.7ab	7073a	1540b	3863ab
T_4	91ab	80.6a	65.0a	6.0cde	5.3dc	5.5bc	2323bc	1556b	1969cde
T_5	107.3a	76.6a	69.5a	6.6bc	7.0abc	6.5ab	4587b	1443b	2627bc
T_6	60.6ab	55.6ab	50.5a	8.3a	8.6a	6.7ab	2165bc	176b	1570dce
T_7	71ab	58.0ab	48.0a	4.7e	4.0ed	4.8c	1567bc	1490b	1321dce
T_8	83.7ab	86.6a	60.5a	5.1ed	5.5bcd	5.6bc	1183bc	1113bc	887de
T_9	20.6b	22.6b	19.3b	2.3f	2.8e	2.1d	493c	393c	462e
SE±	17.69	10.93	12.77	0.36	0.46	0.65	677.1	232.9	697.0

*Means followed by same letter within a column are not significantly different using DMRT at 5%. a.i.= active ingredient, Kg= Kilogram, ha= hectare *Weed cover score: 1 – 10 scale: 1 = uncovered space, 10 = completely covered T_1= Paraquat at 0.8 kg a. i.ha-^{1} followed by Fluazifop at 0.4 kg a.i. ha-^{1} T_2= Paraquat at 0.8 kg a. i. ha-^{1} followed by Pendimethalin at 1.0 kg a. i. ha-^{1} followed by Fluazifop at 0.4 kg a.i. ha-^{1} T_3 = Hoe tilling followed by Paraquat at 0.8 kg a. i. ha-^{1} followed by Fluazifop at 0.4 kg a. i. ha-^{1} T_4 = Ploughing followed by Paraquat at 0.8 kg a.i. ha-^{1} followed by Fluazifop at 0.4 kg a. i ha-^{1} T_5 = Harrowing followed by Paraquat at 0.8 kg a. i ha-^{1} followed by Fluazifop at 0.4 kg a. i. ha-^{1} T_6= Animal Traction followed by supplementary hoe weeding (SHW) T_7= Manual Cultivation followed by SHW T_8 = Harrowing followed by Ridging followed by Pendimethalin at 1.0 kg a. i. ha-^{1} followed by SHW T_9= Ploughing followed by Harrowing followed by Ridging followed by Pendimethalin at 1.0 kg a.i. ha-^{1} followed by SHW*

Table 6. Effect of tillage practices on groundnut stands and kernel yield in 2008, 2009 and 2010 rainy seasons

Treatments	Number of stands ha^{-1}			Kernel yield kg ha^{-1}		
	2008	2009	2010	2008	2009	2010
T_1	16667b	21556b	20889f	146.8b	190.4b	160.2b
T_2	20667b	20889b	22296ef	151.3b	191.3b	161.7b
T_3	24222b	29778ab	29333b-e	183.3b	170.9b	168.1b
T_4	17778b	20222b	25704d-e	153.3b	163.7b	179.8b
T_5	20000b	17778b	23111def	171.6b	186.6b	181.6b
T_6	22667b	39333a	30667a-d	181.8b	178.7b	1104.7a
T_7	20000b	40889a	31630abc	932.2b	1231.7a	1161.3a
T_8	25333b	41556a	35259ab	1038.8a	1108.6a	1010.3a
T_9	35333a	39778a	37852a	1067.7a	1188.3a	1151.3a
SE±	2205.2	4488.1	3548.0	16.81	23.33	25.83

Means followed by same letter(s) within a column are not significantly different using DMRT at 5% a.i.= active ingredient, Kg= Kilogram, ha= hectare T_1= Paraquat at 0.8 kg a. i.ha-^{1} followed by Fluazifop at 0.4 kg a.i. ha-^{1} T_2= Paraquat at 0.8 kg a. i. ha-^{1} followed by Pendimethalin at 1.0 kg a. i. ha-^{1} followed by Fluazifop at 0.4 kg a.i. ha-^{1} T_3 = Hoe tilling followed by Paraquat at 0.8 kg a. i. ha-^{1} followed by Fluazifop at 0.4 kg a. i. ha-^{1} T_4 = Ploughing followed by Paraquat at 0.8 kg a.i. ha-^{1} followed by Fluazifop at 0.4 kg a. i ha-^{1} T_5 = Harrowing followed by Paraquat at 0.8 kg a. i ha-^{1} followed by Fluazifop at 0.4 kg a. i. ha-^{1} T_6= Animal Traction followed by supplementary hoe weeding (SHW) T_7= Manual Cultivation followed by SHW T_8 = Harrowing followed by Ridging followed by Pendimethalin at 1.0 kg a. i. ha-^{1} followed by SHW T_9= Ploughing followed by Harrowing followed by Ridging followed by Pendimethalin at 1.0 kg a.i. ha-^{1} followed by SHW

Table 7. Correlation coefficient of kernel groundnut yield and some weed parameters for three years combined

	Kernel yield	No. of stand at harvest	Weed cover score	Weed density	
Kernel yield	-				
No. of Stand at harvest	0.30348*	-			
Weed cover score	-0.55778**	-0.36496*	-		
Weed density	-0.31514*	0.05568ns	0.59834**	-	
Weed dry weight	-0.47454**	-0.21196ns	0.51042**	0.3236*	-

*=significant at 5%, ** = significant at 1% level of probability*

T1 and T9 had consistently produced the highest and lowest number of weeds per m^2, highest and lowest weed cover score, as well as highest and lowest weed dry weight, respectively. These may likely be due to reduction in number of weeds per m^2, weed cover score and biomass in T9 as well as fragmentation of weeds and their exposure to desiccation due plowing and harrowing. Similar observations were reported by [3]. Highest weed cover score and weed biomass of zero tilled plots (T1), suggested higher weed competition. This was confirmed by high level of weed infestation on these plots and frequency of weed occurrences as indicated in Tables 2 - 5. Higher number of stands and kernel yield were recorded in conventionally tilled plots while the lowest was observed in zero and minimum tilled plots. Highest number of stands in T9 could be attributed to good soil pulverization, which provided an ideal environment for germination and seedling emergence. High infestation by rodents that picked the sown seeds in zero tilled plots could explain the reason for low number of stands recorded in these treatments. Reduced competition from weed for the available environmental resources due to effective weed control in the T9 treatment might have favored higher kernel yield. Similar observations were reported by [3] and [12]. Lower kernel yield was recorded in zero tilled plots probably because of higher weed pressure as indicated by higher weed density, weed cover score and weed dry weight. Similar observations were reported by [6].

Significant but negative correlations were observed between kernel yield and most of weed characters studied. These indicated that as weed characters increased the grain yield decreased due to competition between the crop and the weeds for environmental resources. Similar observation was reported by [13] in cowpea. Although T9 was more effective in controlling weeds as evident by less weed dry weight; less weed cover score; and less weed density; it was however the most uneconomical. A simple cost analysis revealed that about \$29.697 ha^{-1} will be required for T1, as compared to \$41.82 ha^{-1} for T2; \$66.06 ha^{-1} for T3; \$78.18 ha^{-1} for T4; \$66.06 ha^{-1} for T5; \$169.697 ha^{-1} for T6; \$175.757 ha^{-1} for T7; \$193.93 ha^{-1} for T8; and \$264.24 ha^{-1} for T9. The implements required for plowing, harrowing and ridging were also costly and not always available in the study area. Human labour required for supplementary hoe weeding was also very expensive and often not always available at the time of critical need. High cost of manual labour for weeding has been reported by [13] in rainfed and irrigated cowpea.

5. CONCLUSION

From the results obtained, it can be concluded that tillage practices had significant relationship with weed occurrence, as well as kernel yield of groundnut. Although conventional tillage (T8 and T9) had persistently produced better and more desirable effect on yield and weed control, yet some salient issues may be raised against it, especially in terms of cost and time needed for its operation. Zero and minimum tillage may still be at par with conventional system, if time and costs of operation were factored into it. Therefore T9 and T8 can be recommended for weed control in the study area.

COMPETING INTERESTS

Authors have declared that no competing interests exist.

REFERENCES

1. Nigam SN, Dwivedi SL, Gibbons RW. Groundnut breeding, constraints, achievement and future possibilities. Plant Breeding Abst. 1991;61:1127-1136.
2. Akobundu IO. Weed Science in the Tropics: Principles and Practice. John Wiley& Sons; 1987.
3. Adamu S, Abdulrazak H. Effect of tillage on the growth and yield of groundnut (*Arachis hypogaea* L.) at Bauchi in the Northern Guinea Savanna zone of Nigeria. International Journal of Food and Agricultural Research. 2004;48-53.
4. Adeyemi OR, Smith MAK, Ojeniyi SO. Effect of land preparation techniques on weed control effectiveness in okra (*Abelmoschus esculentus*). Nigerian Journal of Weed Science. 2008;21:72-83.
5. Sigurd H. Weeds and weed management on arable crops. An ecological approach. CABi publishing; 2003.
6. Rao VS. Principles of weed science 2nd ed. Oxford: IBH Publishing; 2004.
7. Wirjahdja SO, Pancho JV. Weed survey sampling method and vegetation analysis. Biotech. Bull. 1975;4:20.
8. Dewey DR, Lu KK. A correlation and path coefficient analysis of crested seed production. Agronomy Journal. 1959;51:515-518.

9. Rusu T, Gus P, Bogdan I, Moraru PI, Pop AI, Clapa D, Marin DI, Oroian I, Pop LI. Implications of minimum tillage systems on sustainability of agricultural production and soil conservation. Journal of Food, Agriculture & Environment. 2009;7(2):335-338.
10. Radosevich SR, Hold JS, Ghersa CM. Ecology of weeds and invasive plant relatioship to agriculture and natural resource management 3rd ed. John Willey & Sons; 2007.
11. Olofintoye JA. Effect of tillage and weed control methods on weed growth and performance of upland rice in the Guinea Savanna of Nigeria. Nigerian Journal of Agronomy. 1987;2(3):82-87.
12. Anapurnamma PA, Pratap Kumar R, Madhu Bindu GS, Sreenivasa Raju M. Evaluation of the performance of rice fallow maize under zero-tillage conditions in comparison to conventional planting. Crop Research. 2010;40(1, 2 & 3):7-11.
13. Hussaini MA, Lado A. Influence of weed control methods on yield and yield components of irrigated and rainfed cowpea (*Vigna unguiculata* (L.) Walp). Crop Research. 2010;40(1, 2 &3):76-82.

6

Lactic Acid Bacteria of Potential as a Means of Inhibiting Undesirable Microorganisms in Warm Season Grass Silages

Daniele de Jesus Ferreira[1*], Anderson de Moura Zanine[1], Edson Mauro Santos[2], Juliana Silva de Oliveira[2] and Ricardo Martins Araujo Pinho[2]

[1]Department of Animal Science, Federal University of Mato Grosso, MT, Brazil.
[2]Department of Animal Science, Federal University of Paraíba, PB, Brazil.

Authors' contributions

This work was carried out in collaboration between authors. All authors read and approved the final manuscript.

Editor(s):
(1) Lixiang Cao, Department of Biotechnology, Sun Yat-sen University, Guangzhou, P. R. China.
Reviewers:
(1) Anonymous, Turkey.
(2) Hakan Geren, Department of Field Crops, Ege University, Izmir, Turkey.
(3) Anonymous, Ireland.

ABSTRACT

The presence of some microorganisms in silage besides reducing nutritional value and may represent risks to animal and human health due potentially pathogenic microorganisms. Enterobacteria, bacteria of the genus *Clostridium* spp and bacteria of the genus *Listeria* spp develop in badly fermented silage, in which pH drop is slower. After silos opening, yeasts, fungi and *Bacillus* spp initiate aerobic degradation, leading to pH rising and reappearing of *Clostridium* spp, *Listeria* spp and enterobacteria. Thus, development control those microorganisms by adequate fermentation is extremely important, since besides reducing silage quality, many are pathogenic or produce substances that are harmful to animal and human health.

Keywords: Bacillus; conservation; fungi; Listeria; pathogens.

**Corresponding author: E-mail: dany_dosanjos@yahoo.com.br;*

1. INTRODUCTION

Silage is a metabiosis, which means during the fermentative process a microbial succession occurs, that describes the different stages in silage process. Consequently, many groups of microorganisms develop simultaneously and in succession, as changes occur in redox and the type and amount of substrate [1,2].

Nutrient preservation in silage comes from fermentation by lactobacilli or other lactic acid bacteria (LAB). To obtain effective action from these microorganisms, four conditions are necessary: 1) fermenting material to allow bacterial growth; 2) oxygen absence in the material to favor the growth of anaerobic Lactobacilli; 3) enough number of Lactobacilli so that they are rapidly dominant over other microbial species; and 4) low humidity to avoid the produced acids to dilute favoring butyric fermentation [3,2,7].

The susceptibility of silage's deterioration seems to be ruled more by the fungal population than by the chemical composition of silage [4]. Aerobic microorganisms' breath may be considered one of the main agents that influence silage quality. However, the substrate used for breath depends on the type of microorganism, for example, yeasts consume only soluble chemicals (sugars and fermentation products), while molds degrade a large array of nutrients, including structural carbohydrates and lignin [1,5,6].

The presence of fungi is undesired, not only because they break the sugar and lactic acid by normal breath, but also for they hydrolyze and metabolize cellulose and other cellular wall components. Besides, some molds, mainly the species of the genus *Aspergillus*, *Fusarium* and *Penicillum*, grow in silage where there is air penetration and produce toxins that are harmful to animals and humans [2,7,18].

The best model for fermentative process is the one in which the lactic bacteria become dominant over the groups of undesired microorganisms.

Thus, the object of this review is to describe the effects of inoculation of lactic bacteria over the microbiological quality and silage stability, considering the importance and control of each microbial group involved in the silage process separately.

2. DEVELOPMENT

2.1 Enterobacteria

Enterobacteria are the group of microorganisms most widely studied. Among the reasons, stands out the medical importance and economic impacts, how easy they are to isolate and grow, rapid breeding time and easy genetic manipulation. They are found in the water, on the ground, in animals and humans' intestines and many vegetal tissues [1].

Enterobacteria are Gram-negative, oxidase-negative, do not grow spores, are shaped into short bacilli (0.3-1.0 x 1.0-6.0 p.m.). They move by peritrich flagello. They are not halophyilic and are facultative anaerobic. They are chemoorganotrophic and show respiratory and fermentative metabolism, growing well in temperatures between 22 and 35°C. They are catalase-positive and reduce nitrate to nitrite [2].

Enterobacteria are divided according to the fermentation end products, mixed acid production, but anediol and tri-methylene glycol producers. Glucose catabolism occurs both by EMP (Embden-Meyerhof-Parnas), in which energetic yield of glycolytic way occurs, metabolic Phospho-dihydroxy-Acetone way and pyruvic acid metabolic way, as by HMP. *Escherichia coli*, Serratia, *Salmonella*, Klebsiellia, Aerobacter, Paracolobacterium, Erwinia, Proteus have HMP way identified already [5].

In silage, many microorganisms are undesired for two reasons: first due to pathogenicity, and since they will be used as animal food, these microorganisms should be absent in silage; second, the development of these bacteria results in great nutrient losses, since the metabolic was employed result in high substrate consumption which is lost in the form of secondary metabolites [1].

Studies have shown that such bacteria develops at the beginning of silage process and have their numbers reduced as the pH decreases. Usually their population decreases as lactic bacteria population growth [8]. Verified, in silage of *Digitaria eriantha*, that the enterobacteria population reached its maximum level in the first 24 hours of fermentation with values of 7.1 log of colony formation unit (cfu)/g, reducing to 6.3 log of colony formation unit (cfu)/g at 9 days of fermentation. [9], assessing microbial populations in corn silage, observed enterobacteria values varying from 5.1 to 3.9 log cfu/g from the beginning to the tenth day of fermentation.

Penteado et al. [10] observed in guinea grass (*Panicum maximum* Jacq. cultivar Mombasa) silage increase of lactic acid bacteria population and the decrease of enterobacteria occurred in the first days of fermentation, showing that microbial succession occurs very rapidly and in a very definite way. However, the enterobacteria are still present in silage, despite the lower pH.

The explanations for microbacteria population reduction as the lactic bacteria increase is the simple reduction of pH or, as it has been shown by some authors, the production of bacteriocins. In Fig. 1. Can observe that the populations of enterobacteria decrease as the pH gets lower in guinea grasssilage and that such decrease is more severe in inoculated silage.

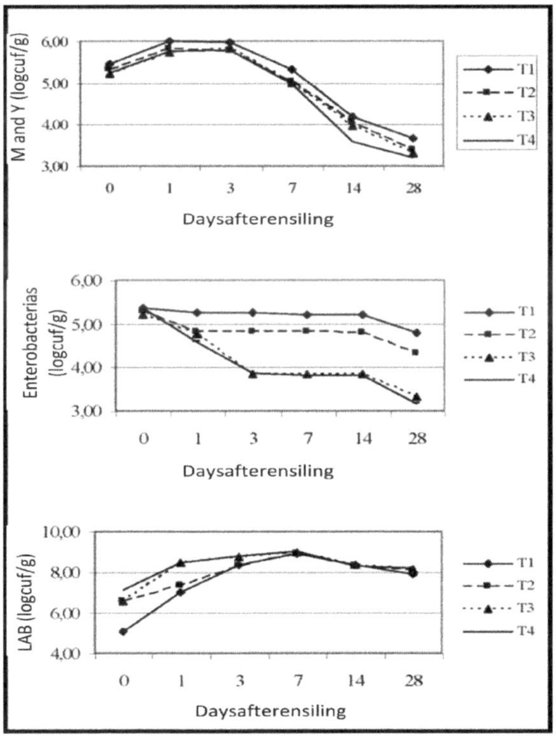

Fig. 1. Populations of molds and yeasts (M and Y), enterobacteria and lactic acidbacteria (LB), in mombasa grass silage without inoculant (T1), inoculated with 10^4CFU/g forage (T2), inoculated with 10^5CFU/g forage (T3) and inoculated with 10^6CFU/g forage of *Lactobacillus plantarum* of epiphytic microbiota

Adapted from [10]

Muck et al. [2] observed a reduction in ammonia concentration and in the population of enterobacteria in guinea grass silage (*Panicum maximum*) inoculated with *Lactobacillus plantarum*, isolated from the epiphytic microflora. Thus the inoculants for silage can facilitate or accelerate the process of ensiling, but they do not replace the fundamental factors (maturity of the plant, dry matter content, exclusion of oxygen), which are essential for producing good quality silage. Among these factors the age of regrowth is the one that influences all the features of the silage, from the fermentation of the silage to the nutritional value, considering the losses.

Penteado et al. [10] evaluated the aerobic stability of *Panicum maximum* cv. Mombasa silage inoculated with two strains of *Lactobacillus buchneri*, one from a commercial inoculant and another isolated from sugar cane (*Saccharum officinarum* L.) silage. It was observed an increase in dry matter content after silo opening, while the carbohydrate ratio did not change due to the low residual concentration, characteristic of grass silage.

Many LAB have antimicrobial peptides, known as bacteriocins, which are responsible for inhibiting growth of related species or species that have similar nutritional requirements. Consequently, bacteriocin production is a form of competition between bacteria that live in a same ecologic niche [2].

Bacteriocin production is a process that consumes a high amount of energy, therefore it is only worthy for the producing microorganism if really necessary. It is known that biosynthesis energy cost is high and, for that reason, it is a process well controlled by molecular regulatory systems with instantaneous catabolic induction and repression. Bacteriocin production is influenced by pH, temperature, environment composition, agents that damage DNA or growth conditions [11].

Antimicrobial activity performed by *Lactobacillus plantarum* was determined by [11,12], who purified and characterized a bacteriocin composed by two peptides, conducted by *Lactococcus lactis* and *Pediococcus pentosaceus*, which was named plantaricin NC8. [13]. Verified that *Lactobacillus* plantarum isolated in cassava (*Manihot esculenta* Grantz) and corn (*Zea mays*) produced a bacteriocin that has shown itself effective in inhibiting bacteria from several geni, including *Staphylococcus, Bacillus, Clostridium, Listeria, Salmonella, Escherichia coli*, besides some lactic bacteria, such as *Streptococcus thermophillus* and *Leuconostoc mesenterioides*.

Silva [14] evaluating the antagonist effect of *Lactobacillus plantarum* isolated from corn verified that there was inhibition of *Eschecrichia coli, Pseudomonas aeruginosa* and *Klebsiella pneumoniae*.

In corn and sorghum silage, [15] showed that enterobacteria do not survive to the end of silage process, fact that could be explained by the high LAB, as well as the rapid and steep drop in silage pH. While [10] verified occurrence of enterobacteria up to the eight fermentation day in mombasa grass silage (Fig. 1).

Thus, in corn and sorghum, inoculation is unnecessary in enterobacteria control, since such do not survive to the low resulting pH at the end of silage process. However, in grass silage, in which many times pH does not reduce enough to eliminate enterobacteria population, the inoculation is necessary and has shown itself effective in controlling enterobacteria growth [8,16,10,17,6,18].

However, [7] observed that the use of inoculant did not improve quality, nutritional and fermentation characteristics of silages *Panicum maximum* cv. Tanzania and cv. Mombasa [6] Reported that there was no effect on the pH value, ammonia, effluent production and recovery rate of MS in Tanzania grass silage, with and without the addition of in oculant.

In Table 1 can be observed the effects of inoculation with *Streptococcus bovis* (strains HC5 and JB1) isolated from rumen, homofermentative lactic bacteria about the development of lactic bacteria and enterobacteria in *Panicum maximum* cv. Mombasa silage. It was observed that in inoculated silage, there was larger development of lactic acidbacteria than enterobacteria. It is important to notice also the persistence of enterobacteria in silage until the twenty eighth day of fermentation, which shows how hard it is to eliminate this group of bacteria in grass silage [17,19,6,18].

2.2 *Clostridium* spp

Bacteria whose final fermentation products are acetate, butyrate, acetone and isopropanol. They are divided into saccharolytic (*Clostridium tyrobutiricum*), that produce butyric acid from sugar fermentation and lactic acid and proteolytic (*Clostridium sporogenes*), that degrade aminoacids, creating ammonia and amines and saccharo-proteolytic (*Clostridium perfringens*), that promote both fermentation and proteolysis [20].

Clostridium forms acetyl CoA or pyruvate acetyl phosphate, with the formation of carbon dioxide and hydrogen, without format (Transferase acetyl phosphate), while enterobacteria use format (hydrogen format lyase).

One of the main problems of the presence of *Clostridium* bacteria in silage is milk contamination. *Clostridium* spores are resistant to cooking temperature of certain cheese, developing and promoting undesired fermentation. The most important are *Clostridium tyrobutiricum* and *Clostridium sporogenes*. *Clostridium tyrobutiricum* ferments lactic acid into butyric acid and CO2 and *Clostridium sporogens* are proteolytic, degrading aminoacids and forming ammonia and amines (histamin, putrescin, cadaverin), which results in putrid smell in cheese [21].

According to [20] the presence of *Clostridium* spp is conditioned to buffering power of ensiled material, since this group of bacteria does not develop in very acidic environments. Thus, in corn and sorghum silage, such bacteria are eliminated by the sudden and extreme pH reduction and, consequently, there is a low accumulation of butyric acid and ammonia in silage.

As for grass silage, the younger and larger amount of humidity, the lower level of soluble carbohydrates, high buffering capacity, more likely the environment will be for the development of bacteria from this genus. For legume, besides the low amount of carbohydrates, the high protein level favors proteolysis, which results in high ammonia production and, consequently, silage with high pH. Such conditions favor butyric fermentation instead of lactic acid [8,9].

As happens with enterobacteria, *Clostridium* spp. bacteria can be inhibited also by the production of bacteriocins. [22] verified that a strain of Streptococcus bovis, called HC5 produces a bacteriocin (bovicin HC5) that inhibits a wide range of microorganisms, including bacteria from the genus Clostridium. Effective bacteriocin when inhibiting bacteria of this genus was also observed in *Lactobacillus plantarum* by [13].

Table 1. Average values of lactic acidbacteria (LAB), enterobacteria (ENT) for treatments throughout the fermentation period of mombasa grass silage

	Openingtime				
Treatments	**0**	**1**	**7**	**14**	**28**
			LAB, CFU/g		
Control	5.39Be	7.48Bd	8.85Aa	8.71Cb	8.27Bc
HC5	5.72Ae	7.81Ad	8.87Ab	9.11Ba	8.57Ac
JB1	5.77Ae	7.82Ad	8.94Ab	9.40Aa	8.69Ac
			ENT, CFU/g		
Control	6.01Aa	5.83Ab	5.94Aa	4.94Ac	4.29Ac
HC5	5.99Aa	5.58Bb	5.08Bc	4.57Bd	3.59Be
JB1	5.93Aa	5.48Bb	4.91Cc	4.53Bd	3.57Be

Averages followed byunlike capital letter in columns and same lowercase letter in rows differ by the Student Newman Keuls test; at 5% significance. Adapted from [17]

Thus, *Streptococcus bovis* inoculation is intended to reduce the development of bacteria from the genus *Clostridium* spp in grass and legume silage, reducing the formation of ammonia, amines and butyric acid and resulting in better quality silage. Besides, the inhibition of the development of this group of bacteria reduces the risks of milk contamination, keeping, thus, the quality of dairy products.

In Table 2 shows the development of lactic acid bacteria and from the genus *Clostridium* spp and the pH of *Digitaria eriantha* silage throughout the period of fermentation inoculated or not with inoculant containing *Enterococcus faecium*. Inoculation resulted in faster and more extreme pH drop, followed by larger development of lactic acid bacteria and lower development of bacteria from the genus *Clostridium* spp. [8].

In Table 3 can be observed the highest concentration of crude protein were observed in silages treated with *Streptococcus bovis* and HC5 and JB1. This may have been associated to the fact that the *Streptococcus bovis* HC5 species releases (bovicine HC5) bacteriocin in themedium that inhibits growth of proteolytic bacteria, such as the enterbacteria or clostridia, and thus decreases the protein nitrogen losses from the inoculated silages [18]. Another fact that may explain the greater concentration of CP in the silages inoculated with *Streptococcus bovis* HC5 and JB1 strains may reflect the common capacity of all the *Streptococcus bovis* strains to synthesize protein from ammonia [22,6].

2.3 *Listeria* spp

The genus *Listeria* spp contains Gram-positive bacillis, non-spores formers, mobile, catalase positive, and are facultative anaerobic. From the seven species found, two are majorly important due its pathogenic effect: *Listeria monocytogenes*, in animals and humans and *Listeria ivanovii*, in animals [23].

Listeriosis is an infectious disease caused by *Listeria monocytogenes*. It affects several animal species, inducing three forms of clinical manifestation: (1) sepsis with abscesses in viscera such as liver and spleen, (2) miscarriage and (3) neurologic disease (meningoencephalitis). The disease is more common in temperate weather regions, where cases occur mostly during winter and beginning of spring. Bad quality silage (pH above 5.5) favors the grown of these bacteria [24,27,28].

Table 2. Values of pH, population of lactic acid bacteria (LAB) and population of bacteria of the genus *Clostridium* spp (CL) in *Digitaria eriantha* with or without inoculant

Period of fermentation	*pH*	*LAB Log CFU/g*	*CL Log CFU/g*
0	6.00	1.60	0.70
1	6.30	5.80	0.50
5	6.10	7.10	1.05
9	5.70	7.20	1.90
	Inoculated		
0	6.00	3.70	1.10
1	5.30	7.70	0.40
5	4.40	8.20	0.20
9	4.30	8.10	0.40

Adapted from [8]

Table 3. Average valuesof pH, ammoniacal nitrogen (N-NH_3) and concentration of dry matter (DM), crude protein (CP), ether extract (EE), neutral detergent fiber (NDF) and acid detergent fiber (ADF) in elephant grass silages without inoculant (control) and inoculated with *Enterococcus* (*Enterococcus faecium*), JB1 (*Streptococcus bovis* JB1) and HC5 (*Streptococcus bovis* HC5)

Treatment	pH	NH_3 (mg/dL)	DM (%)	CP (%MS)	EE (%MS)	NDF (%MS)	ADF (%MS)
Control	4.32a	11.44a	25.53b	6.23b	3.02a	69.74a	38.75a
Enterococcus	4.19b	11.09b	27.12a	6.30b	2.99a	69.37a	37.30a
Streptococcus bovis JB1	3.99c	10.54c	28.12a	6.98a	2.95a	71.71a	36.88a
Streptococcus bovis HC5	4.04c	10.68c	26.93a	7.08a	3.06a	68.09a	37.57a
CV (%)	2.02	2.12	3.87	3.58	5.78	6.76	5.97

Means within a column with unlike lettercase differ by the Tukey test at the level of 5% significance.
Adapted from: [7]

As well as enterobacteria and bacteria from genus *Clostridium* spp and *Listeria* spp. develop better in silage with higher pH and is inhibited in silage with lower pH. [25] verified the presence of *Listeria* spp in 65.6% of the samples at the moment of silage opening of Cynodon sp (tifton-85) and, among them, 10% tested positive for *Listeria monocytogenes*, and all of the assessed silage showed pH levels above 4.70.

Evaluating the occurrence of different ribotypes of *Listeria* spp in corn and grass silage, [24] verified the presence of *Listeria* spp. in 10% of the corn silage samples and 60% in the grass silage samples, showing that grass silage are more susceptible to occurrence of *Listeria* spp. Besides, it was also verified that 83% of the isolates of *Listeria* spp in high quality corn silage were identified as *Listeria monocytogenes*, alerting for the presence of this pathogenic species even in high quality silage such as corn.

Besides the importance of adequate fermentation in controlling *Listeria* in silage, as described for enterobacteria and *Clostridium* bacteria, *Listeria* can be inhibited by LAB that produce bacteriocin, as demonstrated by [26]. In Table 4 There is observed that among the many species inhibited by LAB, to be included *Listeria monocytogenes*. *Listeria* inhibition was also observed by [22], evaluating the effects of bovicine bacteriocin HC% produced by *Streptococcus bovis* HC5.

2.4 Fungi

By the opening silos, occur the oxygen penetration, so aerobic bacteria, fungi and yeasts develop. Such microorganisms use residual sugars and some yeasts and acetic acid bacteria use lactic acid present in the environment, altering the redox potential promoting increased pH. Such conditions favor the reappearance of enterobacteria, *Listeria* spp and *Clostridium* spp., harming even more the microbiological quality silage.

Among the species that develop after silo opening, the fungi are extremely undesired, since, besides causing losses, like microorganisms are mycotoxin producers that can cause harms to animals' health and, in cases of higher intoxication, can be found in animal origin products and this represent risks to human health [8].

In Table 5 are summarized the main mycotoxins found in food and their effects over animal performance.

According to [20] fungi develop at the beginning of fermentative process, using the remaining oxygen among the plant particles and normally reach their maximum value at the first days of fermentation. Such microorganisms produce large amount of spores that are activated when silage is exposed to air by silo opening.

The high residual content of soluble carbohydrates in silage, mainly the ones made of corn, sorghum and sugarcane, favors the aerobic deterioration process by fungi and yeasts, causing losses after the silo opening. However, the organic acids produced by fermentation, mainly acetic acid, have fungicidal effect and can mitigate the deterioration, increasing aerobic stability of the silages [29,30,31,32].

Table 4. Antagonist activity (inhibition halo diameter) of *Lactobacillus* spp. and *Lactococcus* ssp. isolated in cheese

	Producer				
Revealer	***Lactobacillus casei***	***Lactobacillus fermentum***	***Lactobacillus rhamnosus***	***Lactobacillus acidophilus***	***Lactococcus lactis***
Bacillus cerens	87.25	90.00	90.00	71.64	36.84
Staphylococus aureus	51.88	49.54	52.61	33.17	20.16
Salmonella enteric	45.30	43.40	47.24	30.88	22.93
Yersinia enterocolytic	35.33	37.39	46.50	33.73	25.09
Listeria monocytogenes	35.03	47.07	44.80	39.14	22.81
Salmonella enteric	44.98	41.89	39.87	32.61	26.07
Shigella flexneri	79.32	84.18	86.22	90.00	63.60
Pseudonomas aeruginosa	89.12	81.92	86.12	34.64	74.60
Escherichia coli	54.62	41.17	69.77	28.04	54.87

Adapted from: [26]

Different from the positive results about enterobacteria control, *Clostridium* and *Listeria* spp bacteria, inoculants based on homofermentative bacteria are not effective in improving silage stability. In this case, heterofermentative bacteria are more effective, besides reducing the amount of lactic acid and residual carbohydrates in silage, producing more acetic acid inhibiting the development of fungi, yeasts and aerobic bacteria [8,9].

In Table 6 we can observe that in corn and sorghum silage, the development of fungi and yeasts, 5 days after silo opening was higher in silage inoculated with *Lactobacillus plantarum*, while inoculation with *Lactobacillus buchneri* that inhibited such microorganisms growth [27].

[28] also verified in corn silage that the inoculation with *Lactobacillus buchneri* inhibited the development of yeasts and increased aerobic stability of the silage. On the other hand, higher values of pH and N-NH_3 were recorded, showing that yeast control and increase of aerobic stability by means of inoculation with heterofermentative bacteria may occur at the costs of some harmful effects over fermentative parameters.

Another way to improve aerobic stability of the silage is the use of propionic bacteria, which have the ability to transform three mols of lactate into two mols of propionate, one mol of acetate and one mol of CO_2. [29] verified that the inoculation of *Propionibacterium* acidipropionici increased the amount of propionic acid and acetic acid in corn, sorghum and hay silage and decreased the amount of CO_2 produced after silo opening, presenting itself effective in improving stability of such silage.

Considering the aspects of aerobic instability, could conclude that in the composition of certain inoculant for corn and sorghum silage, or other forage species with high level of soluble carbohydrates, there may be present heterofermentative lactic bacteria or propionic bacteria, thus ensuring aerobic stability of silage after silo opening.

Table 5. Mycotoxin most common effects in food over animal performance

Mycotoxins	Effects on animals
Aflatoxin	Liver compromising
Ochratoxin	Weight loss
Deoxynivalenol	Low consumption,food refusing
Thicothene T-2	Low consumption,food refusing
ZearalenoneF-2	reproductive disorders
Slaframin	Diarrhea

Adapted from: [21]

Table 6. Fermentative profile and dry matter loss in silage of corn and sorghum without inoculant (C) or inoculated with *Lactobacillus plantarum* (LP) or *Lactobacillus buchneri* (LB)

		%						
		pH	SC*	Acid lactic	Acid acetic	Ethanol	N-NH3	DM losses**
	C	3,72b	3,15a	4,04c	1,27b	0.47	0,26b	1,65b
Corn	LB	4,13a	0,64b	2,76d	3,89a	0.49	0,28a	3,26a
	LP	3,64b	2,54a	7,94a	0,33c	0.42	0,21c	0,75c
	LB + LP	3,80a	1,08b	5,55b	3,17a	0.45	0,22c	1,14bc
	C	3,87b	6,75a	4,86a	0,96b	0.50	0,28b	1,97b
Sorghum	LB	4,26a	1,36b	2,54d	4,30a	0.53	0,30a	3,49a
	LP	3,75b	5,96a	9,39a	0,62c	0.47	0,24c	0,94c
	LB + LP	3,88b	2,02b	6,18b	3,49a	0.49	0,24c	1,45bc

**Soluble carbohydrates; ** dry matter losses; Adapted from: [27]*

Table 7. Carbohydrate fermentation profile of the isolates EB1, EB2, EB5, and EB6, signal grass plants (*Brachiaria decumbens* cv. Basiliski). + Intense fermentation, - no fermentation; (+) less intense fermentation

	Isolated strain				*Lactobacillus plantarum*
	EB1	EB2	EB5	EB6	
Glycerol	-	-	-	-	-
Erythritol	(+)	(+)	(+)	(+)	-
D-arabinose	-	-	-	-	-
L-arabinose	+	+	+	+	+
Ribose	+	+	+	+	+
D-xylose	-	-	-	-	-
L-xylose	-	-	-	-	-
Adonitol	-	-	-	-	-
β-methyl D-xyloside	-	-	-	-	-
Galactose	+	+	+	+	+
D-glucose	+	+	+	+	+
D-frutose	+	+	+	+	+
D-mannose	+	+	+	+	+
L-sorbose	-	-	-	+	-
Rhamnose	(+)	(+)	(+)	(+)	-
Dulcitol	-	-	-	-	-
Inositol	-	-	-	-	-
Mannitol	+	+	+	+	+
Sorbitol	+	+	+	+	+
α-methyl D-mannose	-	-	-	-	+
α-methyl D-glycoside	-	-	-	-	-
N-acetyl-glucosamine	+	+	+	+	+
Amygdaline	+	+	+	+	+
Arbulin	+	+	+	+	+
Esculin	+	+	+	+	+
Salicin	+	+	+	+	+
Cellobiose	+	+	+	+	+
Maltose	+	+	+	+	+
Lactose	+	+	+	+	+
Melibiose	+	+	+	+	+
Saccharose	+	+	+	+	+
Trehalose	+	+	+	+	+
Inulin	-	-	-	-	-
Melezitose	+	+	+	+	+
D-raffinose	+	+	+	+	+

	Isolated strain				*Lactobacillus plantarum*
	EB1	EB2	EB5	EB6	
Amidon	-	-	-	-	-
Glycogene	-	-	-	-	-
Xylitol	-	-	-	-	-
β-gentibiose	+	+	+	+	+
D-turanose	+	+	+	+	+
L-lyxose	-	-	-	-	-
D-tagatose	-	-	-	-	-
D-fucose	-	-	-	-	-
L-fucose	-	-	-	-	-
D-arabitol	(+)	(+)	(+)	(+)	-
L-arabitol	-	-	-	-	-
Gluconate	+	+	+	+	+
2 Cetogluconate	-	-	-	-	-
5 Cetogluconate	-	-	-	-	-

Adapted from: [30]

Kung et al. [30] conducted a study aiming to characterize and quantify microbial populations in signal grass, harvested at different ages of regrowth. The six strains of lactic acid bacteria isolated from the signal grass were characterized according to the Gram-staining, catalase enzyme reaction, and the form of bacilli, submitted to tests for growth and identification. The identification of the isolates was performed by fermentation of carbohydrates in kit API 50 CH (Bio Meurix-France) (Table 7 above).

3. CONCLUSION

The control of undesirable microorganisms like enterobacteria, *Clostridium* spp. and *Listeria* spp. are performed by adequate fermentation, so that corn and sorghum silage, as well as silage inoculated with homofermentative lactic bacteria are less willing to the development of such microorganisms.

COMPETING INTERESTS

Authors have declared that no competing interests exist.

REFERENCES

1. Mcdonald P, Henderson AR, Heron SJE. The biochemistry of silage. New York: Chalcombe Publications. 1991;339.
2. Muck RE, Spoeltra SF, Wikeselaar PG. Effects of carbon dioxide on fermatation and aerobic stability of maize silage. Journal of the Science of Food and Agriculture. 1992;4(59):405-412.
3. Bughardi SR, Goodrich RD, Meike KC. Evaluation of corn silage treated with microbial additives. Journal of Animal Science. 1980;4(50):729-36.
4. Taylor CC, Ranjit NJ, Mills JA. The effect of treating whole-plant barley with *Lactobacillus buchneri* 40788 on silage fermentation, aerobic stability, and nutritive value for dairy cows. Journal of Dairy Science. 2003;18(85):1793-1800.
5. Muck E, Contreras FE, Mertens DR. Silage inoculant effects on in vitro rumen fermentation. Journal of Animal Science. 2007;6(85):284-299.
6. Ferreira DJ, Zanine AM, Lana RP, Santos EM, Alves GR, Veloso CM. Silage fermentation and chemical composition of elephant grass inoculated with rumen strains of *Streptococcus bovis*. Animal Feed Science and Technology. 2013;2(6): 215-226.
7. Santos EM, Pereira OG, Garcia R, Ferreira, CLLF, Oliveira JS, Silva TC. Effect of regrowth interval and a microbial inoculant on the fermentation profile and dry matter recovery of guinea grass silages. Journal of Dairy Science. 2014;8 (97):4423-4432.
8. Meeske R, Basson HM, Cruywagen CW. The effect of a lactic acid bacterial inoculant with enzymes on the fermentation dynamics, intake and digestibility of *Digitaria eriantha* silage. Animal Feed Science Technology. 1999;8(82):237-248.
9. Meeske R, Basson HM. The effect of a lactic acid bacterial inoculant on maize silage. Animal Feed Science Technology. 1998;3(70):239-274.
10. Penteado DCS, Santos EM, Carvalho, GGP, Oliveira JS, Zanine AM, Pereira OG,

Ferreira CLLF. Inoculação com *Lactobacillus plantarum* da microbiota em silagem de capim mombaça. Archivos Zootecnia. 2007;7(56):191-202.
11. McdonaldP, Henderson AR, Heron SJE, Cruywagen CW. The biochemistry of silage. New York: Chalcombe Publications, 2003;(2), p.458. Maldonado A, Barba JLR, Diaz RJ. Purification and genetic characterization of plantaricin NC8, a novel coculture-inducible two-peptide bacteriocin from *Lactobacillus plantarum* NC8. Applied and Environmental Microbiology. 2003;11 (69):383-389.
12. Ogunbanwo ST, Sanni AI, Onilude AA. Characterization of bacteriocin produced by *Lactobacillus plantarum* F1 and *Lactobacillus brevis*. African Journal of Biotechnology. 2003;2(8):112-123.
13. Oyetayo VO. Phenotypic characterization and assessment of the inhibitory potential of *Lactobacillus* isolates from different sources. African Journal of Biotechnology. 2004;3(7):355-357.
14. Silva AM. Populações microbianas em plantas de milho e sorgo, produtos da fermentação e desempenho de bovinos de corte, suplementados com suas silagens, tratadas com inoculantes microbianos. 2001, 145f. Tese (Doutorado em Zootecnia) - Universidade Federal de Viçosa, Viçosa-MG. Português; 2001.
15. Rocha KD. Silagens de capim-elefante cv. Cameroon, de milho e de sorgo produzidas com inoculantes ênzimo-bacterianos: Populações microbianas, consumo e digestibilidade. 2003, 93f.Dissertação (Mestrado em Zootecnia) - Universidade Federal de Viçosa, Viçosa-MG, 2003. Português
16. Oliveira JS, Santos EM, Zanine, AM, Mantovani HC, Pereira OG, Rosa LO. Populações microbianas e composição química de silagem de capim-mombaça (*Panicum maximum*) inoculado com Streptococcus bovis isolado de rúmen. Archives of Veterinary Science. 2007;2(12):35-40. Português 2007.
17. Ferreira DJ, Zanine AM, Lana RP, Ribeiro MD, Alves GR, Mantovani HC. Chemical composition and nutrient degradability in elephant grass silage inoculated with *Streptococcus bovis* isolated from the rumen. Anais da Academia Brasileira de Ciências. 2014;6(85):425-432.
18. Ferreira DJ, Zanine AM, Lana RP, Ribeiro MD, Alves GR, Mantovani HC. Ingestão e digestibilidade aparente em ovinos alimentados com silagens de capim-elefante inoculadas com *Streptococcus bovis*. Arquivo Brasileiro de Medicina Veterinária e Zootecnia. 2012;6(64):397-402. Português.
19. Mcdonald P. The biochemistry of silage. Chichester: John Wiley & Sons. 1981;218.
20. Jobim CC, Pereira JRA, Santos CT. Sistemas de produção de leite com ênfase na utilização de volumosos conservados. In: Reis, R.A.; Siqueira GR, Bertipaglia LMA, Oliveira AP, Melo GMP, Bernardes TF. Volumosos na produção de ruminantes. Editora Funep- UNESP, Português. 2005;307.
21. Mantovani HC, HUH, Worobo RW, Russel JB. Bovicin HC5, a bacteriocin from *Streptococcus bovis* HC5. Microbiology. 2002;4(148):3347-3352.
22. Ryser ET, Arimi SM, Donnelly CW. Effect of pH on distribuition of *Listeria* ribotypes in cor, hay and grass silage. Applied and Environmental Microbiology. 1997;9(63): 3695-3697.
23. Rissi D, Barros D. Meningoenceflite por Listeria monocytogenes em caprinos. <coralx.ufsm.br/ppgmv/seminarios2005/DanielRicardoRissi.pdf>Acesso em: 5 DEZEMBRO. Português; 2013.
24. Schocken-Iturrino RP, Reis RA,Coan RM, Bernardes TM, Panizzi RC, Pojjati ML, Pedreira MP. Alterações químicas e microbiológicas nas silagens de capim-tifton 85 após a abertura dos silos. Revista Brasileira de Zootecnia. 2005;2(34):464-471, Português.
25. Guedes Neto LG, Pereira JRA, Santos CT. Atividade antimicrobiana de bactérias ácido-lácticas isoladas de queijos de coalho artesanal e industrial frente a microrganismos indicadores. Arquivo Brasileiro de Medicina Veterinária e Zootecnia, 2005;5(57):245-250. Português
26. Filya I. The effect of *lactobacillus buchneri* and *Lactobacillus plantarum* on the fermentation, aerobic stability and ruminal degradability of low dry matter corn and sorgum silage. Journal of Dairy Science. 2003;9(96):3575-3581.
27. Driehuis F, Elferink SJW, Wikselaar PG. *Lactobacillus buchneri* improves aerobic stability and farm scale whole crop maize silage but does not affect feed intake and milk production of dairy cows. In: International Silage Conference. Proceeding Uppsala- Swedish University

of Agricultural Science. 1999;3(48):265-265.

28. Filya I, Sucu E, Karabulut A. The effect of *Propionibacterium* acidipropionici, with or without *Lactobacillus plantarum*, on the fermentation and aerobic stability of wheat, sorghum and maize silages. Journal of Applied Microbiology. 2004;4(97):848-826.
29. Santos EM, Pereira OG, Rasmo G, Ferreira CLLF, Oliveira JS, Silva TC. Microbial populations, fermentation profile and chemical composition of signal grass harvsted of different rgrowth ages. Revista Brasileira de Zootecnia. 2011;4(44):747-755.
30. Kung Jr L, Ranjit N.K. The effect of *Lactobacillus buchneri* and other additives on the fermentation and aerobic stability of barley silage. Journal of Dairy Science. 2001;5(84):1149-1155.
31. Kung Jr L, Stanley R.W. Effect of stage of maturity on the nutritive value of whole-plant sugarcane preserved as silage. Journal of Animal Science. 1982;4(54): 689-696.

7

Vertical Distribution of Available Plant Nutrients in Soils of Mid Central Valley at Odisha Zone, India

Antaryami Mishra[1], Trupti Mayee Pattnaik[2], D. Das[1*] and Mira Das[2]

[1]Department of Soil Science and Agricultural Chemistry, College of Agriculture, Orissa University of Agriculture and Technology (OUAT), Bhubaneswar-751003, India.
[2]Departments of Chemistry, ITER, SOA University, Bhubaneswar, India.

Authors' contributions

This work was carried out in collaboration between all authors. Author AM designed the study, wrote the protocol and wrote the first draft of the manuscript. Author DD reviewed the experimental design, all drafts of the manuscript and performed the statistical analysis. Author TMP managed the analyses of the study and help in draft of the manuscript. Author MD helped in draft preparation. All authors read and approved the final manuscript.

Editor(s):
(1) Mariusz Cycoń, Department and Institute of Microbiology and Virology, School of Pharmacy, Division of Laboratoty Medicine, Medical University of Silesia, Poland.
(2) Anonymous.
Reviewers:
(1) Anonymous, China.
(2) M. R. Olojugba, Dept. of Soil Science and Management, Joseph Ayo Babalola University, Ikeji Arakeji, Osun State, Nigeria.
(3) Celerino Quezada, Soils and Natural Resources, Faculty of Agronomy, University of Concepcion, Chile.

ABSTRACT

Vertical distribution of plant nutrients is most importance for plant production. We always explored nutrients distribution in the top soil. But plants also take nutrients from the sub soil. It is hypothesizing that vertical nutrient distributions are dominated by plant cycling relative to leaching, weathering dissolution and atmospheric deposition. Therefore, four pedons were selected for the study of available plant nutrients with the objective to sustain the better productivity and sustainability. The study area consists of the upper and lower land of Dhenkanal district come under mid central valley at Odisha zone, India which is situated between 20°29' and 21°11' N latitudes and $85^0 58'$ and $86^0 20'$ E longitudes. Soils were analyzed for the status of pH, organic carbon, available nitrogen, phosphorus, potassium, sulphur and boron content in vertical direction. Soil organic carbon, nitrogen, phosphorus and sulphur content decreased from higher elevation to lower elevation while pH, potassium and boron showed reverse trend. Soils of upland are relatively more

**Corresponding author: E-mail: dib_das101@yahoo.co.in;*

fertile with 1200 to 1400 mm annual precipitation and appropriate proportions of soil separates imparts the opportunity to the farmers to grow high value vegetable crops like maize, bottle gourd, okra, ridge gourd or cowpea in kharif whereas tomato, groundnut, sesamum, green gram, horse gram, potato, cabbage, cauliflower, brinjal, French bean or pea in rabi season. Soils of lowland area are suitable for rice cultivation. Long term crop productions planning may be done on the basis of variability of nutrients status in the vertical distribution of soil profile.

Keywords: Vertical distribution; Dhenkanal district; plant nutrients.

1. INTRODUCTION

Knowledge of vertical distribution of plant nutrients in soil is useful as roots of most of the crop plants go beyond the surface layer and draw part of their nutrient requirements from the sub surface layers of the soil [1]. Several workers have studied the horizon wise levels of plant nutrients in soil profiles of different series. However most of the workers have limited their studies on fertility status of surface layer only. Very few have studied layer wise fertility status of sub-surface layer which is very important for effective nutrient application. It is also helpful in understanding the inherent capacity of the soil to supply essential nutrient to the plants. Brar and Sekhon [2] and Pal and Mukhopadhyay [3] observed that the vertical distribution of potassium is of considerable importance as many deep rooted crops known to absorb potassium from the sub surface layers. Therefore, the information of vertical distribution of potassium is importance.

To understand the importance of plants in structuring the vertical distributions of plant nutrients, Jobbagy and Jackson [4] explored nutrient distributions in the top meter of soil, hypothesizing that vertical nutrient distributions are dominated by plant cycling relative to leaching, weathering dissolution, and atmospheric deposition, they examined three predictions that: (1) the nutrients that are most limiting for plants would have the shallowest average distributions across ecosystems, (2) the vertical distribution of a limiting nutrient would be shallower as the nutrient became more scarce, and (3) along a gradient of soil types with increasing weathering-leaching intensity, limiting nutrients would be relatively more abundant due to preferential cycling by plants. They also found that most of the nutrients are dominant in surface soil and decreasing with increasing depth.

As the interface between the atmosphere, biosphere, and lithosphere, soil undergoes an intense vertical exchange of materials resulting in steep chemical and physical gradients from surface to bedrock. Soil stratification is the most visible result of this exchange, and its extensive observation and synthesis from the basis of pedogenetic and taxonomic study [5-8]. The type, thickness, and position of horizons can yield information about soil forming factors such as climate, topography, and vegetation type [6,9,10]. Likewise, the vertical distribution of plant nutrients should yield similar insights into nutrient inputs, outputs, and cycling processes [11,12]. Most knowledge about the role of plant cycling on the distribution of nutrients comes from studies on horizontal nutrient patterns [13].

According to Patil and Patel [14] in the present strategy of high yielding and intensive cropping system, fertilizer application based on soil test assumed great significance for the purpose of fertilizer recommendation. Generally the soil samples are collected from the surface (0-15 cm) layer. The findings of Department of Agricultural Chemistry, Gujurat Agricultural University, Junagada indicate that the sub surface soil fertility also contributes greatly towards the yield of groundnut whose roots have been observed to proliferate to a depth of 45 cm. They also opined that the sub surface soil fertility had greater contribution than that of surface soil towards the production of groundnut. Hence the above study provides evidence that sub surface fertility status of soil has to be taken into account while making fertilizer recommendation of different crops.

Rice and groundnut are the most important crops of Odisha though several deep rooted crops of pulses and other oil seeds, sugarcane besides fruits and vegetables are also grown. In the study of Singh et al. [15] showed the more than 30% of the roots of dry matter to be present in the depth of 25-40 cm even for rice which is normally taken as surface feeder.

With these background, we have conducted a study of vertical distribution of available plant nutrients in the mid central table land agro climatic zone of Odisha with the aim to sustain the better productivity and sustainability.

2. MATERIALS AND METHODS

2. 1 Study Area

Dhenkanal district is located between longitudes 85° 58′ E to 86°20′E and between latitudes 20°29′N to 21°11′N which is come in the mid central table land agro climatic zone of Odisha (Fig. 1). It occupies a central position in the geo-political map of Odisha state in India. The mid central table land agro climatic zone consist of Dhenkanal and Anugul Districts.

2.2 Climate

The climatic condition of the zone is hot and dry in sub humid with a mean rainfall of 1421mm per annum. The climate is hot with high humidity in April and May and cold during December and January. The monsoon generally breaks in the month of June. The mean summer temperature is 38.7°C and mean winter temperature 14°C. The district has 8 numbers of blocks consisting of 1215 numbers of villages covering total area of 4452 km^2 [16]. The main crops of the district are rice, groundnut, sesamum, greengram, horsegram, sugarcane, vegetable and fruits.

2.3 Soils

The district mainly consisting of red and laterite soils through patches of yellow soils are found in some parts of the district. River *Bramhani*, the second biggest river of Odisha flows through this district and alluvial soil are found on the both of the river banks, which are very suitable for vegetable cultivation during summer. So as per older system of classification though red, literate and alluvial soils are present, as per the soil taxonomy these are classified as *Alfisols, Inceptisols and Entisols* [17]. The red colour of the soil is due to the high iron oxides content. The laterite soil have been formed by the process of laterization because of intense leaching of bases due to high rain fall. The alluvial soils are products of the pedogeonic process of illuviation mostly by the river Bramhani and its tributaries. The relief of the district consists of high hills, valleys with dense forest. The pedogenic process of colluvation has also take place in these types of topography. The colluvial deposit are found in the foot slope of most of the hills which are mostly used for cultivation of rice in the Kharif season followed by pulses and vegetables in the Rabi- Summer season. The area has deciduous natural forests and grasses. As the area is irrigated by Brahmani command area rice is grown both in kharif and rabi in different land types.

2.4 Soil sampling

A details soil survey of the area was conducted by using the soil survey manual of USDA [18] and guidelines for soil profile description by Food and Agriculture Organization [19]. Soil profiles of sizes 2m x 1.5m up to a depth of 1.5 to 2 m opened from two different land types (upland and lowland) of 2 blocks in the districts (Figs. 2-3). Nine soil samples from nine different horizons have been collected by spade during summer season. The textural classes analysis to determine the sand, silt, clay was carried out by Boycous Hydrometer method [20]. The pH of 1:2 soil water suspension was determined by means of glass electrode wing digital pH meter.

Organic carbon was determined by Walkley and Black's rapid titration method [21]. Available phosphorous was determined by Olsen's method [22]. Available potassium was determined by neutral normal ammonium acetate method using digital flame photometer [23]. Available boron was done by hot water extraction method and available sulphur was done by 0.15% $CaCl_2$ method.

3. RESULTS AND DISCUSSION

Tables 1-4 described the physic-chemical properties of the four pedons. The soils of the four pedons were sandy clay loam to sandy. Sand was dominating fraction (56.40-90.80%) but sand percent decreased with decreasing height in the soils of all four pedons. An increasing trend of silt and clay fractions down the depth was noted in all four pedons but it is more prominent in the soils of low land due to the process of eluviation and illuviation [24].

Surface soils of the four pedons are acidic in nature (pH<7). pH value increases with decreasing height. Value of pH was relatively higher (pH-8.81) in the soils of low land whereas lower (pH-5.2) in the soils of upland [24].

Organic carbon is comparatively higher (>0.5 g kg^{-1}) in surface horizons of upper land than lower land (Tables 1-4). The higher amount of organic carbon in the surface layers is due to addition of organic matter like farm yard manure and incorporation of stubble left after harvesting of paddy. The irregular amount of organic carbon down the depth in pedon-2 is because of slope wash in the hill bottom lands in different period of times due to the pedogenic process of colluvation.

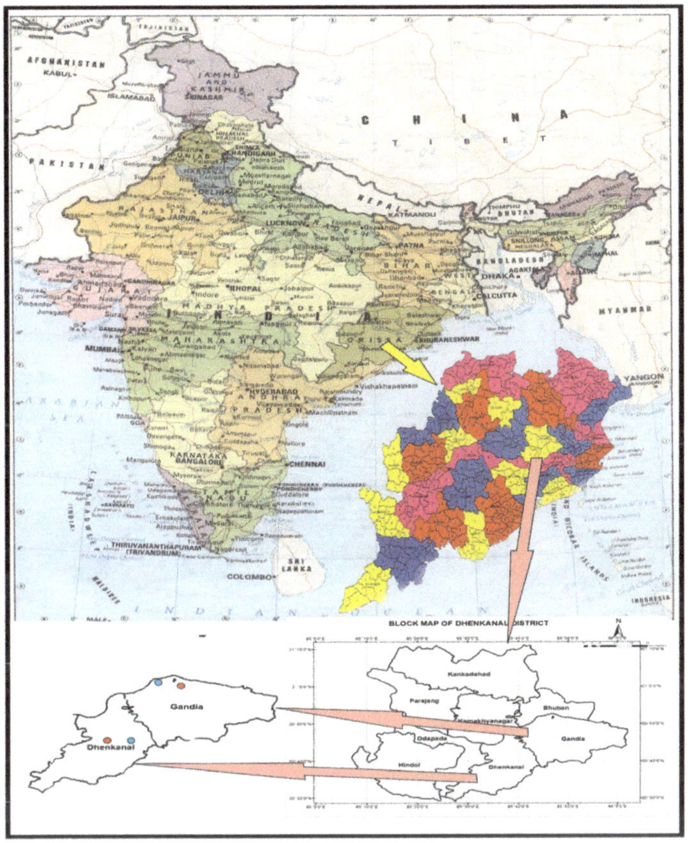

Fig. 1. Study area in the Dhenkanal district, Odisha state, India

Fig. 2. Upper land profile of Tarava village of Dhenkanal Sadar block, Dhenkanal, India

Fig. 3. Upper land profile of Deogaon block of Gondia block, Dhenkanal, India

Table 1. Soil physical and chemicals properties of pedon-1 at upper land (Latitude -20º41.224' Longitude -85º42.355') of Sankhua village, Dhenkanal sadar block in Odisha zone, India

Sl. No.	Sample No.	Depth (Cm)	Horizon	Particle size (%)			Textural class	pH	OC (g Kg^{-1})	Available nutrients				
				Sand	Silt	Clay				N (mg kg^{-1})	P_2O_5 (mg kg^{-1})	K_2O (mg kg^{-1})	S (mg kg^{-1})	B (mg kg^{-1})
1	1	0-13	O	88.00	10.80	1.20	Sandy	5.20	0.521	138.75	5.29	204.29	5.05	2.04
2	2	13-28	Ap	83.00	14.80	2.20	Sandy	5.20	0.212	139.38	4.79	215.71	2.52	1.68
3	3	28-43	A	75.00	18.80	6.20	Loamy sand	5.24	0.308	100.63	4.78	227.14	3.83	1.44
4	4	43-59	AB	74.00	19.80	6.20	Loamy sand	5.43	0.289	100.00	3.66	208.99	1.13	1.44
5	5	59-79	E	69.00	22.80	8.20	Sandy loam	5.79	0.347	109.38	2.53	201.60	3.39	1.38
6	6	79-107	BE	67.60	23.80	8.60	Sandy loam	5.92	0.212	69.38	2.25	214.37	4.70	1.38
7	7	107-132	Bt1	65.60	26.80	7.60	Sandy loam	6.07	0.328	76.88	2.25	230.50	2.26	1.32
8	8	132-162	Bt2	61.60	29.80	8.60	Sandy loam	6.04	0.328	87.50	1.91	303.07	1.74	1.80
9	9	162-180	BC	62.60	28.80	8.60	Sandy loam	6.31	0.193	72.50	1.41	296.35	0.61	2.22

Table 2. Soil physical and chemicals properties of pedon-2 at low land (Latitude -20º41.719' Longitude -85º37.570') of Tarava village, Dhenkanal sadar block in Odisha zone, India.

Sl. No.	Sample No.	Depth (Cm)	Horizon	Particle size (%)			Textural class	pH	OC (g Kg^{-1})	Available nutrients				
				Sand	Silt	Clay				N (mg kg^{-1})	P_2O_5 (mg kg^{-1})	K_2O (mg kg^{-1})	S (mg kg^{-1})	B (mg kg^{-1})
1	1	0-16	Ap	87.80	5.20	7.00	Sandy	6.22	0.328	86.25	5.29	141.79	3.05	0.90
2	2	16-32	A	90.80	4.20	5.00	Sandy	6.24	0.057	84.38	4.22	153.22	3.48	0.72
3	3	32-45	AB	90.80	4.20	5.00	Sandy	6.37	0.057	82.50	2.53	148.51	2.61	0.96
4	4	45-63	E	86.80	5.20	8.00	Sandy	7.34	0.115	71.25	2.25	173.38	4.18	0.96
5	5	63-73	BE	88.80	4.20	7.00	Sandy	7.50	0.057	68.75	1.97	154.56	4.35	1.08
6	6	73-89	Bt1	85.80	5.20	9.00	Sandy	7.63	0.077	71.25	1.97	141.12	3.83	1.14
7	7	89-113	Bt2	78.80	10.00	11.00	Loamy sand	7.72	0.173	65.63	1.69	153.22	5.22	1.26
8	8	113-131	Bt3	75.00	8.20	16.80	Loamy sand	7.73	0.270	73.75	1.06	175.39	4.26	1.32
9	9	131-181	BC	73.00	9.20	17.80	Loamy sand	7.81	0.270	68.75	1.06	181.44	2.96	2.04

Table 3. Soil physical and chemicals properties of pedon-3 at upper land (Latitude -20°42.087' Longitude -85°44.942') of Deogaon village, Gondia block in Odisha zone, India

Sl. No.	Sample No.	Depth (M)	Horizon	Particle size (%)			Textural Class	pH	OC (g Kg^{-1})	Available nutrients				
				Sand	Silt	Clay				N (mg kg^{-1})	P_2O_5 (mg kg^{-1})	K_2O (mg kg^{-1})	S (mg kg^{-1})	B (mg kg^{-1})
1	1	0-15	O	81.40	6.60	12.00	Loamy sand	5.31	0.901	193.13	9.86	34.94	2.87	2.52
2	2	15-30	Ap	80.40	6.60	13.00	Loamy sand	5.39	0.518	167.50	15.21	49.06	3.13	1.50
3	3	30-50	A	73.40	18.60	16.00	Sandy loam	5.72	0.595	169.38	10.14	62.50	4.09	1.68
4	4	50-70	AB	68.40	10.60	21.00	Sandy clay loam	6.10	0.518	171.88	4.51	61.82	3.22	1.62
5	5	70-85	E	64.40	15.60	20.00	Sandy loam	6.43	0.461	163.75	4.79	73.25	3.74	0.60
6	6	85-110	BE	56.40	18.60	25.00	Sandy clay loam	6.79	0.307	173.75	6.76	163.97	4.00	0.18
7	7	110-140	Bt1	60.00	14.60	25.40	Sandy clay loam	6.68	0.326	181.88	4.51	116.93	6.53	0.42
8	8	140-190	Bt2	61.00	13.60	25.40	Sandy clay loam	6.64	0.230	128.75	5.36	96.10	2.52	0.30
9	9	190-220	BC	66.00	6.60	21.40	Sandy clay loam	6.47	0.211	133.75	3.67	100.80	2.18	0.66

Table 4. Soil physical and chemicals properties of pedon-4 at low land (Latitude -20°45.126' Longitude -85°47.350') of Deogaon village, Gondia block in Odisha zone, India

Sl. No.	Sample No.	Depth (M)	Horizon	Particle size (%)			Textural Class	pH	OC (g Kg^{-1})	Available nutrients				
				Sand	Silt	Clay				N (mg kg^{-1})	P_2O_5 (mg kg^{-1})	K_2O (mg kg^{-1})	S (mg kg^{-1})	B (mg kg^{-1})
1	1	0-15	Ap	72.00	10.60	17.40	Sandy loam	6.81	0.345	175.00	12.68	137.76	4.35	0.36
2	2	15-28	A	69.00	11.60	19.40	Sandy loam	7.30	0.369	77.50	9.02	168.00	2.87	1.56
3	3	28-50	AB	74.00	10.60	15.40	Loamy sand	7.90	0.332	58.75	8.73	33.60	3.05	0.48
4	4	50-75	E	68.00	12.60	19.40	Sandy loam	7.87	0.351	56.25	7.33	36.96	2.87	0.36
5	5	75-110	BE	69.00	10.40	20.60	Sandy loam	8.03	0.351	41.25	5.92	40.99	3.13	0.42
6	6	110-125	Bt1	71.00	8.40	20.60	Sandy loam	8.03	0.295	35.63	4.23	32.255	3.48	0.36
7	7	125-165	Bt2	69.00	9.40	21.60	Sandy loam	8.15	0.295	48.63	3.94	26.21	4.79	0.24
8	8	165-180	Bt3	68.00	10.40	21.60	Sandy loam	8.14	0.277	26.88	3.10	16.80	3.74	0.96
9	1	180-215	BC	70.00	10.60	19.40	Sandy loam	8.81	0.271	25.00	2.68	16.20	3.35	0.86

Nitrogen is a part of all living cells and is a necessary part of all proteins, enzymes and metabolic processes involved in the synthesis and transfer of energy [24]. The available nitrogen is comparatively higher in the surface horizons of all the soil profiles, which is because of the higher amount of organic carbon in those horizons [4] (Tables 1-4). A gradual decrease of nitrogen was noted down the depth in all four pedons [24,25]. It is also observed low nitrogen in lowland as compared to upland [26].

Phosphorus (P) is also an essential part of the process of photosynthesis, involved in the formation of all oils, sugars, starches, etc. [24]. The available phosphorus content is higher in the surface horizon and decreases down the depth in the soil profiles [4] (Tables 1-4). It is because of the addition of large quantities of phosphorus for paddy cultivation. The surface horizon of all pedons are ranged from medium to low (7-20 mg kg^{-1}) whereas sub surface horizons are low (<7 mg kg^{-1}) in available phosphorus. A gradual decrease of phosphorus was recorded down the depth in all four pedons [24,25].

Potassium is absorbed by plants in larger amount than any other mineral element except nitrogen and in some cases, calcium [24]. The comparatively higher amount of potassium in the sub surface horizon of all the profiles except pedon 4 is because of the development of the soil from the residual parents material which mostly consist of feldspars and micas which are mostly potash bearing minerals [27] (Tables 1-4). Potassium content of surface horizon of pedon 1 and 2 are high (>140 mg kg^{-1}); rest two are varied between medium and low. It was observed that content of potassium decreased down the depth in upland [24]. It is indicated that silt size particles were the major contributor of potash while clay contributed to a very little extent to crops and it is mainly due to weathering of biotite mica up to the stage of silt fractions [24].

The soluble sulphur content in all the horizons are below the critical limit (<5 mg kg^{-1}) which could be attributed to lower content of organic carbon in these horizons (Tables 1-4). The soluble boron content of the surface horizon of all the soil profile is above the critical limit (>0.5 mg kg^{-1}) except the pedon 2 which is below the critical limit (Tables 1-4). Sulphur is deficient (<5 mg kg^{-1}) in all the sub horizons of all the soil profiles, while boron is deficient (<5 mg kg^{-1}) only in pedon 4.

4. CONCLUSION

It can be concluded that judicious application of phosphorus and sulphur is necessary for higher crop productivity in two blocks of the Dhenkanal district whereas liberal amount of boron should be applied in low lands of Gandia block. Similarly liberal amount of potassium application is necessary in upland of Gandia block. To maintain the soil quality, liberal application of organic matter is required in upland of Central land agro climatic zone of Odisha. The variability in nutrient status in the vertical distribution of soil profile has a long term bearing on the production and productivity of field crops, vegetables, orchard plantation and agro forestry in different blocks of Dhenkanal district. Land use planning can be done on the basis of physico-chemical properties and nutrient status of different horizons of upland and lowland soil of the district.

COMPETING INTERESTS

Authors have declared that no competing interests exist.

REFERENCES

1. Sangwan BS, Singh K. Vertical distribution of Zn, Mn, Cu, and Fe in semi arid soils of Haryana and their relationships with soil properties. J. Indian Soc. Soil Sci. 1993;41:463-467.
2. Brar MS, Sekhon GS. Vertical distribution of potassium in five benchmark soil series in Northern India. J. Indian Soc. Soil Sci. 1987;35:732-735.
3. Pal SK, Mukhopadhyay AK. Distribution of different forms of potassium in profiles of some Entisols. J. Indian Soil. Soil Sci. 1992;40:371-373.
4. Jobbagy E, Jackson R. The distribution of soil nutrients with depth: Global patterns and the imprint of plants, Biogeochemistry. 2001;53:51-77.
5. Hilgard E. Soils, their formation, properties, compositions, and relations to climate and plant growth in the Humid and Arid Regions. Macmillan, New York, US, 1906.
6. Jenny H. Factors of soil Formation. McGraw-Hill, New York, US; 1941.
7. Soil Survey Staff. Soil Taxonomy. USDA, Washington DC, US; 1975.
8. Buol SW, Hole FD, McCracken RJ. Soil genesis and classification, third edition. Iowa State University Press, Ames, Iowa, US; 1989.

9. Marion GM, Schlesinger WH. CALDEP: A regional model for soil CaCO3 (caliche) deposition in the southwestern deserts. Soil Science. 1985;139:468-481.
10. Honeycutt CW, Heil RD, Cole CV. Climatic and topographic relations of three Great Plains soils. I. Soil morphology. Soil Sci. Soc. Am. J. 1990;54:469-475.
11. Smeck NE. Phosphorus: an indicator of pedogenic weathering processes. Soil Science. 1973;115:199-206.
12. Kirby MJ. A basis for soil profile modelling in a geomorphic context. J. Soil Sci. 1985;36:97-121.
13. Noy Meir I. Desert ecosystems: environment and producers. Ann. Rev. Ecol. Syst. 1973;4:25-41.
14. Patil RG, Patel MS. Influences of available nutrients in surface and sub-surface soils on yield and nutrient uptake by Groundnut J. Indian. Soc. Soil Sc. 1983;31:160-161.
15. Singh R, Tripathy RP, Sharma JC. Rooting pattern and yield of rice (*Oryza sativa* L.) as influenced by soil water regimes. J. Indian Soc. Soil Sci. 1997;45:693-697.
16. Mishra Antaryami, Pattanaik Trupti Mayee, Das D and Das Mira. Soil fertility maps preparation using and GPS and GIS in Dhenkanal District, Odisha, India. International Journal for Plant and Soil Sci. 2014;3(8).
17. Sahu GC, Mishra Antaryami. Soils of Orissa and their management. Orissa review. 2005;LXIII(4):56-60.
18. Soil Survey Staff. Soil Survey Manual U.S. Dept. Agr. Handb. 18. Scientific Publishers. P.O.Box 91, Jodhpur-342001, 1995.
19. Food and Agriculture Organization. Guidelines for Soil Profile Description (Second edition). Soil Resources Development and Conv. Service, Land and Water Dev. Divn., Food and Agriculture Organisation of the United Nations, Rome; 1977.
20. Piper CS. Soil and Plant Analysis, Inter Science Publishers, Inc. New York; 1950.
21. Jackson ML. Soil Chemical Analysis. Prentice-Hall of India Private Ltd. New-Delhi; 1973.
22. Olsen SR, Cole CV, Watana FS, Dean L A. Estimation of available phosphorus in soils by extraction with sodium bicarbonate USDA. Circ. 1954;939.
23. Page AL, Miller RH and Keeney DR. Soil Chemical Analysis., Agronomy Series ASA- SSSA Publisher, Madison, Wisconsin, USA, Part-2 (Ed.) No. 9:1982.
24. Sharma RP, Singh, RS, Sharma SS. Vertical distribution of plant nutrients in Alluvial soils of aravalli range and optimization of land use. International Journal of Pharmaceutical and Chemical Sciences. 2013;2(3):1377-1389.
25. Meena HB, Sharma RP, Rawat US. Status of macro and micronutrients in some soils of Tonk district of Rajasthan. J Indian Soc Soil Sci. 2006;54:508-512.
26. Datta M, Saha PK, Choudhary HP. Erodibility characteristics of soils in relation to soil characteristics and topography. J. Indian Soc. Soil Sci. 1990;38:495-498.
27. Sharma RP, Rathore MS, Singh RS, Qureshi FM. Mineralogical Framework of Alluvial Soils Developed on the Aravalli Sediments. J Indian Soc Soil Sci. 2010;58:70-75.

8

Brown Cane Sugar–cattle Production Integration for Rural Economic Development Prospects in South Sulawesi, Indonesia

Rahim Darma[1*], A. Majdah, A. Zain[2] and A. Nixia Tenriawaru[1]

[1]Department of Agricultural Socio-economics, University of Hasanuddin, Makassar, 90245, Indonesia.
[2]Department of Agribusiness, University of Islam Makassar, Makassar, 90245, Indonesia.

Authors' contributions

This work was carried out in collaboration between all authors. Authors RD and AMAZ designed the study; produced data gathering instruments and coordinated the data collection. Authors RD and ANT conducted data gathering and analysis, while authors RD and AMAZ wrote the draft. All authors read and approved the final manuscript.

Editor:
(1) Anthony N. Rezitis, Dept. of Business Administration of Food and Agricultural Enterprises, University of Patras, Greece.
(2) Daniele De Wrachien, State University of Milan, Italy.
Reviewers:
(1) Anonymous, Brazil.
(2) Anonymous, Brazil.
(3) Daniel Staciarini Corrêa, Animal Production Department, Universidade Federal de Goiás, Brazil.
(4) Anonymous, Brazil.
(5) Anonymous, Nigeria.

ABSTRACT

Sugar plays an important role in Indonesian economy; the commodity is one of the major import commodities in recent years. Integration of brown cane sugar (BCS) with cattle production is a potential business opportunity that can be developed for rural economic development. This integration can optimize the use of marginal lands, and facilitate the development of environment friendly organic farming. The research objective was to identify the potential for the development of BCS-cattle production integration based on land resources, technology, culture, labor force, and markets. Data were collected through focus group discussions and interviews of 75 randomly selected farms from three districts (Barru, Wajo, and Bone), all of them in South west Sulawesi, Indonesia. The results showed that there is a huge potential of 302,771 ha's cultivated farm land, comprising of 10,355 ha's rice fields, 86,753 ha's dry land and 205.663 ha's paddock grazing. Cattle

**Corresponding author: E-mail: rdarma@indosat.net.id;*

raising is part of the farming culture in Sulawesi and a major source of income for most rural households. BCS-cattle production integration can generate employment opportunities, promote sugar and meat import substitution, produce biogas as an alternative energy source, increase the use of organic fertilizer, and promote organic-based farming system. For this system to be successful, government support is needed in establishing pilot projects at various locations which are expected to encourage farmers to adapt and develop the integrated farming system for rural economic development.

Keywords: Integrated farm; appropriate technology; organic farm; rural employment; small scale enterprises.

1. RATIONALE

Indonesia is a tropical agricultural country where most agricultural products can be grown, but it imports almost all foods. Currently a lot of agricultural land is not utilized in the form of concession or controlled individually; there are many landless farmers. These conditions make the countryside less attractive and encourage high urbanization. The unemployment rate in Indonesia has increased drastically from 3.2 percent in 2000 to 8.9 percent in 2009 where the unemployment rate in urban areas is higher than in the country side. In 2009, the urban unemployment rate was 11.4 percent, substantially higher than in the country side with *about* 7.6 percent in the rural sector. The same thing happened to women unemployment compared to that of men; amounting to 10.7 and 7.8 percent respectively [1]. Potential areas for sugar cane development in South Sulawesi are approximately 91,108 ha's which consist of 10,355 ha's paddy fields and 86,753 ha's dry land. Pasture land can partly be converted into cane fields for the small scale cane sugar industry development and at the same time be integrated with cattle breeding.

Sugar is a strategic commodity in Indonesia's economy as it belongs to the group of nine-basic needs for daily consumption. In 2010, total national sugar consumption both industrial and household consumption amounted to 4.55 million tons, while domestic production was only 2.44 million tons of sugar, so the shortage was met by imports [2]. Production cannot keep pace with consumption of sugar for several reasons, namely the decline in sugar cane plantations because the land is converted into residential and industrial areas, the decline in yield, continuously declining price of sugar, and decrease in plant efficiency [3].

Cane sugar industry on a household scale is suitable for rural development as it adds several work opportunities for poor farmers including the women. The development of brown cane sugar (BCS) can be integrated with cattle breeding; the shoots and leaves of the sugarcane can be used as feed materials for cattle. Cattle feces and urine waste can be processed to produce organic fertilizer and biogas. Manure as fertilizer is a substitution of chemical fertilizers supporting organic farming. The research objective was to identify potential land resources, technology, farming culture and the market for the development and BCS-cattle production.

Increasing production of BCS serves as import substitution and/or yields an export commodity, which drives regional economic development. It supports an increase in the cattle population by providing feeds for cattle which in turn leads to an increase in animal protein production and meat import substitution. This concept promotes organic fertilizer to substitute chemical fertilizer, encourages organic agriculture and reduces green gas emission. Development of integrated sugar cane with cattle breeding is an appropriate concept for rural economic development because it will generate huge employment opportunities for the rural population, especially for poor households through the use of less productive agricultural land.

2. CONCEPTUAL FRAMEWORK

2.1 Integrated BCS-cattle Development

Markets, natural resources, appropriate technology, and farming culture offer a great potential for the development of BCS and beef cattle production integration. Sugarcane can be developed easily on marginal or unutilized lands. Sugarcane is commonly planted on asphalt sidewalk in both rural and urban areas where it can grow without any further maintenance.

Cultivation of sugar cane has long been known by the Indonesian farmers in rural areas, including the processing into brown sugar or palm sugar substitute such as used as *Saka*, raw vinegar, and sweetener in Japan [4].

The processing of sugar cane into BCS could be regarded as part of the culture of farmers in many rural communities in Indonesia. Brown sugar is widely used for household consumption as a sweetener, flavor and color enhancer. Brown sugar can change the soft food texture other than crystal sugar and it is used as a raw material for small food and beverage industries such as soy and *Tauco* industries using brown sugar as a sweetener [5]. In addition, BCS is one of the alternatives to meet the needs of sugar in Indonesia [6].

Growing cane and cattle simultaneously to produce BCS and beef integrates farming and organic agriculture. Integrated BCS-Cattle farming is an innovation to the farmers which will increase their knowledge and skill for organic farming and sustainable agriculture. The importance of organic farming for the perspective of sustainable agriculture development can be found in several studies [7-10]. A set of agricultural technologies or innovations which are simple, effective and need based are selected and implemented to improve the knowledge, skills and abilities of farmers [11]. The integration of cultivated crops and livestock maintains productivity, protects the environment and the health of farmers and their families and at the same time adopts sustainable agricultural systems [12]. Integration of farming will also provide higher revenues due to reducing the use of external inputs. Integrated farming is a pattern of land use on a regular basis, taking place naturally and not using the external input [13]. The integration of farming also encourages farmers to develop organic farming. Organic agriculture is an integrated farm that aims to reflect the strong linkage between farm biota, production and the environment as a whole [14]. Organic agriculture is an ecological production management system that promotes and enhances biodiversity, biological cycles and soil biological activity. It is based on minimal use of off-farm inputs and on management practices that restore, maintain and enhance ecological harmony [15].

BCS-cattle production integration is a combination of two effective technologies that can be easily developed by farmers. Appropriate technology (AT) is an alternative method to overcome the problems encountered in agriculture, whether in relation to the development, production, development opportunities, or the creation of employment opportunities. AT is a technology appropriate to the needs of the community, to address the problem of society, and to preserve environment that can easily be utilized by the community to generate additional value from the economic and environmental aspects (President Act No. 3/2001). BCS processing technology as an AT is cheap and easy and has huge potential to convert natural resources, reduce unemployment, encourage domestic industry and increase incomes, especially for rural economy.

2.2 Rural Employment Development

Rural workers have low skills and are flexible in entering the employment opportunities available in rural areas, especially in agriculture and in the informal sectors. Therefore, it is necessary to develop a relatively inexpensive technology that can be easily developed locally so that rural unemployment and urbanization can be resolved. The skill level of work needed for the BCS-cattle production integration is generally low matching that of most rural workers. Low human capital, education and experience are dominant characteristics for rural labors [16-18].

The problem of unemployment is not only economic issues but more related to social issues. Types of jobs created in the economy must be in accordance with the capabilities and expectations of the work force [19]. Human resources investment can boost growth through creating agricultural economic activities for which there is a market, for which local resources are available and which are supported by innovations and appropriate technology. In the current era of global trade liberalization specific products that have been produced naturally and environmentally friendly will have a strong bargaining position; besides they can be labeled as super labor intensive involving many members of rural communities [20]. Small scale integrated BSC-Cattle productionis a new technology that can easily be adopted by rural people with low level of education and skill; it can absorb large numbers of family labor, especially female labor. It could also be considered as a training process for local people towards sustainable development.

The BCS development is the reverse of trends towards more capital-intensive processes or to commercial commodities. The small scale industrial development based on BCS-cattle production integration is a concept of labor transformation from urban to rural area. It goes contrary to the concept described by Feridhanusetyawan and Stahlin which the transformation of labor in the global competition is drastically changing from rural to urban areas, from low-technology-based industry to modern technologies, or from low-tech and labor-intensive industries to capital-intensive modern industry [21].

Employment in agriculture and the urban informal sector can be expanded; the labor market and the social structure of the sector are sufficiently flexible to absorb the workers who are not absorbed in the formal sector [22]. At the same time, capacity building is needed through intensive labor investment with appropriate technology as a potential way to improve rural employment and the quality of human resources [23]. So the development of BSC-cattle production integration is an economic development concept for rural economic development. The rural labor development is a key strategy for rural development of agricultural countries [24,25].

2.3 Research Method

The study was conducted in three districts in the center of cattle development South Sulawesi, Indonesia. Barru district is the center of Bali cattle germ plasm development, Wajo district is an area that generated BCS in the 1960s, and Bone district is part of the largest sugar cane planting area in South Sulawesi.

Primary data was collected by method of focus group discussions and interviews with randomly selected samples of 30 farmers in Barru district, 25 farmers in Wajo district, and 20 farmers in Bone district. The data collected include methods of animal husbandry, sugar cane plantation and cultivation of BCS. Secondary data including area of unutilized agricultural land, grazing area, sugar cane planting area, and the BCS development site were taken from South Sulawesi Statistical Bureau. Qualitative analysis was used to describe the development prospects of BCS-cattle integration based on the potential of natural resources, appropriate technology, culture, labor force, and market.

3. RESULTS AND DISCUSSION

3.1 Land Resources Potential

The development of cane and cattle in South Sulawesi has a huge potential. Sugar cane can be developed in the area of non cultivated agricultural land and grazing land, while cattle can be developed by utilizing an abundant unutilized agricultural waste. Farmers can extend their businesses thanks to sufficient labor, paddocks and grazing land that can be converted into cane growing land and be integrated with cattle farming.

Table 1 showed that in 2010 there were about 205.663 ha's of uncultivated agricultural land or 8.44 per cent of the total agricultural land In South Sulawesi which could be classified as potential for cane development. The total area of agricultural land of 2,434,968 ha's in 2010 consisted of 582,444 ha's of rice fields and 852,524 ha's of dry land (not for paddy). Approximately 10,355 ha of rice fields or 1.78 percent of the total paddy fields and 86,753 ha's or 4.68 percent of the total area of dry land are not cultivated.

Table 1. illustrates the decrease in agricultural land at around 13.41 per cent over three years from 237.518 ha in 2007 to 205.663 ha in 2010. Land is widely used to plant corn, cocoa, sugarcane in accordance with the activities of the agricultural development program. There is around 97,108 ha consisting of 10,355 ha of rice fields and 86,753 ha of dry land which has potential for cane development. Pasture land could be converted into sugar cane for the development of household scale sugar cane industry and cattle development in an integrated manner without significant changing the function of pasture land. If 25 percent of that land (205,663 ha) would be used for sugar cane development then there would be about 51.4 thousand hectares with an average productivity of s 70 tons of sugar cane per ha. The conversion factor of sugar cane is 10 percent. The BCS produced could be about 367.5 thousand tons or about 13 percent of the national sugar production or about 20 percent of sugar imports that could be substituted. On top of that an additional 500 thousand heads of cattle can be bred or about 50 percent of the cattle population in the South Sulawesi.

3.2 Human Resource Potential

BSC is a business model that can be developed on a small and medium scale making it very suitable to encourage the rural economy in Indonesia. BSC provides employment opportunities and income throughout the year and reduces the risk of food shortages for poor families. Integration of low cost BSC production and cattle breeding can be a development strategy through which farming can provide higher revenues [26].

Population growth and less optimal use of agricultural resources generate an increase in unemployment in the agricultural sector. In contrast, the unemployment rate in the non-agricultural sector is decreasing. The population of South Sulawesi in 2010 was 8,034,776 of which approximately 3,272,365 people were employed and 298,952 unemployed, including open unemployment about 0.14% of total the population. There were an increase in employment and a decrease in unemployment of 4.12 percent and -21.30 percent respectively during the period of 2010-2012 [27,28]. In contrast, the number of people employed in the agricultural sector reduced with about 123 thousand workers, i.e. from 1,592,299 workers in 2011 to 1,469,325 workers in 2012. At the same time other sectors increased: by about 2.2 thousand in the processing industry and around 60 thousand in trade, restaurant, accommodation, and services. GDRP (Gross Domestic Regional Product) for the processing sector increased by about 6.10 percent and for the trade sector, hotels and restaurants by 10.07 percent per year during the period 2005-2010.

Employment opportunities in the agricultural sector (including farming, forestry, and fisheries) increased by 50.9 thousand workers from 1,569,316 workers in 2009 to 1,620,177 workers in 2010. BPS also reported that the total female workforce in South Sulawesi 1,07 million people or 88 percent of the total population in 2009 which was increased from 71.10 percent in 2005. The proportion of the female workforce was 47.9 percent in agriculture, 29.9 percent in the commercial sector, restaurants and accommodation services, and 12.3 percent in the service sector, community, social and personal [29]. The idle female workforce is a potential for the development of BCS-cattle production integration because many types of jobs are matched by female labor. This potential can be seen from the great absorption of female labor in the agricultural sector (45.21%).

In urban areas, the majority of women workers are employed in the large commercial, retail, restaurants and hotels (47.33%) and social service sector (30.06%). Of particular interest is that the proportion of female workers absorbed in the processing industry tends to be higher than that of male employees, respectively 7.69 percent and 6.05 percent in both urban and rural areas. Based on these data, there is a potential workforce to develop the BCS-cattle integration. Temporary employment opportunities that can be

Table 1. The potential land cultivated and non-cultivated for cane development (ha)

No.	Kinds of Land	2007(ha)	2010(ha)	Changes (ha)	Percentage (%)
1.	Unutilized rainfed	7,527	7,368	-159	-2.11
2.	Temporary Unutilized rainfed	4,194	2,987	-1,207	-28.78
3.	Uncultivated Dry Land	114,297	86,753	-27,544	-24.10
4.	Paddock Grazing	111,500	108,555	-2,945	-2.64
	Total	237,518	205,663	-31,855	-13.41

Source: South sulawesi provincial office of food and horticulture; 2011

Table 2. Production, import, and domestic consumption of sugar (in million ton)

Year	Production	Import	Total Supply	Household consumption	Industrial consumption	Total demand
2005	2.24	2.37	4.61	2.78	1.21	3.99
2006	2.31	1.71	4.02	3.08	1.22	4.30
2007	2.95	2.84	5.79	3.39	1.31	4.70
2008	2.57	2.04	4.61	3.83	1.51	5.34
2009	2.30	2.75	5.05	2.97	1.57	4.54
2010	2.24	2.91	5.15	2.86	1.69	4.55

Source: National symposium of sugar, Surabaya, Indonesia; 2012

created with the expansion model are at least 150 thousand. Cane growing-cattle breeding integration is an excellent economic activity because it can optimally utilize the unproductive agricultural land, develop livestock population, support the development of organic agriculture, create employment opportunities of around 2-3 people per hectare, reduce the level of poverty in rural, and empower women workers [30].

3.3 Market Potential of BCS and Beef

Sugar is a strategic commodity in Indonesia's economy as it belongs to the group of nine basic needs for daily consumption. In 2010, total national sugar consumption both industrial and domestic amounted to 4.55 million tons, while production was only 2.44 million tons of sugar [31]. Shortage of sugar supply is met by importing sugar. In Table 2, it appears that national production is unable to meet the needs of overall consumption and so the government had to import sugar. Sugar consumption cannot be matched by domestic sugar production due to several reasons, such as the decrease in sugar cane plantations because of land being converted to residential and industrial areas, the decline in yield, continuous decline in sugar prices and decreased sugar plant efficiency [32].

Sugar supply shortage will be met from brown sugar as one of the alternatives that can meet the needs of sugar in Indonesia [33]. The BCS market potential is magnified by the food industry such as in East Java with a need of 30-40 thousand tons per year, while the BCS industries can only produce about 5 thousand tons per year [34]. BCS has also export potential with the greatest demand from Canada, USA, Belgium, Australia, and Europe as the amount can reach 500 tones per month while the current BCS supply is only 30 to 50 tons per month [35]. East Java province has a sugar manufactory, owned by Ahmad Rubai, who has become the sole producer and exporter of raw sugar to Japan since 1995. Rubai exports BCS as much as 300 tons per year with a turnover of Rp 15 Billion per year. Japan uses BCS as raw material for industrial syrup, ketchup, and cake [36].

Annual beef consumption per capita of Indonesia has reached 1.87 kg. This figure is low when compared to other countries in Southeast Asia. Low consumption even then, Indonesia needs at least 448,000 tons of beef equivalent to 4.48 million cows weighing 100 Kgs on average. Of these, only about 85 percent can be met by domestic beef production and the remaining, approximately 15 percent, was derived from imports. Based on these data, the number of additional cattle needed is about 672 thousand cows which can be met from the increase of cattle population in South Sulawesi and import. Cattle population in South Sulawesi can be increased with about 500 thousand heads through the development of BCS-cattle integration of production equaling approximately 74.4 percent of the total shortage of meat in Indonesia.

3.4 Development of Brown Cane Sugar

Processing of cane into BCS can be considered as a new economic activity for farmers, especially in the area of cane sugar factories in South Sulawesi. Farmers are presently interested in planting sugar cane not only for selling it to sugar cane factories but also for processing it to BSC integrated with cattle farming. There is renewed interest in reintegrating crops and livestock because of concerns about natural resource degradation, the profitability and stability of farm income, long-term sustainability, and fostering diversified cropping systems [37-39].

In general, cane plantation is developed in South Sulawesi on marginal land with limited water availability. BSC-cattle production integration is one form of farm management than can utilize marginal lands with limited water resources. Improving the effectiveness of land resources with limited water resources requires knowledge of good water management practices [40]. Sugar cane in South Sulawesi has huge potential to develop based on the availability of land, suitable climate and sugar cane cultivation techniques that are well known by the farmers. Sugar cane has long been cultivated in the cane planting areas of three factories in South Sulawesi. Over the last two years, the total cane area of these three sugar cane mills have increased by about 5 percent. The increase occurred mainly in *PG-Takalar* (Takalar Sugar Factory while both *PG-Bone* (Bone Sugar Factory) and *PG-Camming* (Camming Sugar Factory) decreased respectively 0.60 and 5.43 percent. This is because the three sugar mills suffered losses.

The small-holder planting area of sugar cane at the three plants showed a significant increase of around 17.84% from 2010 to 2011 due to government programs to increase the domestic production of sugar and reduce sugar imports.

Table 3. shows that the total area of cane for the three plants in South Sulawesi increased from 10,196.42 ha's in 2010 to12,900 ha's in 2011; 28.43 percent and 17.84 percent of the sugar cane factories concessions and the farmers' consessions respectively. The development of sugarcane area owned by farmers is financed by National Development Budget (*APBN-Anggaran dan Pendapatan Belanja Negara*), the credits of food security and energy (*KKPE-Kredit Ketahanan Pangan dan Energy*) from BRI Bank and the Partnership Program for Environmental Development (*PKBL-Program Kemitraan Bina Lingkungan*) from (state owned) PTPN-X company. Although there was an increase in sugarcane acreage, the milling period of the three sugarcane factories was brief due to limited availability of cane as raw materials. Sugarcane acreage is around 13 thousand hectares with an average productivity of 60 tonnes of cane per hectare and milling capacity per factory of 2,500 TCD (tonnes cane per day), so that the production processing period of the three plants is only about 70-90 days per year, which should normally be around 150-180 days per year. Development of BCS could encourage farmers to grow sugar cane and then produce BCS themselves or sell the cane to sugar factories.

3.5 Cattle Breeding Potential

Raising cattle is a way of life for most the farmers in South Sulawesi and serves as a main or as an additional sources of income. Various innovations are being made by farmers, such as the development of collective cages to reduce labor requirements improve the security of livestock, and simultaneously collects the waste that can be converted into organic fertilizer and biogas [41]. Integrated BCS-cattle production is one of the important farm models, generating benefits, creating employment opportunities in rural areas, attracting the rural labor force to work in the agricultural sector where urban-oriented job seeking is predominant. Some forms of employment in the non-agricultural sector are becoming less attractive, while agriculture and other rural businesses have turned into a promising business [42-44].

Table 3.The cane area on the region of three sugar cane factories in South Sulawesi, 2010-2011 (in ha basis)

No.	Factories /Acquisition land	2010	2011	Change	%
I.	**PG-takalar**	**3,276.68**	**4,186.03**	**909.35**	27.75
	- Cane of Factory	2,376.68	3,118.49	741.81	31.21
	- Cane of Farmer	900.00	1,067.54	167.54	18.62
II.	**PG-bone**	**3,822.30**	**3,799.19**	**-23.11**	-0.60
	- Cane of Factory	3,822.30	3,743.49	-78.81	-2.06
	- Cane of Farmer	0.00	55.7	55.70	100.00
III.	**PG-camming**	**5,197.44**	**4,914.99**	**-282.45**	-5.43
	- Cane of Factory	4,253.22	3,864.99	-388.23	-9.13
	- Cane of Farmer	944.22	1,050.00	105.78	11.20
IV.	**Total**	**10,196.42**	**12,900.21**	**2,703.79**	26.52
	- Cane of Factory	8,352.20	10,726.97	2,374.77	28.43
	- Cane of Farmer	1,844.22	2,173.24	329.02	17.84

Sources: South Sulawesi Provincial Office of Plantation; 2012

Table 4. Cow population development in South Sulawesi (2007-2010)

No.	Year	Beef cattle	Dairy cows	Total
1.	2007	666,662	1,867	668,529
2.	2008	703,303	1,919	705,222
3.	2009	769,066	1,826	770,892
4.	2010	848,916	2,198	851,114
5.	2011	1,021,110	1,894	1,023,004
	Growth (%)	11.38	1.12	11.35

Source: South sulawesi statistical livestock; 2011

Table 5. Number of respondents (advanced farmers) in three districts based on rice farming and livestock cultivated

No.	Farm Classifications	District of barru (n=30)	District of wajo (n=25)	District of bone (n=20)
I		**Commodity**		
	Paddy	76.6 %	60.0 %	90.0 %
	Crops	20.0 %	60.0 %	75.0 %
	Bulrush	36.6 %	-	62.5 %
	Cane	-	30.0%	82.5 %
	Cattle	86.3%	25.0%	62.5%
II		**Cattle Farming**		
	Range of Ownership (head)	2-14	2-25	5-25
	Average of Ownership (head)	6.5	8.8	7.0
	Farmers as Agricultural Waste Users	80.0%	15.0%	62.5%
	Intensive Cage	3.3%	5.0%	37.5%
	Seller of Manure	10.0%	25.0%	-

Cattle breeding have the potential and opportunities to develop as a competitive advantage in South Sulawesi. It is supported by the labor input cost components which are relatively cheap, plenty non cultivated agricultural land, and a culture of cattle breeding. The capacity of the dry matter crop waste suitable for animal feed amounted to 25.8 million heads of livestock in South Sulawesi on 2006 [45].

Table 4 showed that over the last five years (2007-2011), the beef population growth in South Sulawesi increased about 11.35 percent per year. This happened because the cow is a commodity and cow breeding is one of the development priorities of the provincial government of South Sulawesi. Various forms of financial assistance were provided by the government, such as the financing for the purchase of cattle which shall be returned to the participating banks for revolving to other farmers. Revolving fund financing is given to sugarcane farmers through Cane Farmer Group (*KPT-Kelompok Petani Tebu Rakyat*) distributed by BRI bank.

BPS reported that the population of cattle in South Sulawesi has reached one million head in 2011 which is a fantastic figure, as it is faster than planned, with a target of reaching one million in 2013 [46]. The districts of Bone, Wajo, and Barru are cattle development centers in South Sulawesi. The proportion of the cattle population of these three districts by about 40 percent of the total population of cattle in South Sulawesi, while 60 percent of the cattle population is spread in other 19 districts in South Sulawesi. Table 5 illustrates that The population density in the two districts of Bone and Barru is very high, respectively 20.73 and 22.68 heads per hectare, while the average cattle density for South Sulawesi was 5.23 heads per hectare and in other areas only 3.74 heads per hectare. This figure shows that the development of an integrated BCS-cattle production in the two districts should be promoted, because the cattle is a main business for most of the famer community in both regions.

Cattle breeding, with the utilization of agricultural waste, and the intensive stall system, are a form of farming activity that can support the development of integrated sugar cane-cattle farming system. Cattle breeding in South Sulawesi has grown from the traditional grazing system to the system supported by intensive stalls with specific grass being grown as fudder. Some farmers use agricultural waste from grain and fodder crops as forage source, but in small quantities.

On all three locations of the research the farmers cultivated paddy and other crops and kept cattle, but forage and sugar cane cultivation was not done in all areas. Cattle management systems differ among the three areas. Cultivation of animal feed and agricultural waste utilization were correlated with cattle management development in the districts of Bone and Barru.

There is great potential area for the development of an integrated BCS-cattle system in the three study sites, Barru, Wajo, and Bone districts. Communities in Wajo district were the first to know or to process cane into brown sugar, but this business was not growing because it was not supported by processing technology. Cattle business is also done by some farmers, but more in the form of grazing system. Semi-intensive or intensive systems have been introduced by the government, but were not well responded by the community. The farmers do not use specific fodder or agricultural waste, so that the cattle business is still a secondary enterprise, even though Wajo district is one of the Bali cattle development centers in South Sulawesi.

Cattle breeding, especially Bali cows is the traditional business or culture for most farmers and growers as well as the main business. Barru district is one of Bali cattle development centers in South Sulawesi as well as the center for germplasm preservation of Bali cattle in Indonesia. Cows are breed by semi-intensive and intensive systems. Semi-intensive is done by using natural sources of feed in the form of forage that grows wild and agricultural waste such as straw, corn stalks, and other crops, while intensive culture is practiced by using a source of food from the cultivation of elephant grass *(orchard grass)* or hay, and by manure and urine processing for organic fertilizers. Farmers in Barru district are not familiar with the cane plant that can be processed into sugar and be used as a forage source for cattle. The average farmer in Barru is raising between 5-10 heads of cattle. It can be said that every 5 cows can provide employment opportunity for around 1 person. Cow manure and biogas are utilized in this area. Feces as manure is sold at Rp 500 /kg, while the urine is processed to liquid fertilizer product, but the concentration and effectiveness of urine use is not known yet by farmers.

Cultivation of sugar cane for crystal sugar has been well known by farmers in the district of Bone since the 1960s, when *PG. Bone* was built. However, the BCS processing technique has not been widely known by sugar cane growers in this area. The top of the cane has not been used extensively as a feed source for livestock. This happens because the sugar harvest is conducted once a year during the milling period of the sugar factory causing the availability of sugar cane tops as cattle feed source to be very short, so it was not a concern to the farmers.

The district of Bone is also a center for cattle development in South Sulawesi with the largest cattle population among districts in South Sulawesi. Cattle, especially Bali cows, is a traditional livestock bred by most farmers and growers as the main business; the cows are bred in semi-intensive and intensive systems. Semi-intensive is done by using natural sources of food in the form of forage that grows wild and agricultural waste such as straw, corn stalks, and other crops, while intensive culture is conducted by using food sources of grass farming (orchard grass) and hay.

3.6 APPROPRIATE TECHNOLOGY FOR BCS-CATTLE INTEGRATION

Cattle, agricultural waste utilization, and intensive stall system is a form of agricultural activity that can support the integration of sugarcane farm and livestock.BCS could be produced easily by routinely harvesting sugarcane after which new shoots will grow that can be harvested after nine months. Sugarcane can be harvested every day during 7-8 months per year outside the rainy season. Household farmers can harvest 0.5 to 1 ton per day depending on the cultivated sugar cane area. Taking in account the nature of this crop, sugar cane shoot leaves can be used as fodder. In addition, there is molasses as a by-product in the process of cooking cane juice into BCS. Molasses is considered as an excellent mixture of cow's feed.

The processing of BCS is an appropriate technology (AT), using a sugar cane juice press that can be manufactured locally, powered by an engine of 8-12 HP (Horse Power), and complemented with a cooking stove using 4-6 pans in line for each stove. The processing capacity is about 2 tons of cane per day. AT-BCS is fuelled efficiently because the bagasses produced from sugar cane juices are enough to cook the sap of sugar cane into BCS. BCS-cattle breeding integration can be developed locally with a small amount of investment. It supports the development of intensive stall integrated with bio digester which is an AT that has been well known to the rural community, but which has not been developed because it is not supported by the availability of adequate forage to grow cows. Economies of scale of integration of both ATs can be reached easily and provide enormous impacts to the economic, social, and environmental aspects. BCS increases employment opportunities in the agricultural sector and farmers' income and simultaneously boosts employment opportunities outside the agricultural sector. The development of employment opportunities outside the agricultural sector associated with BSC will encourage rural and regional economic development. The creation of new rural agricultural employment has a positive influence on the growing of non-agricultural sector [47]. This contrasts with other regions, where the business linkages of agriculture and non-agricultural sectors lowers farm incomes, especially for small-scale farmers who, because of lack of land or capital, are engaging in non-farm employment to supplement farm incomes [48].

BCS-cattle breeding integration will encourage the development of the cattle population growth both in grassland-based and in fattening system breeding. The Balicow's body weight increases around 0.25 percent by feeding 10 percent body weight of forage or 3 percent of dry matter/concentrate of body weight. The average potential of Bali cattle weight gain is thus 0.25 kg per day based own 100 kg cow weight. This weight gain is lower the 0.2-0.3 kg /day with a traditional culture, 0.35 kg/day by supplementing forage, and 0.42 kg/day with forage and concentrate as shown in other research [49]. Intuitively, the farmer could produce best feed from the mixture of the old leaves, shoot leaves, molasses, and sugarcane juice foam. If the price of beef is Rp 80,000 /kg or equivalent to Rp 48,000/kg of price live-weight (60% of the price of meat), it means that farmers can earn Rp 12,000/day of additional cow weight (0.25 kg x Rp 48,000/kg live weight).

Integrated BSC-cattle farming is a system that is highly efficient and environmentally friendly. This system can reduce nutrient leaching in the soil because tillage activities are not carried out, while older leaves of sugarcane can be used as mulch to reduce nutrient leaching during the rainy season. Leaching of N in sugarcane is difficult to avoid given the absence of cover crops, especially during the rainy season [50]. Potential cane shoots are 30 percent of the production of sugarcane and old leaves make up 10 percent totalling to about 40 percent of the biomass. The production of 80-90 tones of sugarcane equals to a biomass production of 125-140 tonnes/ha/year consisting of sugar cane, shoots leaves, and old leaves. Potential old and shoots leaves are available for harvesting (200 days) at around 150-200 kgs per day meeting the needs of 8-10 cows heads per hectare during harvest period. BCS and cattle breeding integration could provide employment for about 3 people per hectare. However, behind this promising potential there is the need to provide at least 20% additional feed and the additional labor needed for bagasse fermentation to overcome lignin content.

It is recommended that the use of sugarcane shoots and leaves should not exceed 8 per cent of the live weight of a cow and a high content of lignin must be fermented [51]. Another problem also occurs because the BCS-cattle production integration requires investments to build processing capacity and sugar cane plantations; these investments are relatively large for most farmers, not only in South Sulawesi, but also in Indonesia in general. This concept is difficult to develop quickly if there is no serious support from the government.

4. CONCLUSION AND RECOMMENDATIONS

When BCS-cattle integration scenarios are developed the scale of household BCS industries have potential for further growth with prospects such as 1) brown sugar can be marketed in the form of brown crystal sugar (*palm suiker*) which is often found in hotels and as raw materials for food industries; 2) BCS can also be developed as an export commodity for which there is a great demand abroad because it is perceived to have less or no chemical content; 3) The BCS-cattle production integration will promote the utilization of agricultural wastes as alternative sources of income; and 4) it is expected in future there will be a processing technology that converts BCS to white crystal sugar, so that the BCS produced in big amounts by cane farmers becomes the raw material for white crystal sugar factories. Thus the white crystal sugar factories could be operated all year long as they will not be depending on the cane harvesting season like conventional crystal sugar factories. BCS-cattle production integration is an alternative concept for rural development and sustainable agriculture in South Sulawesi. There is a great potential for cultivated farm land of around 302,771 ha that can be used for the development and BCS-cattle integration. This occurs because cow breeding is part of the farmer culture and a major source of income for most residents in rural areas which can be improved through the integration of processing BCS. Meanwhile, the development of BCS-cattle integration creates employment opportunities for 3 people per hectare, allows for sugar and meat import substitution, enables biogas as an alternative household energy source, as well as organic fertilizers that encourage environmentally friendly farming. The government support is expected to make pilots in each area involving skilled farmers considered as early adaptors to develop BCS-cattle production integration which will encourage other farmers to be followers.

ACKNOWLEDGEMENTS

The authors would like to express their thanks to Mr. Ad de Leeuw, Director siSinga Consultancy and former director of Netherlands Education Centre Jakarta for editing and valuable suggestions.

COMPETING INTERESTS

Authors have declared that no competing interests exist.

REFERENCES

1. BPS. South Sulawesi in figure 2011. South Sulawesi Statistical Bureau, Indonesia; 2011.
2. DGI. Production, consumption, and import of Indonesian sugar 2005-2013. Indonesian Sugar Board; 2013. Accessed 10 December 2013. Available:http://e-journal.uajy.ac.id/3956/2/1EP18014.pdf
3. Ariesa FN, Tinaprilla N. The comparative study on several countries for sugarcane agribusiness development in Indonesia. Paper presented on National Symposium of Economic Sugar on January12, 2012, Surabaya-Indonesia; 2012.
4. Astuti RS. Rubai as a brown cane sugar exporter, Kompas, Wenedsday, 15 April 2009.
5. Akzar R. Feasibility analysis of brown cane sugar processing of Julu Atia company, Sub district of West Polongbangkeng Selatan, South Sulawesi. Thesis (unpublished). Departement of Management, Bogor Agricultural University-IPB. Bogor; 2012.
6. Sinaga D. Feasibilty Study of Business in Gobal Economy. Mitra Wacana Media. Jakarta; 2009.
7. Eghball B, Ginting D, Gilley JE. Residual effects of manure and compost applications on corn production and soil properties. Agronomy Journal. 2004;96 (March-April 2004):442-447.
8. Kaur K, Kapoor KK, Gupta AP. Impact of organic manures with and without mineral fertilizers on soil chemical and biological properties under tropical conditions. Journal of Plant Nutrition Soil Science. 2005;168(1):117-122.
9. Kihanda FM, Warren GP, Micheni AN. Effect of manure application on crop yield and soil chemical properties in a long-term field trial of semi-arid Kenya. Nutrient Cycling in. Agroecosystem. 2016;76(2-3): 341-354.
10. Yaduvanshi NPS. Substitution of inorganic fertilizers by organic manures and the effect on soil fertility in a rice-wheat rotation on reclaimed sodic soil in India. Journal of Agriculture Science. 2003; 140(2):161-168.
11. Kalyani KS, Rao CC, and Murthy TGK. Impact analysis of empowerment programmes of tribal community in East Godavari, Andhra Pradesh, India. Asian Journal of Agricultural Extension, Economics & Sociology. 2013;2(2),128-139.
12. Hesterman OB, Thorburn TL. A comprehensive approach to sustainable agriculture: W.K. Kellogg's integrated farming systems initiative. Journal of Production Agriculture. 1994;7(1):133-142.
13. El Titi A. Integrated farming: an ecological farming approach in European agriculture. Outlook on Agriculture.1992;21(1):33–39.
14. Mannion AM. Agriculture and environmental change. Temporal and Spatial Dimensions. Wiley, Sussex, UK; 1995.
15. IFOAM. Definitions of organic agriculture. International Federation of Organic Agriculture Movements (IFOAM). 2007. Accessed 2 January 2014. Available:http://www.ifoam.org/organic_facts/doa/index.html
16. Renes G, Ridder G. Are Women Qualified? Labor Economics. 1995;2(1):3-18.
17. Gorgens T. Reservation wages and working hours for recently unemployed US women. Labor Economics. 2002;9(1):92-123.
18. Wolf E. Lower wage rate for fewer hours: A Simultaneous Wage-hours Model for Germany. Labor Economics. 2002;9(5): 643-663.
19. Suryadarma D, Suryahadi A, Sumarto S. Reducing Unemployment in Indonesia: Results from a Growth-Employment Elasticity Model. SMERU Working Paper, Jakarta; 2007.
20. Darma R, Tenriawaru N, Sallatu A. Brown sugar cane-cattle integration as an economic mover. Paper presented on the

national seminar for creative economy based agricultural products, 26 January 2012, Solo, Indonesia.

21. Feridhanusetyawan T, Stahl CW. The impact of trade liberalization on labor markets in the Asian Pacific Region. Asia Pacific Journal of Economics and Business. 2001;5(1):26-60.
22. Wie TK. Economic Growth and Poverty during the New Order and After the Asian Economic Crisis: A Brief Overview. Introductory for Chapter 1 of SMERU Book on Poverty and Social Protection in Indonesia. Jakarta; 2007.
23. Darma R, Arsyad M. Determinants of unemployments in Indonesia: A study on development program of rural employment in Pinrang District, South Sulawesi. Journal of Economics and Development. Ryukoku University, Nagoya. Japan. 2010;5(1):147-159
24. Ellis, F. Household strategies and rural livelihood diversification. Journal of Development Studies. 1998;35(1):1-38.
25. Foster A, Rosenzweig M. Learning by doing and learning from other: Human capital and farm household change in agriculture. Journal of Political Economy. 1995;103(6):1176-1209.
26. Blank SC, Erickson KW. Moss CB. Nehring R. Agricultural profits and farm household wealth. American Journal of Agricultural Economics. 2004;86(5):1299–1307.
27. BPS, South Sulawesi statistical livestock 2010. *South Sulawesi Statistical Bureau*, Indonesia; 2011.
28. BPS. South Sulawesi in figure 2013. South Sulawesi Statistical Bureau, Indonesia; 2013.
29. BPS, South Sulawesi statistical livestock 2010. South Sulawesi *Statistical Bureau*, Indonesia; 2011.
30. Darma R, A. Majdah AZ, Amandaria R. Zakat. Local Social Organization, and Social Capital in Rural Economic Development. Journal of Sociology Study. 2012;2(3):189-197.
31. DGI. Production, consumption, and import of Indonesian sugar 2005-2013. Indonesian Sugar Board; 2013. Accessed 10 December 2013. Available:http://e-journal.uajy.ac.id/3956/2/1EP18014.pdf
32. Ariesa FN, Tinaprilla N. The comparative study on several countries for sugarcane agribusiness development in Indonesia. Paper presented on National Symposium of Economic Sugar on January12, 2012, Surabaya-Indonesia; 2012.
33. Akzar R. Feasibility analysis of brown cane sugar processing of Julu Atia company, Sub district of West Polongbangkeng Selatan, South Sulawesi. Thesis (unpublished). Department of Management, Bogor Agricultural University-IPB. Bogor; 2012.
34. Rosdiansyah. The great demand of brown sugar for foreign market (*Permintaan Gula Merah keLuar Negeri Sangat Besar*); 2012. Accessed 20 December 2013. Avalibale:http://www.lensaindonesia.com/2012/01/24permintaan-gula-merah-ke-luar-negeri.html
35. Akzar R. Feasibility analysis of brown cane sugar processing of Julu Atia company, Sub district of West Polongbangkeng Selatan, South Sulawesi. Thesis (unpublished). Department of Management, Bogor Agricultural University-IPB. Bogor; 2012.
36. Astuti RS. Rubai as a brown cane sugar exporter, Kompas, Wenedsday, 15 April 2009.
37. Russelle MP, Entz MH,Franluebbers AJ. Reconsindering Integrated Crop-Livestock Sysytem in North America, Agronomy Journal. 2007;99(2):325-334.
38. Crews TE, Mohler CL, Power AG. Energetics and ecosystem integrity: The defining principles of sustainable agriculture. American Journal of Alternative Agriculture. 1991;6(3):146-149.
39. Cooper N, Gamon J. Sustainable ariculture what does it mean? The Agricultural Education Magazine. 1991;63(8):12-22.
40. Woods R, Sassenrath GF, Halloran J, Whittaker W. The Economic Feasibility of Implementing Irrigation in Small, Limited Resource Farming Systems. American Journal of Experimental Agriculture. 2013;3(4):664-684.
41. Darma R, A. Majdah AZ, Amandaria R. Zakat. Local Social Organization, and Social Capital in Rural Economic Development. Journal of Sociology Study. 2012;2(3):189-197.
42. Darma R, Arsyad M. Determinants of unemployments in Indonesia: A study on development program of rural employment in Pinrang District, South Sulawesi. Journal of Economics and Development.

Ryukoku University, Nagoya. Japan. 2010;5(1):147-159

43. Ellis F. Rural livelihoods and diversity in developing countries. Oxford: Oxford University Press; 2000.
44. BPS, South Sulawesi statistical livestock 2010. *South Sulawesi Statistical Bureau*, Indonesia; 2011.
45. BPS, South Sulawesi statistical livestock 2010. *South Sulawesi Statistical Bureau*, Indonesia; 2011.
46. Brazil La T. Bukenya JO. Analysis of non-farm employment growth in Alabama. American Journal of Experimental Agriculture. 2012;2(3):288-305
47. ADB. Technical assistance for rural, urban, and subregional linkages in the Mekong region: A holistic approach to development and poverty reduction. 2003; Accessed 20 December; 2013. Available:http://www.adb.org/Documents/TARs/OTH/tar_oth_37156.pdf.
48. Sugama IN, Budiari NG.Rice straw as an alternative feed for Bali heifers. Agriculture Scientific Magazine. 2012;5(1):21-25.
49. Ishikawa S, Ando S, and Yoshikawa S. Effects of planting sugarcane and napier grass on N leaching from lysimeters under high application of cattle manure. American Journal of Experimental Agriculture. 2014;4(5):497-514.
50. Andajani T. General Guidelines for Livestock Integration. Directorate of Feed, Ministry of Agriculture, Republic of Indonesia. Jakarta; 2011.

9

Effects of Plant Spacing on the Growth, Yield and Yield Components of Okra *(Abelmoschus esculentus L.)* in Botswana

Mogapi E. Madisa[1*], Thembinkosi Mathowa[1], Christopher Mpofu[1] and Thato A. Oganne[1]

[1]Department of Crop Science and Production, Botswana College of Agriculture, Private Bag 0027, Gaborone, Botswana.

Authors' contributions

This work was carried out in collaboration between the authors. Authors MEM and TM designed the study and authors TAO and CM carried out field and laboratory work. Authors MEM and TM managed the literature searches and author TM managed statistical analysis. Authors TM and MEM managed the manuscript. All authors read and approved the final manuscript.

Editor(s):
(1) Lanzhuang Chen, Laboratory of Plant Biotechnology, Minami Kyushu University, Japan.
Reviewers:
(1) Ouoba Kondia Honoré, UFR-Sciences and Technology, Polytechnic University of Bobo, Burkina Faso.
(2) Anonymous, Pakistan.

ABSTRACT

Okra is a newly cultivated crop in Botswana. Plant population has been identified as one of the factors that contribute to poor plant development and lower yields. The effects of various intra-row spacing on yield and yield components of okra variety; Clemson Spineless was evaluated at Botswana College of Agriculture in Sebele. The treatments consisted of five intra-row spacings of 30, 45, 60, 75 and 90 cm for treatments 1-5 respectively, each of the treatment was replicated three times in a Randomized Complete Block Design (RCBD). Yield and yield components were determined on five pre-determined plants from each plot and the raw data was subjected to analysis of variance (ANOVA). Generally, a significant treatment effect was revealed for plant height with narrower plant spacing of 30 cm significantly increasing the plant height. Wider plant spacing of 90 cm significantly increased the plant weight, number of branches and leaves. A non-significant treatment effect was observed for stem diameter, fruit length and diameter, number of flowers and fruits. Based on the results wider intra-row spacing of 90 cm is recommended for okra production.

**Corresponding author: E-mail: madisamogapie@yahoo.co, memadisa@bca.bw;*

Keywords: Plant spacing; clemson spineless; vegetative development; A. esculentus; Botswana.

1. INTRODUCTION

Okra (*A. esculentus*) is a member of the mallow family Malveceae and genus Abelmoschus [1-5]. It is a vigorous, half-woody, herbaceous, semi-fibrous, warm season annual and dicotyledonous plant [6]. Okra thrives well in most soil types from sandy loam, loam and or clay soils with a pH range of 5.8-8.0 [7], it is native to an area extending from Ethiopia to the Sudan and was introduced to Egypt in the seventh century [1,2]. It was then carried through North Africa and areas bordering the Mediterranean and eastward [1,2]. Okra is also known elsewhere as okro, gumbo, ochro, bhindee, lady`s finger and guimbabo [3]. There are numerous types which differ in fruiting time and shape, colour of leaves and stem length [8,3,4]. The mostly cultivated varieties are *Abelmoschus caillei* (*A. Chev*) and *Abelmoschus esculentus* (L.) Moench. *A. caillei* is an unconventional okra type which grows naturally in many parts of West Africa whereas *A. esculentus* is a conventional type which is a native of Asia [4]. The successful exotic cultivar to Botswana is *A. esculentus* commonly referred to as Clemson Spineless which is a uniformly spineless, medium dark green, angular pods 12-15 cm long. It grows up to 1.2-1.5 m in height and takes about 55-58 days to fruit [8]. The immature fruits are ready for fresh harvest between three and seven days after flowering [9].

Total world production of okra was estimated at 4.99 million metric tons, cultivated from an area of 0.78 million hectares with an average yield of 6.39 t ha^{-1} [5]. West and Central Africa region accounts for more than 75% of okra produced in Africa, however the average productivity is very low at 2.5 tha^{-1} [10]. Production of okra is all year round but more abundant during the rainy season. However, specific local cultivars are grown in specific areas depending upon their photoperiodic requirements. During dry season okra is grown under irrigation or in valley bottoms or riverbanks using residual moisture [11]. According to Majanbu et al. [12] fresh pod is highly valuable and can be found in almost every African market. Moreover, fresh pods are usually marketed in open streets markets or supermarkets without any kind of temperature or humidity control [13]. Although okra is a new crop in Botswana, it is of economic importance in the country because of the influx of foreign nationals from countries where it is widely consumed.

One of the major aspects of crop ecology, production and management which limit crop production is improper crop spacing in the field [1,14-17] and to some extent fluctuating environmental factors [4]. A spacing of 60-90cm×20-90cm is used depending mostly on the growth habit of the cultivar [14,17]. In addition, wider row spacing is said to be more productive since management practices like weeding and other practices can be easily carried out [1]. Fatokun and Chheda [18] investigated the response of two high yielding okra cultivars to different population densities in Ibadan (Nigeria). It is reported that number and weight of fruits plant^{-1} as well as the vegetative branches plant^{-1} decreased significantly with increase in population density from 61×30.5 cm to 61×91.5 cm. However, with late planting in the season two plants stand^{-1} were found to increase yield for the two cultivars over the single plant stand. Norman [19] reported in studies of four plant population density in Makurdi (Nigeria) that a spacing of 40×40 cm produced not only highest number of productive branches and fruits plant^{-1}, but also the highest fruit weight plant^{-1} and together with spacing of 40×30 cm they produced the highest fruit yield ha^{-1}. Fruit size for spacing 40×30 cm, 40×35 cm and 40 cm × 40 cm which was essentially similar was larger than that of 40×25 cm spacing which produced the lowest number of productive branches and fruits plant^{-1} and the lowest fruit weight plant^{-1} and fruit yield ha^{-1}. Similar results were obtained by [20,6]. Many local farmers practice broadcasting system in Botswana. This wastes seeds planted ha^{-1} and promotes over-crowding, thus increasing competition among plants and making weeding and other farm operations difficult. There has not been substantial research done on okra by the Department of Agriculture Research (DAR) in Botswana, as such information on the effects of plant spacing on the growth and yield of okra is not sufficient. The aim of this study was to evaluate the effect of intra-row spacing on commonly grown okra variety in Botswana.

2. MATERIALS AND METHODS

2.1 Location and Climate

The field experiment was conducted using variety Clemson Spineless at Botswana College of Agriculture (BCA) garden in Sebele from January to April 2010. The garden is located 24º33`S and longitude 25º54`E at elevation of

994 m above sea level, 10 km from Gaborone City center, along Gaborone-Francistown highway. The climate of the study area is semi-arid with an average annual rainfall (30 year mean) of 538 mm [21,22]. Most of rainfall is received in the summer months, starting in late October, continuing to March/April (Fig. 1). Soils are predominantly sandy loams (76% sand, 10% silt and 14% clay) with low water holding capacity, low cation exchange capacity (1.2 meq/100 g) and pH of 6.3.

2.2 Experimental Design and Establishment

The experiment was laid out in a Randomized Complete Block Design (RCBD) with five intra-row spacing of 30 cm, 45 cm, 60 cm, 75 cm and 90 cm being treatments 1-5, thus R_{30}, R_{45}, R_{60}, R_{75} and R_{90}, respectively. The inter-row spacing was kept constant at 50 cm. Each treatment was replicated three times. This gave a total of fifteen (15) plots of 2.5 m by 5.0 m, each plot with six rows including guard rows.

2.3 Cultural Practices

The experiment was established on piece of land previously ploughed using a tractor. The land was leveled using hand tools. Compound fertilizer NPK 2:3:2 (22) was applied once at a rate of 80 kg ha^{-1} before planting. Two Clemson Spineless seeds were sown per hole and thinning out to one seedling per stand at three to four leaf-stage was done. Watering was done once or twice per week depending on the weather conditions to keep the soil moist. Weeds were removed whenever they appeared in the experimental site; the most common and problematic weeds species identified during the experiment were; *Argemone mexicana, Chenopodium carinatum, Datura ferox* and *Tribullus teristeris.*

2.4 Data Collection

Developmental measurements were carried out on five pre-determined plants from each plot. Plants were tagged and the following data was recorded every fortnight from 6 weeks after emergence (WAE); plant height measured using a meter ruler from base to the terminal bud, stem diameter measured using a digital calibrated vernier caliper at about 10 cm above the soil surface, number of leaves and plant braches both determined quantitatively by counting. Number of flowers and fruits produced were both determined quantitatively by counting and were recorded every seven days from 8 and 9 WAE, respectively. A sample of five fruits from each plot was selected at random and their fresh weight measured using an electric balance (PGW 4502e), length measured using a 30 cm ruler, fruit diameter measured using a digital calibrated vernier caliper and seed number determined quantitatively by counting were recorded at termination of the study.

2.5 Data Analysis

The data collected was subjected to analysis of variance (ANOVA) using the STATISTIX-8 program. Treatment means were separated using Tukey's Studentized Range (HSD) Test at P=.05.

3. RESULTS AND DISCUSSION

3.1 Branch-lets and Leaves

Treatments had no effect on okra branching in weeks 6 and 8 and number of leaves in week twelve. However, wider spacing treatments (R_{90}, R_{75}, R_{60} and R_{45}) significantly increased branching in weeks 10 and 12 and leaf production in weeks 6, 8 and 10 as compared to (R_{30}) narrower spacing (Table 1). The non-significant treatment effect observed in weeks 6 and 8 was expected since the plants were still small hence no intra plant competition. While the significant treatment effect observed in weeks 10 and 12 is attributed to less intra plant competition for light and nutrients. Wider spacing helped the okra plant to utilize its energy properly in the production of leaves and branching because there was less competition for light, nutrients nor was overlapping from adjacent okra plants within the row [23]. It was also observed that the taller okra plants had fewer branches and hence fewer number of leaves (Tables 1 and 2). Leaf numbers also decreased over time (Table 1), possibly due to unfavourable weather conditions and senescence. Similar result was obtained by [20,6] where taller okra plants and fewer leaves were observed at closer spacing. Ibeawuchi et al. [1] observed that okra height decreased over time as row spacing increased. The study also showed that okra spaced at 30 cm within the row was significantly taller than other okra plants spaced otherwise from 2-8 weeks after planting. Ijoyah et al. [20] also reported that tallest okra plants were produced from the intra-row spacing of 25 cm which was significantly (P<.05) greater than that produced from the wider intra-row spacing. The number of branches plant^{-1} and leaf area also decreased as intra-row spacing reduced. This may be attributed to competition for light because plants at closer spacing get less amount of sunlight and other growth resources hence their tendency to grow upright instead of lateral and producing branches. Wider spacing helped the plants to utilize its energy properly in branching because there was not much competition for light nor was overlapping from adjacent okra plants within the row.

3.2 Number of Flowers and Fruits

Spacing treatments had no significant (P>.05) effect on the number of flowers and fruits (Table 1). The flower production across the treatments was affected by rainfall that rained every day of the week with an average rainfall amount of 113.9 mm recorded in April (Fig. 1). It was expected that treatment R_{90}, R_{75} and R_{60} respectively would have high flower production but most of the flowers were destroyed and dropped by the rains. Fruit production was also indirectly affected by the loss of flowers. Table 2 shows that treatments R_{90}, R_{75} and R_{60} had notably higher fruit production. This could be attributed to the increase in productive node of okra that increased with row spacing; R_{90} had 77%, while R_{75} and R_{60} had 69% and 66% respectively which affected yield as reported by [1]. However, the differences in means of the fruits produced were not significantly different for all row spacing. Ijoyah et al. [20] reported that the number of okra pods per plant decreased as intra-row spacing reduced.

3.3 Seeds

Treatment R_{60} significantly (P<.01) produced the highest number of seeds per fruit (84.3) though similar to R_{45} (Table 1). Although there was no significant differences in both fruit length and diameter, intra plant spacing of 45 cm and 60 cm had superior values (Table 2). This might have influenced significant differences in number of seeds observed (Table 1). This could be attributed to the fact that at lower plant spacing there was intense competition among the plants while at wider spacing vegetative production was stimulated at the expense of seed production. The findings suggest that 45 cm to 60 cm gives the highest number of seeds. A similar trend was observed by Norman [19], Ijoyah et al. [20] and Maurya [6] for fruit weight and diameter which positively correlates with number of seeds. Moniruzzaman et al. [24] found out that at the spacing of 60×30 cm the plants grew relatively taller and increase seed yield per hectare but reduced number of mature fruits per plant, length and diameter of mature fruit and number of seeds per fruit. The highest number of seeds fruit^{-1} (62.2) was recorded in the widest spacing of (60×60 cm) identically followed by 60×50 cm spacing. Similarly, the highest seed yield plant^{-1} was obtained from the widest spacing (60×60 cm) followed by 60×50 cm. However, the widest spacing of (60×60 cm) gave the lowest seed yield ha^{-1} (1.96 tons).

3.4 Plant Height (cm)

Generally, effect of spacing seemed to be negatively correlated on okra plant height with plants spaced at 30 cm revealing superior absolute numbers compared to the rest. However, the effect was non-significant in week 6, a highly significant and significant (P<.05) effects were observed in weeks 8 and 10, respectively (Table 2). Moreover, a non-significant effect was observed for week twelve (Table 2). This may be attributed to the close spacing of 30 cm which resulted in tall plants, possibly because of intra competition for light. However, both number of branches and leaves increased with increase in plant spacing (Table 1), possibly because of less intra plant completion for other resources such as nutrients. Similar results were reported by [25,17] in green pepper where closer spacing produced the tallest plants and shortest plants were obtained from the widest spacing. Plants spaced at 45, 60, 75 and 90 cm respectively are of approximately the same height possibly because there was less competition for light. Maya et al. [25] and Islam et al. [17] also reported maximum average number of branches per plant was recorded from plants of the widest spacing and the lowest number of branches from closest spacing. This might be due to the fact that the plants of wider spacing receive more light, nutrients and other resources than of close spacing.

3.5 Stem Diameter (cm)

Spacing treatments had no significant (P>.05) effect on okra stem diameter for the entire period of the study (Table 2). However, wide row spacing with lesser plant population revealed superior stem diameter. On contrary, study by Islam et al. [17] on green pepper found out that stem girth was statistically significant due to different plant spacing. The widest spacing produced the maximum stem girth and it gradually decreased with decreasing spacing. The plants planted at widest spacing produced profuse branching without lodging [20]. From the present study, wider row spacing tended to produce stronger plants that did not require any support unlike at closer spacing where the plants were thinner and weaker.

3.6 Fresh Fruit Weight (g)

Treatments effect seemed to be positively correlated on okra fresh fruit weight with 90 cm spacing significantly ($P<.01$) increasing fruit fresh weight for week 8 and a significant ($P<.05$) effect was observed for weeks 9, 10 and 11, respectively (Table 2). Plant spacing influenced the individual fruit weight. The maximum fruit weight was obtained in the widest spacing (90 cm). The widely spaced plants produced heavier fruits as they had stronger plants than at closer spacing. The result are in agreement with Ijoyah et al. [20], Islam et al. [17] and Maurya [6] who reported maximum yield at wide spacing because wider spacing facilitated the plants to develop properly with less inter and intra plant competition for utilizing the available resources resulting in higher yield plant^{-1}.

3.7 Fruit Size (Length and Diameter)

Mean fruit length and diameter were not significantly different at $P>.05$ across the five treatments evaluated (Table 2). However, treatment R_{45} and R_{60} had the largest fruit compared to the rest. In a study by Norman [19], narrower plant spacing of 40×30 cm and 40×40 cm produced the largest fruit as compared to wider plant spacing, this finding partly agrees with the current study for the wider spacing which had smaller fruits. Ijoyah et al. [20] and Maurya [6] obtained significantly higher fruit weight and fruit diameter of okra at spacing of 60×40 cm higher than at wider spacing. The non-significant treatment effect between wider and narrower spacings could be attributed to the effect of rainfall that affected the number of flowers resulting in less competition among fruits at lower plant spacing. The present results seems to suggest that over spacing of plant does not necessarily result in corresponding increase in fruit size because excessive spacing had no impact on fruit size. Instead it leads to underutilization of the land hence lower productivity. Therefore, it can be inferred that intra plant spacing of 45 cm to 60 cm gives the optimum fruit size.

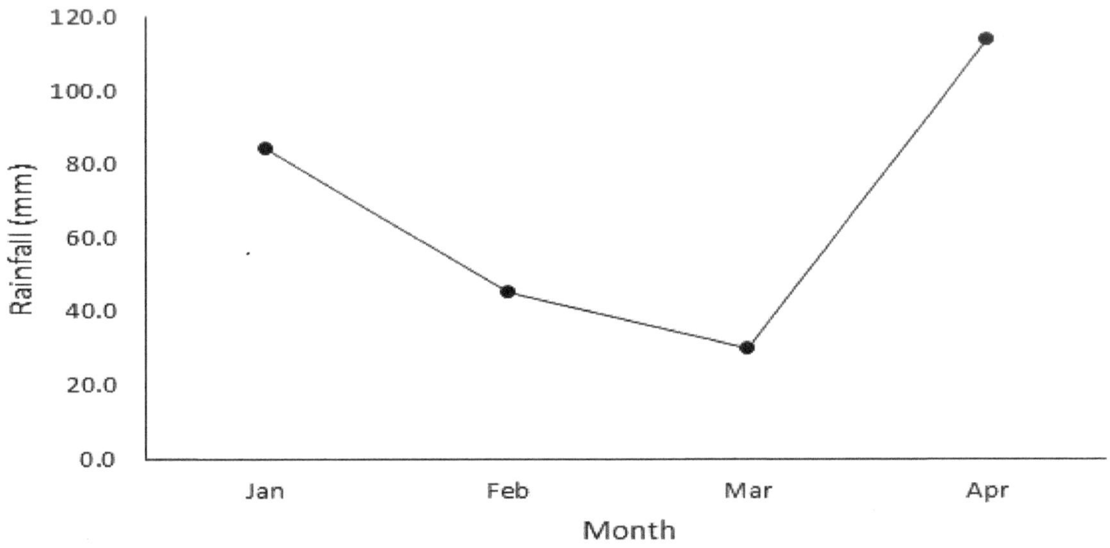

Fig. 1. Mean monthly rainfall recorded during the study

Table 1. Mean branch, leaf, flower, fruit and seed number of okra as influenced by plant spacing

Treatments	Branch				Leaves				Flowers				Fruits			Seeds
	Week 6	Week 8	Week 10	Week 12	Week 6	Week 8	Week 10	Week 12	Week 8	Week 9	Week 10	Week 11	Week 9	Week 10	Week 11	
R_{30}	1.0	1.0	1.0b	1.0b	12.3b	21.7b	22.7b	15.7	1.3	3.0	3.0	2.3	1.0	1.7	1.7	65.3bc
R_{45}	1.0	1.0	1.0b	1.3b	21.7ab	31.3ab	27.7ab	18.7	1.0	2.7	2.7	2.3	1.0	2.0	2.0	79.3ab
R_{60}	1.0	1.3	1.3ab	2.7a	21.7ab	31.0ab	29.7ab	18.7	1.0	2.7	3.0	3.0	0.3	1.7	2.7	84.3a
R_{75}	1.3	1.3	1.7ab	2.7a	22.7ab	35.0a	32.7a	18.7	1.0	3.0	3.0	3.0	0.6	2.7	2.7	59.7c
R_{90}	1.3	1.3	2.0a	3.0a	28.3a	34.7a	32.7a	19.7	1.0	2.7	2.7	2.7	1.0	2.7	2.7	63.7c
Significance	ns	ns	*	**	*	*	*	ns	ns	ns	ns	ns	ns	ns	ns	**
HSD	ns	ns	0.96	1.03	10.85	12.12	8.77	ns	ns	ns	ns	ns	ns	ns	ns	14.39
CV (%)	27.90	26.35	24.40	17.12	18.06	14.01	10.71	20.99	24.21	15.97	13.51	16.06	42.70	24.21	22.13	7.25

*** Highly significant at P<.01, * significant at P<.05, ns non-significant at P>.05. Means separated using tukey's studentized range (HSD) test at P≤.05, means within columns followed by the same letters are not significantly different. Where R_{30}-R_{90} are intra-row spacing of 30, 45, 60, 75 and 90 cm*

Table 2. Mean plant height, stem diameter, fresh fruit weight, fruit length and fruit diameter of okra as influenced by plant spacing

Treatments	Plant height (cm)				Stem diameter (cm)				Weight (g)				Fruit length (cm)	Fruit diameter (cm)
	Week 6	Week 8	Week 10	Week 12	Week 6	Week 8	Week 10	Week 12	Week 8	Week 9	Week 10	Week 11		
R_{30}	256.7	569.7a	759.7a	861.5	1.4	2.0	2.3	2.6	60.8b	55.6b	58.7b	54.5b	13.4	2.7
R_{45}	236.9	527.8b	683.6ab	800.8	1.5	2.7	2.5	2.7	64.1b	62.3ab	59.3b	62.6ab	15.0	3.0
R_{60}	232.7	516.3b	683.7ab	790.7	1.6	2.5	2.6	2.8	65.7b	66.3ab	62.1ab	61.9ab	15.3	2.9
R_{75}	243.5	516.7b	660.7ab	774.3	1.6	2.5	2.5	2.7	69.7ab	72.5ab	76.2a	72.8ab	13.7	2.8
R_{90}	230.5	511.5b	611.2b	722.4	1.7	2.5	2.9	3.1	78.2a	80.6a	71.1ab	80.2a	13.9	2.8
Significance	ns	**	*	ns	ns	Ns	ns	ns	**	*	*	*	ns	ns
HSD	ns	38.73	129.12	ns	ns	Ns	ns	ns	10.06	23.50	15.00	24.60	ns	ns
CV (%)	13.88	2.60	6.75	12.07	18.61	13.44	27.08	17.86	5.28	12.37	8.13	13.16	16.76	9.33

*** Highly significant at P<.01, * significant at P<.05, ns non-significant at P>.05. Means separated using tukey's studentized range (HSD) test at P≤.05, means within columns followed by the same letters are not significantly different. Where R_{30}-R_{90} are intra-row spacing of 30, 45, 60, 75 and 90 cm*

4. CONCLUSION

This study showed that intra-row spacing of 90 cm significantly increased the plant weight, number of branches and leaves. The narrower spacing of 30 cm produced taller and weaker plants, while intermediate intra-row spacings revealed superior values for non-significant parameters. On the basis of these results, wider intra-row spacing of 90 cm is recommended for production of okra. However, further studies need to be conducted to verify the findings.

ACKNOWLEDGEMENTS

We are grateful to the Ministry of Education, Skills and Development for providing a scholarship to Thato A. Oganne. We thank Botswana College of Agriculture for providing resources including, amongst other things, field and laboratory.

COMPETING INTERESTS

Authors have declared that no competing interests exist.

REFERENCES

1. Ibeawuchi II, Obieefuna JC, Ofoh MC. Effect of row spacing on yield and yield of okra (*Abelmoschus esculantum*) and mixture groundnut (*Archish ypogaea*). Journal of Agronomy. 2005;4(4):304-307.
2. Dhankhar BS, Singh R, (Eds). Okra handbook: Global production, processing and crop improvement. HNB Publishing, New York; 2009.
3. Osawaru ME, Dania-Ogbe FM, Chime AO. Epidermal morphology of West African okra *Abelmoschus caillei* (A. Chev.) stevels from South Western Nigeria. Science World Journal. 2011;6(3):5-19.
4. Ade Oluwa OO, Kehinde OB. Genetic variability studies in West African okra (*Abelmoschus caillei*). Agriculture and Biology Journal of North America. 2011;2(10):1326-1335.
5. Konyeha S, Alatise MO. Yield and water use of okra (*A. esculentus* L. Moench) under water management strategies in Akure, South-Western city of Nigeria. Inter. J. Emerging Technology and Advance Engineering. 2013;3(9):8-12.
6. Maurya RP, Bailey JA, St. A. Chandler J. Impact of plant spacing and picking intervals on the growth, fruit quality and yield of okra (*Abelmoschus esculuntus* L. Moench). American Journal of Agriculture and Forestry. 2013;1(4):48-54.
7. Gajete TD. Technoguide for okra production. In: Technoguides for agricultural production and livelihood projects, CLSU Research Office, RET, Science City of Munoz, Nueva Ecija; 2004.
8. Ellersieck M. Growth of okra and fruiting pattern as affected by growth regulators. Hortscience. 1990;25(4):431-433.
9. Cantwell M, Suslow T. Okra: Recommendations for maintaining postharvest quality. Department of Plant Sciences, University of California, Davis; 2013. Available:http://postharvest.ucdavis.edu/pfvegetable/Okra, accessed 02/09/2014
10. Kumar S, Dagnoko S, Haougui A, Ratnadass A, Pasternak D, Kouame C. Okra (*Abelmoschus* spp.) in West and Central Africa: Potential and progress on its improvement. African Journal of Agricultural Research; 2010.
11. Abd El-Kader AA, Shaaban SM, Abd El-Fattah MS. Effect of irrigation and organic compost on okra plants (*Abelmoschus esculuntus*) grown in sandy calcareous soil. Agric. and Bio. J. of North America. 2010;1(3):225-231.
12. Majanbu IS, Ogunlela VB, Ahmad MK, Olarewaju JD. Response of two okra (*Abelmoschus esculentus* (L) varieties to fertilizer: Yield and yield components as influenced by nitrogen and phosphorus application. Fertilizer Research. 1985;6(3):257-267.
13. Babarinde GO, Fabunmi OA. Effects of packing materials and storage temperature on quality of fresh okra (*Abelmoschus esculentus*) fruit. Agricultura Tropica Et Subtropica. 2009;42(4):151-156.
14. Aslam M, Naseer A, Khan M, Mirza S, Naeem Ullah. Effect of different row and planting spacings on soybean yield and its components. Pakistan J. Agric. Res. 1993;14(2&3):143-148.
15. Conley SP, Binning LK, Connell TR. Effect of cultivar, row spacing and weed management on weed biomass, potato yield and net crop value. Amer. J. of Potato Res. 2001;78(1):31-37.
16. Njoku DN, Muoneke CO. Effect of cowpea planting density on growth, yield and productivity of component crops in cowpea/cassava intercropping system.

Journal of Tropical Agriculture, Food, Environment and Extension. 2008;7(2):106-113.

17. Islam M, Saha S, Hasanuzzaman Akand MD, Abudur Rahim MD. Effect of spacing on the growth and yield of sweet pepper (*Capsicum annum* L.). J. Central European Agric. 2011;12(2):328-335.
18. Fatokun CA, Chheda HR. The influence of population density on yield and yield components of okra (*Abelmoschus esculentus* (L.) Moench). Acta Horticulturae. 1983;123:273-281.
19. Norman JC. Observations on the response of okra (*Abelmoschus esculentus* (L.) Moench) to leaf harvest and intra-row spacing. Swazi. J. Sci. Technol. 1988;9(2):1-6.
20. Ijoyah MO, Unah PO, Fanen FT. Response of okra (*Abelmoschus esculentus* L. Moench) to intra-row spacing in Makurdi, Nigeria. Agric. Biol J. N. Am. 2010;1(6):1328-1332.
21. Bekker RP, De Wit PV. Contribution to the soil and vegetation classification of Botswana, Field Document No. 34 FAO/UNDP/Government of Botswana. Soil Mapping and Advisory Services Project; 1991.
22. Legwaila GM, Marokane TK, Mojeremane W. Effects of intercropping on the performance of maize and cowpeas in Botswana. International Journal of Agriculture and Forestry. 2012;2(6):307-310.
23. Anonymous. Vegetables research and development in the 1990s-A strategic plan. AVRDC Publication; 1991.
24. Moniruzzaman M, Uddin MZ, Choudhury AK. Response of okra seed crop to sowing time and plant spacing in South Eastern hilly region of Bangladesh. Bangladesh J. Agril. Res. 2007;32(3):393-402.
25. Maya P, Natarajan S, Thamburaj S. Effect of spacing, N and P on growth and yield of sweet pepper cv. California wonder. South India Hort. 1997;45:16-18.

10

Effect of Cow Dung, NPK and Their Combinations on Soil Properties and Performance of Sweet Potato (*Ipomoea batatas* L.) in Sudan Savanna, Nigeria

M. Haliru[1*], M. Audu[1], S. A Lukman[1], M. M Sauwa[1], I. Aliyu[1] and N. G Hayatu[1]

[1]Department of Soil Science and Agricultural Egineering, Usmanu Danfodiyo University, Sokoto, Nigeria.

Authors' contributions

This work was carried out in collaboration between all authors. Author MH invented the idea, designed and make the overall supervision of the research. Authors MA and SAL conducted and reviewed the statistical analysis respectively. Authors MMS, IA and NGH conducted both the field and laboratory work of the research.

Editor(s):
(1) Alejandro E. Ferrari, Department of Science and Technology, Universidad Nacional de Quilmes, Argentina.
Reviewers:
(1) Deepak Bhattacharya, Head-Oddisi Research Laboratory, C/O Sri Radha Krishna, Bhubaneswar, India.
(2) Tunira Bhadauria, Department of Zoology, Feroz Gandhi College, Kanpur University, India.
(3) Anonymous, USA.
(4) Chukwuma Emmanuel Chibundo, Agricultural and Bioresources Engineering, Nnamdi Azikiwe University Awka, Nigeria.

ABSTRACT

In Nigeria, one of the major problems confronting agriculture is poor soil fertility replenishment strategy that could allow sustainable production. Two field trials were conducted simultaneously in Sokoto (UDUS, Vegetable Research farm) and Zamfara (Bakalori Irrigation Project) both in Sudan-Savanna agro-ecological zone of Nigeria, with the aim of studying the effect of different rates of cow dung, nitrogen-phosphorus-potassium fertilizer and their combinations on soil properties and performance of sweet potato. The research was conducted during the 2011/2012 dry season using ten treatments: 3, 2, and 1tons per hectare of cow dung, 400, 300, and 200 kilogram per hectare of nitrogen-phosphorus-potassium fertilizer, 2 tons of cow dung + 100 kilogram nitrogen-phosphorus-potassium fertilizer, 1.5 ton of cow dung + 150 kilogram of nitrogen-phosphorus-potassium fertilizer, 1 ton of cow dung + 200 kilogram of nitrogen-phosphorus-potassium fertilizer per hectare and no fertilizer application (control). The result obtained was consistent with regard to locations indicating a numerical increased in soil organic carbon, cation exchange capacity(CEC), total nitrogen, available phosphorus, exchangeable calcium and potassium due to the treatments

*Corresponding author: E-mail: halirumuazu@gmail.com;

application as compared to the initial values. However, the effect was not significant (p>0.05) among treatments. While on sweet potato yield the treatments effect was significant (p<0.05) and application of 300 kilogram of nitrogen-phosphorus-potassium fertilizer per hectare gave the highest sweet potato yield at both locations. This research therefore, recommended that treatments used are important materials for soil nutrients replenishment and application of 300 kilogram per hectare of nitrogen-phosphorus-potassium fertilizer is recommended for better sweet potato yield in the study areas.

Keywords: Fertilizer; rate; soil properties; sweet potato.

1. INTRODUCTION

Plant nutrients are essential for the production ofhigh quality crops to providenutrient requirement for the world's expanding population. Plant nutrients are therefore, a vital component of sustainable agriculture. Increased crop production largely relies on the type of fertilizer used to supplement essential nutrients for plants. The nature and the characteristics of nutrients released by mineral and organic fertilizer are different and each type of fertilizer has its advantages and disadvantages with regard to crop growth and soil fertility. The sound management of fertilization must attempt to ensure both an enhanced crop productivity and safeguarded environment. Therefore, a balanced fertilization strategy that combines the use of mineral and organic fertilizers must be developed and evaluated [1].

Sweet potato (*Ipomoea batatas* L.) Belongs to the family *Convolvulaceae*, the family has about 45 genera and 1000 species but only sweet potato (*Ipomoea batatas* L.) Is of economic importance as food, it is an important food and vegetable crop that is widely distributed throughout the world particularly in the Tropics. It originated from Tropical America and has been cultivated probably from B.C. Today, the crop remains one of the three most important root crops in the world, following potato (*Solanum tuberosum*) and cassava (*Manihot esculenta*) [2].

In an effort to reduce over and under application as well as establishing appropriate types of fertilizer that will enhance both nutrient availability to plant and the soil fertility, the need therefore, arises for assessing the comparative effect of different rates of organic, mineral and organo-mineral fertilizers on the soil, growth and yield of crops. This research was therefore, aimed at assessing the effect of different rates of cow dung, nitrogen-phosphorus-potassium fertilizer(NPK) and their combinations on soil properties, growth and yield of sweet potato in Sudan savanna agro ecological zone of Nigeria.

2. MATERIALS AND METHODS

The trials were conducted in Sokoto and Zamfara States, simultaneously. For Sokoto (L1); the trial was conducted at the Usmanu Danfodiyo University, Sokoto Vegetable Research Farm located at Kwakwalawa village in Dundaye district of Sokoto State. Sokoto is located between latitudes 120 and 130 5'N and longitudes 40 8' and 60 4'E at an altitude of 350m above Sea Level (ASL) [3]. In Zamfara (L2); the trial was conducted at the Bakalori Irrigation Project. The area lies on both sides of the Sokoto river between latitude 120 30'-120 50'N and longitude 50 59'-60' 20'E near Talata Mafara, about 110km from Sokoto on the Sokoto – Gusau road in Zamfara State. The treatments consisted of three rates of cow dung, three rates of nitrogen-phosphorus-potassium fertilizer (NPK 15:15:15), three rates of cow dung and nitrogen-phosphorus-potassium fertilizer combinations and control (no fertilizer application) the treatments were coded as:

Treatment	Code
3 tons per hectare of cow dung	CD(H)
2 tons per hectare of cow dung	CD(M)
1ton per hectare of cow dung	CD (L)
400 kilogram per hectare of nitrogen-phosphorus-potassium Fertilizer (NPK)	NPK (H)
300 kilogram per hectare of nitrogen-phosphorus-potassium Fertilizer (NPK)	NPK(M)
200 kilogram per hectare of nitrogen-phosphorus-potassium Fertilizer (NPK)	NPK (L)
2 tons per hectare of cow dung plus 100 kilogram per hectare of nitrogen-phosphorus-potassium fertilizer	CD+NPK (H)
1.5 ton per hectare of cow dung plus 150 kilogram per hectare of nitrogen-phosphorus-potassium fertilizer	CD+NPK (M)
1 ton per hectare of cow dung plus 200 kilogram per hectare of nitrogen-phosphorus-potassium fertilizer	CD+NPK (L)
No fertilizer application	Control

The treatments were arranged in a randomized complete block design (RCBD) replicated three times. Cow dung was applied in full dose and incorporated into the soil one week before transplanting; NPK was applied in split dose: first dose at two weeks after transplanting and the remaining dose at 50 days after the first application, using broadcasting method.

Data was collected for the following growth and yield parameters; Number of leaves and vines per plant, vine length, number of tubers per plant, tuber weight and tuber yield. Soil samples were collected using soil auger from 0 to 30 cm depth before and afterthe experiment and analyzed for soil pH, organic carbon, cation exchange capacity (CEC), total nitrogen and available phosphorus. Data was analyzed using analysis of variance (ANOVA) and Duncan's New Multiple Range Test (DNMRT) for mean separation.

3. RESULTS AND DISCUSSION

3.1 Soil Analysis after the Experiment

The chemical compositions of cow dung and soil before the experiment is presented in Table 1. Result shows that the soil of the two locations was almost the same and cow dung used was rich in nutrients.

Effect of cow dung, NPK and their combinations on soil chemical properties is presented in Table 2. The result indicated that treatments haveno significant effect ($p>0.05$) on soil pH, organic carbon, CEC, total nitrogen and available phosphorus at both locations. This result could be attributed to the time frame at which the experiment was conducted as related to the rate of decomposition, mineralization and crop quality. However, as compared to the initial values, the soil pH declined whereas the concentration of organic carbon, CEC, total nitrogen and available phosphorus increased. With regard to drop in pH, this could be due to formation of organic acid during decomposition of organic materials and or accumulation of nitric acid from nitrification process with application of mineral fertilizer. Decrease in soil pH could be linked to accumulation of plant products which produces various organic acids [4]. Similar observation was reported in a study on the effect of continuous use of chemical fertilizers and manures on soil properties for five years and reported that, the organic carbon level of the soil Increased to 0.24 from 0.19 percent by manure application [5]. The long term effect of farm yard manure on soil properties and yield in cotton – wheat rotation at Northwest Rajasthan. The observation made was, incorporation of 16 tons of farmyard manure accounted for the highest increase in organic carbon content of soil. The increase in CEC was associated with rise in organic matter content [6].Application of farmyard manure resulted in a significant increase in CEC [7]. An increased in CEC through manure application and the lowest CEC of the soil was found in the treatment which received only N. Increase in total nitrogen could be attributed to direct addition of nitrogen through organic and inorganic fertilizers [5]. Continuous addition of manure for 20 years increased the soil total and available nitrogen content significantly from 0.05 to 0.083 percent, while increase was only 0.05 percent in N fertilizer applied plots as large portion of N was removed by the crops [8]. An increase in phosphorus with incorporation of wheat crop residue and farmyard manures was observed while inorganic fertilizer alone decreased available phosphorus when compared with initial status [9]. It was similarly, observed that, farmyard manure treated plots showed more increase in available phosphorus than inorganic fertilizer which was due to coating of sesquioxides by organic materials that reduced phosphorus fixing capacity of the soil [10].

3.2 Sweet Potato Performance

Effect of cow dung, NPK and their combinations on growth parameters of sweet potato is presented in Table 3. The result indicated that treatments had significant effect ($p<0.05$) on all the considered growth parameters at both locations. Where NPK(H) recorded the best sweet potato growth performance while the least performance is in the control, this could be attributed to the nutrient availability throughout the crop vegetative period. This was similarly reported that, vegetative growth parameters of sweet potato tend to increase with increasing application of mineral fertilizer and different forms of organic manure [11]. Shoot dry matter, leaf area and plant height showed highly significant difference in relation to the rate of nitrogen fertilizer, as they increased significantly and linearly with increasing rate of nitrogen application. Hence, the result of this research could be due to high rate and availability of nitrogen in the best treatment NPK(H) [12]. It was reported by many researchers [13-19] that, increased concentration of nitrogen fertilizer can increase nitrogen uptake and this increase has a positive effect on chlorophyll concentration, the photosynthetic rates and leaf expansion.

Table 1. Initial soil analysis and chemical composition of cow dung used

Parameters	UDUS (kwalkwalawa)	Bakalori	cow dung
pH (H_2O) 1:1	5.63	5.9	8.07
Organic carbon (g/kg)	0.33	0.45	0.34
Total nitrogen (%)	0.16	0.17	0.66
Avail. Phosphorus (mg/kg)	4.74	4.67	5.06
CEC (cmol/kg)	3.41	3.94	
Exchangeable Ca (cmol/kg)	0.29	0.32	
Exchangeable Mg (cmol/kg)	0.52	0.72	
Exchangeable K (cmol/kg)	0.35	0.40	0.50

Table 2. Effect of cow dung, NPK and their combinations on soil ph, organic carbon and cec at udus and bakalori in 2011/2012 dry season

	UDUS					BAKALORI				
Treatment	pH	Organic carbon (%)	CEC (cmol/ kg)	Total nitrogen (%)	Avai. P (mg/ kg)	pH	Organic carbon (%)	CEC (cmol/ kg)	Total nitro gen (%)	Avai . P (mg/ kg)
CD (H)	5.3	0.66	5.03	0.23	4.60	5.8	0.61	4.76	0.18	4.36
CD (M)	5.3	0.57	5.16	0.18	4.94	5.8	0.55	5.88	0.20	4.89
CD (L)	5.2	0.34	4.70	0.16	4.22	5.8	0.47	3.70	0.21	4.52
NPK (H)	4.9	0.43	5.00	0.19	4.76	5.0	0.35	5.40	0.18	4.42
NPK (M)	5.1	0.49	4.93	0.15	4.66	5.5	0.27	5.35	0.16	4.71
NPK (L)	5.12	0.55	4.73	0.15	4.56	5.4	0.39	4.83	0.23	4.30
CD+NPK(H)	5.38	0.68	4.83	0.19	4.93	5.6	0.46	4.20	0.20	4.72
CD+NPK(M)	5.23	0.41	4.73	0.2	4.80	5.8	0.54	5.46	0.20	4.36
CD+NPK (L)	5.26	0.47	5.06	0.35	4.54	5.7	0.41	4.90	0.20	4.64
CONTROL	5.10	0.38	4.73	0.16	4.78	5.7	0.41	3.43	0.12	4.62
SE	0.05	0.05	0.08	0.01	0.06	0.03	0.26	0.02	0.06	0.05
level of sig	ns	Ns	ns	Ns	ns	ns	Ns	ns	Ns	Ns

ns = Not significant at 5% level of probability; Avai. P = available phosphorus

Table 3. Effects of cow dung, NPK and their combinations on growth parameters of sweet potato at 14 wat in udus and bakalori during 2011/2012 dry season

	UDUS			BAKALORI		
Treatment	No. of leaves /plant	No. of vines/plant	Vine length(cm)	No. of leaves /plant	No. of vines/plant	Vine length(cm)
CD (H)	197.6c	9.3c	48.4b	231.4cde	12.3cd	49.0cd
CD (M)	209.1c	9.7c	34.7c	221.0def	11.2d	38.0e
CD (L)	172.4e	9.4c	27.5d	183.1f	9.1e	53.6c
NPK (H)	329.4a	16.5a	65.0a	300.9ab	16.6a	71.5a
NPK (M)	187.5cde	11.4b	49.8b	327.3a	15.3b	64.2b
NPK (L)	180.9de	12.7b	48.7c	190.9ef	11.3d	63.0b
CD+NPK (H)	181.7de	9.3c	37.8c	212.8def	11.4d	43.6de
CD+NPK(M)	235.7b	10.7bc	51.0b	277.3bc	13.0c	62.0b
CD+NPK (L)	206.0cd	12.3b	49.5b	241.2cd	11.0d	48.3cd
CONTROL	167.4e	9.6e	35.6c	237.0cde	9.3e	37.6e
SE	7.17	0.63	1.14	14.00	0.52	3.38
level of sig	*	*	*	*	*	*

*Mean followed by the same letter(s) within the same raw are statistically the same at 5% level of probability; * = Significant at 5% level of probability; WAT= Weeks after transplanting*

Table 4. Effects of cow dung, NPK and their combinations on yield and yield components of sweet potato at udus and bakalori during 2011/2012 dry season

Treatment	UDUS			BAKALORI		
	No. of Tubers/plant	Tuber weight/plant	Tuber yield/ha	No. of Tubers/plant	Tuber weight/plant	Tuber yield/ha
CD (H)	7.1abc	0.8bc	14.4c	5.9a	0.8b	18.7b
CD (M)	6.1abc	0.4de	8.2ef	5.7a	0.4de	8.5d
CD (L)	6.1abc	0.5d	10.7de	5.8a	0.3de	6.5d
NPK (H)	7.1abc	0.7c	15.2c	6.4a	1.1a	25.9a
NPK (M)	8.0a	1.3a	27.3a	6.5a	0.9b	22.7a
NPK (L)	6.2abc	0.4de	12.7cd	5.3a	0.5cd	12.5c
CD+NPK(H)	5.9bc	0.9bc	21.1b	5.0a	0.4d	12.5c
CD+NPK(M)	7.6ab	0.9b	19.4b	6.2a	0.6c	14.9c
CD+NPK (L)	5.5c	0.3e	9.6ef	5.0b	0.3de	9.2cd
CONTROL	6.3abc	0.5d	8.0f	3.8b	0.2e	6.3abc
SE	0.58	0.03	0.64	0.05	0.05	1.05
level of sig	*	*	*	*	*	*

Mean followed by the same letter(s) within the same raware statlstlcally the samo at 5% level of significant;
** = Significant at 5% level of probability*

Effect of cow dung, NPK and their combinations on yield and yield component of sweet potato is presented in Table 4 (above). The result indicated that, treatments have significant effect ($p<0.05$) on the yield and yield component of sweet potato at both locations. Where NPK (M) recorded the highest tuber yield at both locations while the least tuber yield was in plot where no fertilizer was applied (control). Increased in yield due to NPK (M) application could be as a result of the readily available N P and K in the best recorded treatment. Mineral fertilizer was reported to have increased shoot weight, leaf area, plant height and subsequently total yield [20-29,12]. Other researches showed that, leaf area, shoot weight, plant height and tuber yield increased by organic fertilization [30,17]. The findings of this research was similarly reported by [31] who reported significant effect on the total tuber yield of sweet potato with application of different rates of Chicken manure.

4. CONCLUSION

This research revealed that, application of different rates of cow dung, NPK and their combination on soil under sweet potato production have no significant effect on soil pH, organic carbon, CEC, total nitrogen and available phosphorus. While performance of sweet potato was significantly influenced by rates of cow dung, NPK and their combination where application of 300 kg/ha of nitrogen-phosphorus-potassium fertilizer gave the best sweet potato yield and therefore, recommended.

ACKNOWLEDGMENT

Special appreciations to the entire staffs of the department of soil science and agricultural engineering, UDUS.

COMPETING INTERESTS

Authors have declared that no competing interests exist.

REFERENCES

1. Chen J. The combined use of chemical and organic fertilizer and/or biofertilizer for crop growth and soil fertility. International workshop on sustained management of the soil rhizosphere system for efficient crop production and fertilizer use. 16-20 October 2006. Level Development Department Bangkok Thailand; 2006.
2. Nwauzor EC, Emehute JKU, Okorocha EAO, Njoku AO, Afuape J, Korieocha D. National root crops and Research Institute, Umudike; 2010. Available:http://www.nreri.org
3. Mamman AB, Oyebanji JO, Peter SW. Nigeria: A people united, a future assured. Survey of states. Gobumo Publishing Company Limited. Calabar, Nigeria. 2000;1(12).
4. Lynch JM. Soil biotechnology. Microbiology factor in crop protection. London, U.K Blackwell. 1983;981.
5. Gattani PD, Jain SV, Seth SP. Effect of continuous use of chemical properties.

Journal of the Indian Society of Soil Science. 1976;24:284-289.
6. Mathur GM. Effect of long term application of fertilizer ad manure on soil properties and yield under cotton wheat rotation in North-West Rayasthan. Journal of the Indian Society of Soil Science. 1997;45:288–292.
7. Yaduvanshi HS, Tripathi BR, Kanwar BS. Effect of continuous manuring on soi roperties of an alfisol. Journal of the Indian Society Of Soil Science. 1985;33:700–703.
8. Anderson FN, Peterson GA. Effective contionous corn (*zea mays* L) manuring fertilization on yield and protein content of the grain and on the soil N content. Agronomy Journal. 1973;65:697–700.
9. Somani LL, Saxena SN. Effect of some organic matter sources on nutrient availability in humus buildup, soil physical properties and wheat yield under field condition. Anals of Arid Zone. 1975;14:149-158.
10. Bharandary V, Omanwar PK. Long–term effect of continuous rational cropping and fertilization on crop yield and soil properties – 11. Effect on EC, pH, organic matter and available nutrient of soil. Journal of Indian Society of Soil Science. 1994;42:387–392.
11. El-Glamry AM. Soil fertility and potato production as affected by conventional and organic systems. Journal of Soil Science and Agricultural Engineering. 2011;2(2):141-156.
12. Najim AA, Hay SeyedHadi MR, Fazali F, TaghiDarzi M, Shamorady R. Effect of utilization of organic and inorganic nitrogen source on the potato shoots, dry matter, leaf area index and plant height, during middle stage of growth. International Journal of Agriculture and Biological Science. 2010;1:1.
13. Tam RK, Magustad OC. Relationship between nitrogen fertilization and chlorophyll content in pineapple leaves, Plantphysiology. 1935;10:159-168.
14. Chapman SC, Barreto HJ. Using a chlorophyll meter to estimate specific leaf nitrogen of tropical maize during vegetative growth. Agronemy Journal. 1997;89:557-562.
15. Vos J, Van Der Putten PEL. Effect of nitrogen supply on leaf growth, leaf nitrogen economy and photosynthetic capacity of potato. Field Crops Research. 1998;59:63-72.
16. Janat M. Efficiency of nitrogen fertilizer for potato under fertigation utilization a nitrogen traler techniques. Communication in Soil Science and Plant Analyses. 2001;38:2401-2422.
17. Abou–Hussein SD, Shorbagy TEL, Abou-Hadid AF, El-Behairy U. Effect of cattle and chicken manure with or without mineral fertilizers on tuber quality and yield of potato crop. ISHS ACta Horticulture. 2003;698:95-100.
18. Rashid MY, Voroney P, Parkin G. Predicting nitrogen fertilizers requirement for corbby chrolophyll metter under different N availability condition. Canadian Journal of Soil Science. 2005;85:149-159.
19. Zebarth BJ, Rosen CJ. Research perspective on nitrogen BMP development for potato. American Journal of Potato Research. 2009;84:3-1.
20. Bradley GA, Pratt AJ. The effect of different combination of soil moisture and nitrogen level on early plant development tuber set of the potato. American Potato Journal. 1995;32:254-258.
21. Watson DJ, Wilson JH. An analysis of the effects of infection with leaf roll virus on the growth and yield of potato plants, and of its interation with nutrient supply and shading. Annuals of Applied Biology. 1956;44:390-396
22. Logan BJ. Effect of irrigation and nitrogen on yield and quality of potatoes growth in semi-arid westerm. N.W.S ISHSACT A Horticulture. 1989;247:243-247.
23. Biemond H, Vos J. Effect of nitrogen on the developmet and growth of the potatos plant. 2. The partitioning of dry matter nitrogen and nitrate. Annals of Botany; 1992.
24. Vos J. The nitrogen response of potato (*Solanun tuberosum* L) in the field: Nitrogen uptake and yield, harvest index and soil nitrogen response to water and nitrogen. Agronomy Journal. 1997;90:420–429.
25. Meyer RD, Marcum DB. Potato yield, petoide nitrogen, and nitrogen response to water nitrogen. Agronomy Journa. 1998;90:420–429.
26. Vos J, Van Der Putten PEL. Effect of nitrogen supply on leaf growth, leaf nitrogen economy and photosynthetic capacity of potato. Field Crops Research. 1998;59:63-72.
27. Zvomuya F, Rosen CJ, Miller JC. Response of russet nortokah clonal

selections to nitrogen fertilization. American Journal of Potato Research. 2002;79:231-239.

28. Kumar P, Pandey SK, Singh BP, Singh SV, Kumar D. Effect of nitrogen rate on growth, yield, economic and crop quality of Indian potato processing cultivars. Potato Research. 2007;50:143-155.
29. Zelalem A, Ekalign TT, Nigussies D. Response of potato (*Solanum tuberosum* L.) to different rate of nitrogen and phosphorus fertilization on vertisols at Debre Berhan, in the central highlands of Ethiopia. African Journal of Plant Science. 2009;3:16-24.
30. Stoner KA, Ferrandino FJ, Gent MPN, Elmer WH, Lamendia JA. Effect of straw mulch, spent musroom compost, and fumigation on the density of Colorada potato be bttbee the (*Coleopteran*: *Chrysomelidae*) in potatoes. Journal of Economic Entomology. 1996;89:1265-1280.
31. Magagula NEM, Ossom EM, Rhykered RL, Rhykerd CL. Effect of chicken manure on soil properties under sweet potato (*Ipomoea batatas* (L) Lam) culture in Swaziland. American – Eurasian Journal of Agronomy. 2010;3(2):36-43.

11

A Rapid Morphological Screening Procedure for Pea (*Pisum sativum* L.) under Drought Stress in Greenhouse Settings

Eliane T. Bodah[1*], Kurt Braunwart[1], Brian W. Bodah[2] and Alcindo Neckel[3]

[1]Division of Breeding, ProGene Plant Research LLC, 860 S. Crestline, Othello, WA 99344, USA.
[2]WSU Extension, Pierce County Extension, Washington State University, 3602 Pacific Ave, Tacoma, WA 98418, USA.
[3]Department of Geography, Faculdade Meridional University, Passo Fundo, RS 99100, Brazil.

Authors' contributions

This work was carried out in collaboration between all authors. Authors ETB and BWB designed the study, wrote the protocol and wrote the first draft of the manuscript. Authors KB and AN reviewed the experimental design and all drafts of the manuscript. Author ETB managed the analyses of the study. Author BWB identified the plants. Author ETB performed the statistical analysis. All authors read and approved the final manuscript.

Editor(s):
(1) Masayuki Fujita, Department of Plant Sciences, Faculty of Agriculture, Kagawa University, Japan.
Reviewers:
(1) Inge Gazendam, Agricultural Research Council-Roodeplaat, Vegetable and Ornamental Plant Institute, Department of Crop Protection, South Africa.
(2) Edward Missanjo, Malawi College of Forestry and Wildlife, Private Bag 6, Dedza, Malawi.
(3) Klára Kosová, Laboratory of Plant Stress Biology and Biotechnology, Division of Crop Genetics and Breeding, Crop Research Institute, Prague, Czech Republic.
(4) Anonymous, Hungary.

ABSTRACT

Aims: This work proposes a rapid morphological screening procedure for peas grown under drought stress aiming to provide phenotypic evaluations for further research.
Materials and Methods: The plant yield and morphological changes of two widely grown pea cultivars ('Aragorn' and 'Banner') to drought stress was evaluated in this work. A total of eighty seeds of each pea variety were grown in a randomized completed block with four replicates in a greenhouse located in Moscow, ID. Plants were subjected to optimal irrigation (1400 ml as a control) and two treatments that reduced optimal irrigation rates by 40% (840 ml, moderate drought) and

Corresponding author: E-mail: eliane.bodah@email.wsu.edu;

60% (560 ml, severe drought) to induce drought stress. The test was repeated.
Results: Varieties significantly ($P<0.05$) differed in their response to water deficiency. The variety 'Banner' appeared to be the most drought tolerant than 'Aragorn' with high values at control, moderate and severe as (85.00±4.08) cm, (87.21±3.26) cm and (66.02±2.92) cm respectively for total plant height, while the values for total dry weight were (3.65±0.20) g, (2.18±0.42) g and (1.26±0.10) g respectively. Similarly, there were significant ($P<0.001$) differences among treatments on growth parameters, with highest values recorded at optimal irrigation (1400 ml) as (21.52±2.42) cm, (62.18±4.68) cm, (0.24±0.04) g, (2.86±0.38) g and (2.30±0.18) seeds/pod for root length, shoot height, root weight, shoot weight and number of seeds respectively. The growth parameters decreased with a decrease of irrigation.
Conclusion: As drought and stress conditions are expected to increase as global climate change progresses, breeding for drought is a promising area. Therefore, the variety 'Banner' has been identified as a potential parental material to be used in breeding for drought tolerance.

Keywords: Drought; greenhouse plants; agronomic methods; stress tolerance; peas.

1. INTRODUCTION

Over the past several decades, drought stress has become one of the biggest concerns in agriculture worldwide. Adaptation to abiotic stresses is an important requirement of agriculturally relevant crops and development of enhanced drought tolerance plays an essential role in plant breeding [1].

The development of drought tolerant plants is of increasing priority and relevance as the area affected by significant drought is expected to markedly increase in the future due to global climate change [2]. While progressive water withdrawal has been studied in the greenhouse for beans, it generally does not mimic actual field conditions [3].

While limited studies and techniques for simulating field drought stress in a greenhouse environment have been conducted [4], more studies are needed to adequately formulate a reliable protocol for legumes, especially peas. It is the aim of this research to provide a rapid morphological screening procedure for pea under drought stress in greenhouse settings.

The use of peas as a model crop for genetic studies has a very rich history, beginning with Mendel in the 1860s. *Pisum sativum* is adapted to cool, semi-arid to sub-humid growing conditions [5]. However, pea has been grown in diverse environments across the globe and the demand for production continues to increase [6]. Drought is of great concern in pea and it has been reported to result in a loss of between 38.5 and 43% production under different stress levels [7].

2. MATERIALS AND METHODS

Two dry pea cultivars, 'Aragorn' and 'Banner,' commonly grown in the U.S. and New Zealand and developed by ProGene Plant Research LLC in conjunction with NZ Plant Research LTD were utilized for this experiment. 80 seeds of each genotype were prepared by sterilization utilizing a 10% bleach aqueous solution, rinsed in sterile, distilled H_2O and air dried for a period of 24 hours.

Five seeds of each variety were planted in a 3.75 liter high density polyethylene/polypropylene polycan container (Anderson Pots, Oregon, USA) filled with 690 g of Sunshine Mix #1 growing media (Sun Gro Horticulture, Agawam, MA) and arranged in a complete randomized design with four replicates per cultivar. The experiment was replicated in triplicate.

The peas were grown under controlled greenhouse conditions with temperatures ranging from 12 to 18°C at night and 21 to 27°C during daylight hours (Micro Grow Greenouse Systems Growmate (Temecula, CA) utilizing 400 Philips HID Ceramalux High Pressure Sodium bulbs). The plants were subjected to a 12-hour photoperiod per 24 hours.

Three different irrigation regimes were tested: control, moderate drought and severe drought. Field capacity of the growing media served as the experimental control. Field capacity was determined by wetting 690 g of media until saturation was reached. The media was then allowed to drain for a period of 24 hours and oven dried to determine maximum moisture holding capacity (field capacity).

Previous field trials conducted by Pro Gene Plant Research [8] have shown that approximate reductions in irrigation of 40% and 60% caused visual drought stress symptoms on plots of field peas. Control plants were irrigated to the full field capacity of the growing media (1400 ml) and irrigated once every 7 days until flowering and then once every 3.5 days from flowering until maturity. Drought stress test pots received amounts of water decreased by 40% (840 mL) and 60% (560 mL) of field capacity, respectively, at the same watering interval.

Phenotyping methods consisted of the collection of the following agronomic notes: emergence and flowering dates, plant survival, shoot height and root length, shoot and root dry weight, number of pods per plant, pod length and width, and number of viable seeds produced per plant. Statistical analysis of the data, including the Fisher's LSD method, was performed using Minitab 17 Statistical Software (Minitab, State College, PA) and Agrobase Generation II® (Agronomix Software Inc., Winnipeg, Manitoba).

3. RESULTS AND DISCUSSION

As shown in previous research, emergence and flowering dates were not affected by drought stress treatments [9]. All 'Banner' treatments (including control) flowered 52 days after planting, while all 'Aragorn' treatments (including control) flowered 54 days after planting. However, varieties differed in their response to water deficiency and plant survival was jeopardized. The 40% reduction from field capacity trial yielded a 20% rate of mortality for both varieties tested. The 60% reduction from field capacity trial resulted in 20% mortality in 'Aragorn' and 27% in 'Banner.' For all morphological measurements: shoot height, shoot dry weight and number of seeds per pod showed a consistent decrease due to drought for both varieties (Table 1).

A statistically significant ($P<0.001$) difference in shoot height was observed in both variety and treatment ($P<0.001$) comparisons. 'Banner' was taller than 'Aragorn' in all treatments and both varieties suffered from increasing stunt with relation to decreasing irrigation (Fig. 1).

'Banner' and 'Aragorn' exhibited a 6.3% and 9.8% reduction in plant shoot height, respectively, in the moderate drought treatment. Both had the least shoot height, at only 70% of their potential, when compared to the control plants, in the severe drought treatment. Shoot dry weight was also impacted by drought, being significantly different ($P<0.001$) among treatment and between varieties. For the moderate and severe drought stress treatments, 'Banner' shoot dry weight (Fig. 2) was reduced by 44% and 70%, respectively, while 'Aragorn' was reduced by 55% and 69%, respectively.

Growth reduction was one of the major drought stress treatments results. It was also observed that leaf area was visually reduced. In a similar study involving wheat cultivars, leaf area was also reduced but root development was retarded in only one cultivar [10]. A reduction in plant height, lower leaf and stem dry weights were also shown, in addition to the production of a lower overall aerial biomass among drought-stressed plants grown in a greenhouse environment [4].

The maintenance of root growth as observed in 'Banner' under drought stress condition can be an important character of drought tolerant plants. Root length did not significantly $P>0.05$) differ among treatments and between varieties. 'Aragorn' roots were not consistently affected on the moderate drought treatment but had an important reduction of 32% on root length on the severe drought; while 'Banner' roots were not consistently affected between the treatments. Root dry weight did not differ ($P>0.05$) significantly among treatments and between varieties. Overall, 'Aragorn' produced more roots than 'Banner' in all trials.

The number of pods per plant were highly significantly ($P<0.001$) different among treatments and also showed a significant ($P<0.05$) difference between varieties. In all treatments, 'Banner' had more pods than 'Aragorn' but both varieties exhibited a reduction in the number of pods with increasing drought. 'Banner' displayed a 38.5% and 41.5% reduction in the overall number of pods as drought increased, respectively; while 'Aragorn' had a reduction of 42.7% and 56%, respectively.

In addition, pods also became shorter and thinner with decreasing irrigation when compared to the control (Table 1). 'Aragorn' did not produce any viable seeds in either drought treatment while 'Banner' was able to produce viable seed throughout all treatments. 'Banner' suffered a reduction in the number of viable seeds by 19% in the moderate drought treatment and 27% in the severe drought treatment.

Yield measured by yield components (number of pods, pod length, width, and number of viable seeds per pod) was also reduced under simulated drought conditions, with simulated severe drought underperforming all other treatments. In a similar study, analysis of variance has indicated that there were significant differences between genotypes regarding agronomic and morphological traits of wheat under drought [11].

When both varieties are combined for the analysis in each treatment it is possible to observe the overall effect of drought on the investigated plant material. Significant differences can be then observed across treatments for all the evaluated parameters (Table 2).

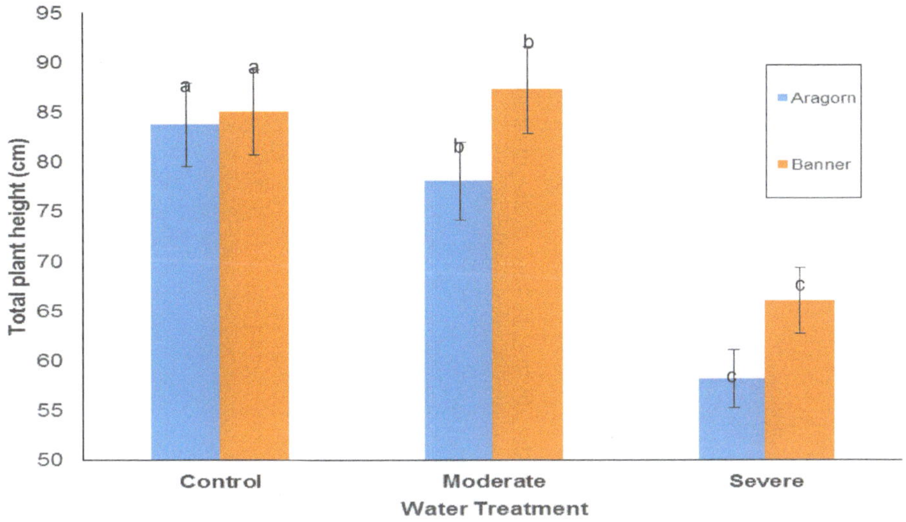

Fig. 1. Plant total height (cm) of 'Banner' and 'Aragorn' across water treatments - control (1400 ml), moderate drought (840 ml) and severe drought (560 ml) stress. Total height included shoot and root measurements. The letters a, b, and c represent the grouping information for the treatments using Fisher LSD method and 95% confidence. Means that do not share a letter are significantly (P<0.05) different

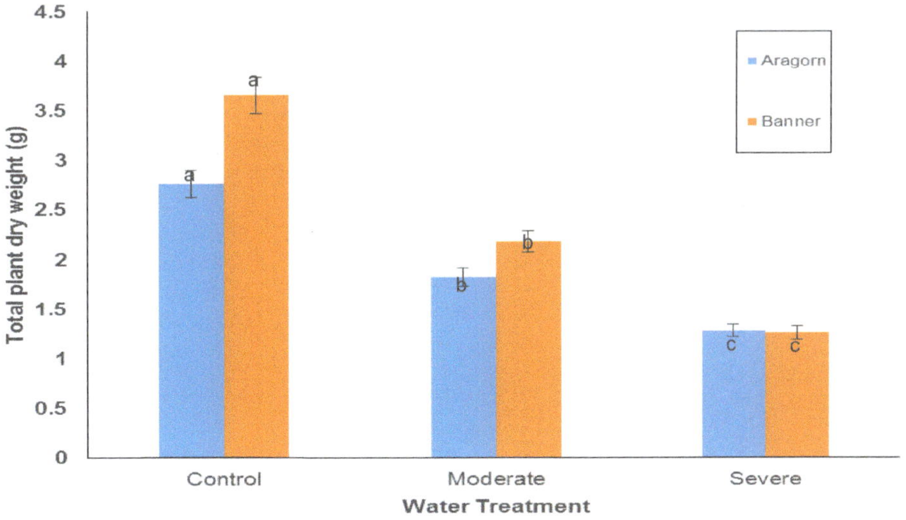

Fig. 2. Total plant dry weight (g) of 'Banner' and 'Aragorn' across water treatments: control (1400 ml), moderate drought (840 ml) and severe drought (560 ml) stress. Total dry weight included shoot and root measurements. The letters a, b, and c represent the grouping information using Fisher LSD method and 95% confidence for the treatments. Means that do not share a letter are significantly (P<0.05) different

Table 1. Morphological data means collected in each treatment (control, moderate, and severe drought), showing: varieties, root length, shoot height, shoot weight, root weight, pod length, pod width, number of pods per seed, number of seeds per pod, and mortality

Irrigation treatment	Variety name	Root length (cm)	Shoot height (cm)	Root weight (g)	Shoot weight (g)	Pod length (cm)	Pod width (cm)	Number of pods/ plant	Number of seeds/pod	Mortality %
Control	Aragorn	22.54±2.1	61.21±2.6	0.32±0.04	2.44±0.2	2.99±O.67	0.29±0.07	2±0.22	0.60±0.4	0
	Banner	17.57±1.2	67.43±3.9	0.18±0.02	3.47±0.33	4.79±0.24	0.59±0.01	2.6±0.23	2.75±0.45	0
Moderate	Aragorn	22.79±2.58	55.27±1.6	0.27±0.03	1.55±0.14	1.55±0.03	0.067±0.01	1.08±0.23	0.00	20
	Banner	24.02±2.56	63.19±2.03	0.23±0.03	1.95±0.13	4.22±0.3.2	0.55±0.04	1.5±0.23	2.23±0.37	20
Severe	Aragorn	15.25±1.5	42.93±2.03	0.19±0.031	1.09±0.13	1.91±0.46	0.12±0.05	0.83±0.27	0.00	26
	Banner	19.48±1.48	46.54±2.52	0.19±0.02	1.07±0.1	3.92±0.39	0.59±0.02	1.07±0.19	2.00±0.33	27

Table 2. Morphological data means across treatments (control, moderate and severe drought) combining both varieties

Treatment/parameters	Control*		Moderate*		Severe*	
Root length	21.52±2.42	A	19.57±3.65	A	15.57±2.11	B
Shoot height	62.18±4.68	A	54.37±2.56	B	40.22±3.20	C
Root weight	0.24±0.04	A	0.23±0.04	AB	0.17±0.04	B
Shoot weight	2.86±0.38	A	1.58±0.19	B	0.96±0.13	C
Number of pods	2.30±0.18	A	1.29±.17	B	0.92±16	C

* *Grouping information using Fisher LSD method and 95% confidence. Means that do not share a letter within a row are significantly ($P<0.05$) different*

Our data suggests that plant characteristics can influence drought tolerance. Comparing the two varieties used in our study, 'Banner,' the variety that performed better under drought stress, has good early vigor emerging five to six days earlier than 'Aragorn' [8].

There are many metabolic changes that can enable a plant to withstand drought, including specific proteins such as chaperonins and dehydrins that are induced in certain plants under drought [12]. Water stress also induced the accumulation of soluble sugars in epicotyls between 2.8- and 5.1-fold in another study and the osmotic adjustment varied greatly depending on cultivar [13]. Moreover, temperature can contribute to drought stress and can induce significant changes in plant cell gene expression [14].

4. CONCLUSION

Our study has proposed a simple, rapid morphological screening procedure for pea under drought stress in greenhouse settings and provided evaluation of two commercial pea cultivars. 'Banner' appeared to be the most drought tolerant, which was reflected by both growth and yield parameters. 'Aragorn' had shown many drought stress symptoms even in the moderate drought treatment, such as very low yield. Our work also identified potential parental material to be used in breeding. As drought and stress conditions are only expected to increase as global climate change progresses, breeding for drought is a promising area.

COMPETING INTERESTS

Authors have declared that no competing interests exist.

REFERENCES

1. Witt S, Galicia L, Lisec J, Cairns J, Tiessen A, Araus JL, Fernie AR. Metabolic and phenotypic responses of greenhouse-grown maize hybrids to experimentally controlled drought stress. Mol Plant. 2012; 5(2):401-17. DOI: 10.1093/mp/ssr102.
2. de Paiva Rolla AA, Carvalho JDFC, Fuganti-Pagliarini R, Engels C, do Rio A, Marin SRR, Oliveira MCN, Beneventi MA, Corrêa F, Guimarães M, Farias JBC, Neumaier N, Nakashima K, Yamaguchi-Shinozaki K, Nepomuceno AL. Phenotyping soybean plants transformed with rd29A: AtDREB1A for drought tolerance in the greenhouse and field. Transgenic Research. 2014;23(1):75-87.
3. Kavar T, Maras M, Kidric M, Sustar-Vozlic J, Meglic V. Identification of genes involved in the response of leaves of *Phaseolus vulgaris* to drought stress. Mol Breeding. 2008;1(21):159-172.
4. Pennypacker BW, Leath KT, Stout WL, Hill RR. Technique for simulating field drought stress in the greenhouse. Agronomy Journal. 1990;82(5):951-957.
5. Coyne CJ, McGee RJ, Redden RJ, Ambrose MJ, Furman BJ, Miles CA. Genetic Adjustment to Changing Climates: Pea. Wiley Blackwell: Chichester, UK. 2011;1:238-250.
6. EMBRAPA – Empresa Brasileira de Pesquisa Agropecuária. Centro Nacional de Pesquisa de Hortaliças. Ervilha para produção de grãos verdes; 2008. Available:www.cnph.embrapa.br/cultivares/ervilha
7. Carvalho JA, Rezende, FC, Aquino RF, Freitas WA, Oliveira EC. Produção da ervilha cultivada em ambiente protegido sob diferentes tensões de água no solo. Revista Brasileira de Engenharia Agrícola e Ambiental, Campina Grande. 2012; 1(16):44-50.
8. Braunwart K, Wood M. Drought stress on pea crops. [Oral communication]. ProGene Plant Research. Othello, WA, US; 2014.
9. Muchow RC. Phenology, seed yield and water use of grain legumes grown under different water regimes in a semi-arid tropical environment. Field Crops Research. 1985;11(1):81-87.
10. Klamkowski K, Treder WW. Response to drought stress of three strawberry cultivars grown under greenhouse conditions. J. Fruit Ornam. Plant Res. 2008;180(16):179-188.
11. Ahmadizadeh M, Nori A, Shahbazi H, Aharizad S. Correlated response of morpho-physiological traits of grain yield in durum wheat under normal irrigation and drought stress conditions in greenhouse. African Journal of Biotechnology. 2014; 10(85):19771-19779.
12. Vaseca I, Akiscan Y, Demirevska K, Anders I, Feller U. Drought stress tolerance of red and white clover: comparative analysis of some chaperonins and dehydrins. Scientia Horticulturae, Canada. 2011;4(130):653-659.

13. Sánchez FJ, Andrés EF, Tenorio J, Ayerbe L. Growth of epicotyls, turgor maintenance and osmotic adjustment in pea plants (*Pisum sativum* L.) subjected to water stress. Field Crops Research. 2014;2(86): 81-90.

14. Talalaiev O, Korduym E. Expression of small heat shock protein (sHSP) genes in the garden pea (*Pisum sativum*) under slow horizontal clinorotation. Plant signaling & behavior. Plant Signaling & Behavior; 2014. DOI:10.4161/psb.29035.

12

Influence of Pelleted and Unpelleted Composted Organic Waste Materials on Growth, Dry Matter Accumulation and Yield of Three Varieties of Cucumber (*Cucumis sativus*) in the Greenhouse

E. E. Ikenganyia[1], U. M. Ndubuaku[1*], C. C. Onyeonagu[1] and U. Ukonze[2]

[1]Department of Crop Science, Faculty of Agriculture, University of Nigeria, Nsukka, Nigeria.
[2]Department of Science Education, University of Nigeria, Nsukka, Nigeria.

Authors' contributions

This work was carried out in collaboration between all authors. All authors read and approved the final manuscript.

Editor(s):
(1) Peter A. Roussos, Lab. Pomology, Agricultural University of Athens, Greece.
(2) Masayuki Fujita, Department of Plant Sciences, Kagawa University, Japan.
Reviewers:
(1) Robert Kalbarczyk, Institute of Landscape Architecture, Wrocław University of Environmental and Life Sciences, Poland.
(2) Anonymous, Nigeria.
(3) Anonymous, Egypt.

ABSTRACT

The study was conducted in a greenhouse to determine the influence of pelleted and unpelleted composted organic waste materials on the morphological growth, dry matter accumulation and yield of three varieties of cucumber (*Cucumis sativus*). The experiment was a 3 × 13 factorial trial in a completely randomized design (CRD) with three replications. The trial consisted of one hundred and seventeen polythene pots laid out in the greenhouse of the Teaching and Research Farm of Department of Crop Science, University of Nigeria, Nsukka. The treatments were pelleted and unpelleted composted rice husks (100%), moringa pod husks (100%), maize cobs (100%), rice husks + poultry manure (75%:25%, volume to volume; v/v), moringa pod husks + poultry manure (75%:25%,v/v), maize cobs + poultry manure (75%:25%,v/v), control and three varieties of cucumber (Poinsett, Marketer and Supermarketer) giving a total of thirty-nine treatment combinations. Cucumber plants grown in soil amended with unpelleted composted rice husks + poultry (75%:25%, v/v) had significantly (P = .05) the highest values of plant height, leaf area per

**Corresponding author: E-mail: uchemay@yahoo.com;*

plant, number of leaves per plant, stem girth, dry matter accumulation in the leaves and stems while those grown in the pelleted rice husks + poultry manure significantly (P = .05) gave the highest values of dry matter accumulation in the roots of the three cucumber varieties after six weeks of planting (the peak of vegetative growth). The unpelleted rice husks + poultry manure significantly (P = .05) also gave the highest values of number of fruits and fruit weight per plant. Supermarketer gave the highest values of the morphological growth traits, dry matter accumulation and yield in all the treatments followed by Marketer, then Poinsett.

Keywords: Pelleted; unpelleted; compost; poinsett; marketer; supermarketer.

1. INTRODUCTION

One of the challenges facing crop production in the tropics is the inherent low concentration of essential nutrients in the soil for crop growth and development. Essential nutrients are those required by plants to complete their life cycle such as nitrogen (N), phosphorus (P), potassium (K) known as primary elements; calcium (Ca), magnesium (Mg) and sulphur (S) known as secondary elements; and micronutrients like boron (B), copper(Cu), manganese (Mn), iron (Fe), molybdenum (Mb) zinc (Zn) etc.[1]. These nutrients can be provided to the soil through the use of organic or inorganic fertilizers. The inorganic fertilizers are associated with increased soil acidity, leaching and nutrient imbalance [2]. The nutrients in organic manures are released more slowly and are stored in the soil for a long time. Manure application also increases soil porosity, aggregate stability, soil water infiltration and holding capacity. It also enhances soil organic matter content, pH, cation exchange capacity and nutrient availability [3, 4 and 5]. However, the main defect in the use of organic fertilizers (manures) in crop production is the limited nutrient supply and bulkiness. Composting is a management practice in agriculture to improve manure quality and efficiency. Well managed compost has good agronomic properties such as good water holding capacity, light weight, small particle size. It is cheap and can easily supply nutrients to the crop [6].

Cucumber (*Cucumis sativus*) is a vegetable crop eaten by most families in the Tropics. It is one of the most popular members of *Cucurbitaceae* (vine crop) family. It is cultivated for fresh fruit which is locally consumed or exported to increase national income. The crop is cultivated in most parts of Northern Nigeria and some parts of Eastern Nigeria by peasant farmers who lack information on some important cultural practices [7]. Cucumber is a vital ingredient in vegetable salads. The production of this crop requires soil with high levels of organic matter content to provide the essential nutrients and soil physical conditions needed for proper growth and development of the crop. However, there is paucity of information on the use of pelleted and unpelleted composted organic materials for the production of cucumber crop. The aim of this study was to determine the influence of pelleted and unpelleted composted organic materials on the morphological growth, dry matter accumulation and yield of three varieties of cucumber (*Cucumis sativus*) in the greenhouse.

2. MATERIALS AND METHODS

2.1 Experimental Site

The experiment was carried out at the Teaching and Research Farm of the Department of Crop Science, University of Nigeria, Nsukka. Nsukka, is located on latitude 06° 52'North, longitude 07^0 24'East and altitude 447.26 meters above sea level (m.a.s.l). Rainfall distribution pattern in this region is bimodal with peaks in July and September and a short dry spell around mid-August. The soil of the experimental site is a reddish-brown clay loamy Ultisol (Oxicpaleustult) belonging to the Nsukka series [8]. The experimental site was on a flat land and, therefore, did not need any drainage system. Nsukka is located in a derived savanna vegetation zone with mean annual minimum and maximum temperatures of 25 and 32°C respectively [9].

2.2 Composting and Pelleting of the Organic Waste Materials

Organic waste materials such as rice husks, moringa pod husks, maize cobs and poultry manure obtained from Songhai Farm Adani, Nsukka and Telefood Research Project, University of Nigeria, Nsukka (Hujii-UNN) were composted. Table 1 below shows the organic materials and their mixing ratios during composting.

Table 1. Organic waste materials and their mixing ratios during composting

Symbol	Organic waste materials	Mixing ratio (v/v)
1. RH	Rice husks	100%
2 .MPH	Moringa pod husks	100%
3. MC	Maize cobs	100%
4. RH3:PM2	Rice husks + poultry manure	75%:25%
5. MPH3:PM2	Moringa pod husks + poultry manure	75%:25%
6. MC3:PM2	Maize cobs + poultry manure	75%:25%

Before composting, the organic materials were air-dried for seven days. The moringa pod husks and maize cobs were shredded into smaller coarse particle sizes (not mesh sizes) for quick degradation during composting. Composting was done aerobically in perforated black plastic buckets with lids for three months. Turning and watering were done whenever necessary throughout the period of composting. After composting, the organic materials were air-dried for seven days to get rid of moisture before pelleting them. The dried composted organic materials were ground at the Project Development Institute (PRODA), Enugu using the institute's milling machine after which they were passed through a sieve with two millimeter (2 mm) standard mesh size. Corn starch obtained from Farm Associates Nigeria Limited, Enugu was used as a binder during the pelleting operation. The milled organic materials and corn starch were homogenized and watered to form a coarse paste. Pelleting was done by using small electrically operated screw pelleting machine. The average diameter of a pelleted material was 6 mm. Tables 2 and 3 show the physical and chemical properties of the pelleted and unpelleted compost materials.

2.3 Experimental Design

One hundred and seventeen black perforated polythene pots of 23 cm diameter and 17 cm depth were filled with topsoil/compost manure mixtures mixed in the ratio of 3:2 (v/v) and laid out in the greenhouse of the Faculty of Agriculture, University of Nigeria, Nsukka, in a completely randomized design (CRD). The compost manure consisted of pelleted and unpelleted rice husks (100%), moringa pod husks (100%), maize cobs (100%), rice husks + poultry manure (75%:25%,v/v), moringa pod husks + poultry manure (75%:25%,v/v), maize cobs + poultry manure (75%:25%,v/v), control and three varieties of cucumber (*Poinsett, Marketer and Supermarketer*). A total of thirty-nine treatment combinations were used, each replicated three times.

Table 2. The physical properties of pelleted and unpelleted composted organic materials

Variable	Bulk density (g cm^{-3})	Total porosity (%)	Available water holding capacity (%)
RH	0.62	76.60	111.68
MPH	0.65	75.47	109.35
MC	0.69	73.96	107.40
RH3:PM2	0.71	73.21	103.11
MPH3:PM2	0.72	72.83	98.32
MC3:PM2	0.78	70.57	81.68
*RH	0.68	74.34	108.24
*MPH	0.73	72.45	92.53
*MC	0.77	70.94	87.37
*RH3:PM2	0.81	69.43	78.32
*MPH3:PM2	0.84	68.30	74.36
*MC3:PM2	1.07	59.62	53.31
CONTROL	1.21	54.34	45.59
F-LSD $_{(.05)}$	0.03	0.15	0.02

*RH = rice husks (100%), MPH = moringa pod husks (100%), MC = maize cobs (100%), RH3:PM2 = rice husks + poultry manure (75%:25%, v/v), MPH3:PM2 = moringa pod husks + poultry manure (75%:25%, v/v), MC3:PM2 = maize cobs + poultry manure (75%:25%, v/v), F-LSD $_{(.05)}$ = Fishers least significant differences at P = .05, * = pelleted*

Topsoil and the composted materials were thoroughly mixed two weeks before planting. Three varieties of cucumber (*Poinsett, Marketer and Supermarketer*) were planted to the depth 2 cm. Planting was done at the rate of two seeds per hole and thinned down to one plant after seedling emergence. Vertical staking was done by tying a twine at the base of the polythene pots to the roof of the greenhouse.

Prophylactic application of 15 mls of Karate in five liters of water was done at five days of seedling emergence to avert pest incidence. Routine watering was done every other day and manual weeding whenever necessary. The experiment was carried out between December, 2012 and March, 2013.

Table 3. The chemical properties of pelleted and unpelleted composted organic materials

Variable	TN (%)	OM (%)	OC (%)	C:N	pH (water)
RH	0.88	17.72	10.28	11.68	6.50
MPH	0.81	23.10	13.40	16.54	6.60
MC	0.54	26.48	15.36	28.44	6.80
RH3:PM2	1.03	11.60	6.73	6.53	7.10
MPH3:PM2	0.98	14.02	8.13	8.30	7.83
MC3:PM2	0.83	17.02	9.87	13.52	7.90
*RH	0.62	17.71	10.27	16.57	6.70
*MPH	0.59	22.19	12.87	21.81	6.74
*MC	0.51	26.07	15.12	29.65	6.90
*RH3:PM2	0.96	10.52	6.10	6.35	7.02
*MPH3:PM2	0.92	12.66	7.34	8.00	8.10
*MC3:PM2	0.80	15.55	9.02	11.28	8.13
CONTROL	0.04	1.85	1.07	25.48	5.00
F-LSD $_{(.05)}$	0.03	0.03	0.02	0.03	0.02

*RH = rice husks (100%), MPH = moringa pod husks (100%), MC = maize cobs (100%), RH3:PM2 = rice husks + poultry manure (75%:25%, v/v), MPH3:PM2 = moringa pod husks + poultry manure (75%:25%, v/v), MC3:PM2 = maize cobs + poultry manure (75%:25%, v/v), OC = organic carbon, OM - organic matter, TN - total nitrogen, C:N - carbon nitrogen ratio, F-LSD $_{(.05)}$ = Fishers least significant differences at P = .05, * = pelleted*

2.4 Data Collection and Analysis

Morphological growth parameters were measured biweekly for six weeks (peak of vegetative growth). Plant height was determined by measuring the length of the plant from the soil level to the shoot tip using a measuring tape. Leaf area per plant was estimated as leaf length (L) x width (W) x 0.85 as described by Blanco et al. 10. Number of leaves per plant was determined by counting. Stem diameter of the main vine was measured with vernier calipers and converted to girth using the following formula:

$$SG = SD\pi = SD \times 22/7$$

Where SG is stem girth and SD is the stem diameter.

The dry weights of the leaves, stems and roots were also measured at six weeks of planting. Five plants from each treatment combination/replicate were taken from the greenhouse and their roots carefully removed from the soil medium by eroding the soil around the root system in a bucket of water. The leaves, stems and roots were carefully separated and dried in an electric oven at 80°C until constant weights were obtained. The yield parameters i.e. fruit length, fruit diameter, fruit circumference, number of fruits/plant and total fruit fresh weight/plant were measured at harvest. Data collected were subjected to analysis of variance (ANOVA) as outlined by Obi [11]. Significant means were separated using Fishers least significant difference (F-LSD) at 5% probability level.

3. RESULTS AND DISCUSSION

The results of the morphological characteristics are presented on Tables 4 and 5. At At six weeks of planting, cucumber plants grown in unpelleted composted rice husks + poultry manure significantly (P = .05) had the highest values of plant height, stem girth, leaf area and number of leaves with *Supermarketer* (P3) variety having the greatest values followed by *Marketer* (P2) and *Poinsett* (P1). The interaction effects of the sowing media x cucumber varieties on the morphological characteristics were also significant.

The pelleted composted rice husks + poultry manure significantly (P = .05) gave the highest dry matter accumulation in the roots while the unpelleted rice husks + poultry manure significantly gave the highest values in the stems and leaves. The *Supermarketer* variety had the highest values of the dry matter accumulation in the roots, stems and leaves followed by *Marketer* and *Poinsett* in that order.

Table 4. The effects of pelleted and unpelleted composted organic materials on the plant height (cm) and stem girth (cm) of three varieties of cucumber (*Cucumis sativus*) at two and six weeks after planting (WAP) in the greenhouse

Organic materials	Plant height (cm)							Mean	Stem girth (cm)							
	2WAP				6WAP				2WAP				6WAP			
	P1	P2	P3	Mean	P1	P2	P3		P1	P2	P3	Mean	P1	P2	P3	Mean
RH	15.30	26.90	27.40	23.20	149.00	174.76	184.97	169.58	0.80	1.00	1.30	1.03	3.00	3.62	4.70	3.77
MPH	13.58	24.39	25.13	21.03	143.30	172.38	180.20	165.29	0.60	0.80	1.00	0.80	2.51	3.60	4.70	3.60
MC	9.10	22.89	24.02	18.67	139.03	169.30	179.01	162.45	0.60	0.80	0.90	0.77	2.51	3.52	4.63	3.55
RH3:PM2	23.30	30.00	37.20	30.17	167.91	187.29	204.57	186.59	2.10	3.00	4.20	3.10	3.49	4.60	5.67	4.89
MPH3:PM2	20.16	28.42	35.30	27.96	161.37	183.07	201.00	181.81	2.00	2.00	3.10	2.37	3.33	4.55	5.50	4.46
MC3:PM2	19.47	26.80	34.12	26.80	159.00	178.83	198.31	178.71	1.00	2.00	2.00	1.67	3.31	4.20	5.31	4.27
*RH	13.50	23.52	24.70	20.57	136.74	168.97	175.20	160.30	0.50	0.60	0.80	0.63	2.30	3.40	4.12	3.27
*MPH	10.62	20.17	23.60	18.13	135.63	162.40	173.10	157.04	0.50	0.50	0.80	0.60	2.15	3.12	4.00	3.09
*MC	8.69	19.34	21.45	16.49	134.40	160.53	170.69	155.21	0.40	0.50	0.70	0.53	1.91	3.10	3.90	2.97
*RH3:PM2	19.80	28.30	34.00	27.37	158.71	178.24	196.23	177.73	1.00	2.08	3.30	2.13	3.09	4.17	5.30	4.19
*MPH3:PM2	17.82	26.50	32.27	25.53	154.13	175.05	194.13	174.44	1.00	1.20	2.20	1.47	3.09	4.00	5.00	4.03
*MC3:PM2	17.03	24.63	31.03	24.23	151.94	173.41	189.37	171.57	0.90	1.00	1.10	1.00	3.04	4.00	5.00	4.01
CONTROL	6.10	15.03	19.31	13.48	106.00	124.23	134.73	121.65	0.70	0.90	1.00	0.87	0.90	1.34	2.94	1.73
Treatment mean (T)	14.96	24.38	28.43	22.59	145.94	169.88	183.19	166.34	0.93	1.26	1.72	1.31	2.66	3.63	4.68	3.66

	Plant height		Stem girth per plant	
	2WAP	6WAP	2WAP	6WAP
F-LSD $_{(.05)}$ for 2 varieties means (V)	.72	.95	.04	.32
F-LSD $_{(.05)}$ for 2 organic materials means (AT)	1.50	1.97	.07	.67
F-LSD $_{(.05)}$ for 2 V × T means	2.59	3.41	.13	1.16

*RH = rice husks (100%), MPH = moringa pod husks (100%), MC = maize cobs (100%), RH3:PM2 = rice husks + poultry manure (75%:25%, v/v), MPH3:PM2 = moringa pod husks + poultry manure (75%:25%, v/v), MC3:PM2 = maize cobs + poultry manure (75%:25%, v/v), * = pelleted, P1 – Poinsett, P2 - Marketer, P3 - Supermarketerand F-LSD $_{(.05)}$ = Fishers least significant difference at P = .05*

Table 5. The effects of pelleted and unpelleted composted organic materials on the leaf area (cm^2) and number of leaves per plant of three varieties of cucumber (*Cucumis sativus*) at two and six weeks after planting (WAP) in the greenhouse

Organic materials	Leaf area per plant (cm^2)								Number of leaves per plant							
	2WAP				6WAP				2WAP				6WAP			
	P1	P2	P3	Mean (V)	P1	P2	P3	Mean (V)	P1	P2	P3	Mean (V)	P1	P2	P3	Mean (V)
RH	39.49	66.60	86.68	64.26	232.52	357.49	406.30	332.10	3.00	6.00	5.67	4.90	10.33	17.00	17.00	14.78
MPH	36.80	61.55	81.42	59.92	212.71	349.48	398.30	320.16	2.67	5.67	5.33	4.56	10.33	16.00	16.67	14.33
MC	32.21	58.93	77.89	56.34	204.90	342.06	384.45	310.47	2.00	5.33	5.00	4.11	9.67	15.67	16.00	13.78
RH3:PM2	57.72	97.84	120.67	92.08	388.34	545.26	648.34	527.32	4.00	7.00	8.33	6.44	15.00	20.00	21.00	18.67
MPH3:PM2	54.61	76.22	113.08	81.31	374.52	479.99	555.26	469.92	3.67	6.67	7.33	5.89	14.67	20.00	20.00	18.22
MC3:PM2	52.97	71.57	107.57	77.37	343.59	431.05	506.60	427.08	3.67	6.33	7.00	5.67	14.67	19.00	19.67	17.78
*RH	32.62	63.54	70.79	55.65	205.89	345.10	381.93	310.97	2.00	5.33	5.00	4.11	9.00	15.00	16.00	13.33
*MPH	29.53	55.71	66.22	50.49	198.09	301.61	367.15	291.95	2.00	5.00	4.67	3.89	8.67	14.33	15.33	12.78
*MC	27.81	52.05	62.30	47.39	186.22	290.26	354.26	276.91	2.00	5.00	4.33	3.78	7.33	14.00	15.00	12.11
*RH3:PM2	49.84	76.68	105.49	77.34	310.54	436.31	509.24	418.70	3.00	6.33	7.00	5.44	13.33	18.33	18.67	16.78
*MPH3:PM2	45.01	70.03	99.33	71.46	294.27	406.49	430.49	377.08	3.00	6.33	6.33	5.22	13.00	18.00	18.33	14.78
*MC3:PM2	43.45	65.14	96.97	68.52	287.43	387.47	406.20	360.37	3.00	6.00	6.00	5.00	13.00	17.67	18.00	16.22
CONTROL	17.30	25.78	44.57	29.22	99.12	248.29	311.05	219.49	2.00	2.67	3.67	2.78	6.00	11.00	13.33	10.11
Treatment mean (T)	39.95	64.74	87.15	63.95	256.78	379.22	435.35	357.12	2.77	5.67	5.82	4.75	11.15	16.62	17.31	15.03

	Leaf area per plant		Number of leaves per plant	
	2WAP	6WAP	2WAP	6WAP
F-LSD $_{(.05)}$ for 2 varieties means (V)	.86	.31	.20	.34
F-LSD $_{(.05)}$ for 2 organic materials means (T)	1.79	.64	.41	.71
F-LSD $_{(.05)}$ for 2 V × T means	3.11	1.10	.71	1.22

The interaction effects of media x varieties on dry matter accumulation were also significant (P = .05) (Table 4). The fruit length, fruit diameter, fruit circumference, number of fruits per plant and total fruit fresh weight per plant were significantly (P = .05) greatest in unpelleted rice husks + poultry manure compared to the other sowing media. The *Supermarketer* variety significantly (P = .05) also had the highest values of the yield traits followed by *Marketer* and *Poinsett* in that order. There were significant (P = .05) media x variety interaction effects on all the yield attributes (Tables 6, 7, 8 and 9).

Generally, the plants grown with unpelleted composts performed better than those grown with pelleted ones in terms of morphological growth characteristics, dry matter accumulation and yield traits.

The consistent superiority of unpelleted composts over the pelleted ones in terms of morphological characteristics, dry matter accumulation and yield attributes could be due to the coarse and fragmented nature of the particles which would have increased soil aeration, drainage and rate of nutrient release for plants absorption. The compact nature of the pelleted composts would have contributed to slow nutrient release and relatively poor performance of the plants in the media. The observed best performance of the plants in unpelleted rice husks + poultry manure could be attributed to the high protein content of rice husks in addition to that in the poultry manure. This agrees with the earlier findings of Ndubuaku et al. [12] who obtained outstanding performance of okra plants in the liquid fraction (biol) of fermented rice husks + topsoil. Rice bran contains a lot of protein. During rice grain processing, a lot of the protein is removed with the bran and the husks. This can be one of the reasons for high level of protein in the rice husks. Mixtures of other organic materials i.e. rice husks, moringa pod husks, maize cobs with poultry manure gave higher values of dry matter accumulation, morphological growth and yield attributes compared with the single organic material treatments. This result reflects the substantial contribution of poultry manure in enriching soil nutrient contents. Poultry manure is rich in nitrogen which enhances crop development especially vegetative growth.

Table 6. The effects of pelleted and unpelleted composted organic materials on the root dry weight (g/plant), stem dry weight (g/plant) and leaf dry weight (g/plant) of three varieties of cucumber *(Cucumis sativus)* at six weeks after planting in the greenhouse

Organic materials	Root dry weight				Stem dry weight				Leaf dry weight			
	P1	P2	P3	Mean (V)	P1	P2	P3	Mean (V)	P1	P2	P3	Mean (V)
RH	0.80	1.00	1.29	1.03	6.52	9.14	12.00	9.22	16.42	19.56	21.63	19.20
MPH	0.65	0.90	1.10	0.88	6.14	9.10	11.78	9.01	16.02	19.00	20.52	18.51
MC	0.65	0.90	1.00	0.85	6.00	9.00	11.69	8.90	15.41	18.71	20.00	18.04
RH3:PM2	1.00	1.25	1.43	1.23	9.89	12.70	14.70	12.43	20.20	23.60	25.17	22.99
MPH3:PM2	0.85	1.24	1.40	1.16	9.60	12.27	14.60	12.16	20.00	23.35	24.20	22.52
MC3:PM2	0.85	1.20	1.40	1.15	9.13	12.05	14.50	11.89	19.34	23.10	24.00	22.15
*RH	1.20	1.40	1.75	1.45	5.14	8.03	10.67	7.95	15.00	17.87	19.73	17.53
*MPH	1.18	1.40	1.31	1.30	5.09	7.81	10.63	7.84	14.34	17.60	19.32	17.09
*MC	1.02	1.40	1.60	1.34	5.00	7.55	10.00	7.52	13.80	17.00	19..00	16.60
*RH3:PM2	1.36	1.63	1.99	1.66	8.63	11.40	13.70	11.24	18.05	21.73	23.61	21.13
*MPH3:PM2	1.24	1.60	1.92	1.59	8.51	11.25	13.10	10.95	17.71	21.40	23.03	20.71
*MC3:PM2	1.20	1.60	1.90	1.57	8.43	11.10	12.50	10.68	17.30	21.00	23.00	20.43
CONTROL	0.20	0.40	0.70	0.43	2.99	5.00	8.50	5.50	11.50	13.00	16.10	13.53
Treatment mean (T)	0.94	1.23	1.50	1.20	7.01	9.72	12.18	9.64	16.55	19.76	21.49	19.27

	Root dry weight	Stem dry weight	Leaf dry weight
F-LSD $_{(.05)}$ for 2 varieties means (V)	.11	.27	.32
F-LSD $_{(.05)}$ for 2 organic materials means (T)	.22	.55	.66
F-LSD $_{(.05)}$ for 2 V × T means	.39	.96	1.14

Table 7. The effects of pelleted and unpelleted composted organic materials on the fruit length (cm) and fruit width (cm) of three varieties of cucumber *(Cucumis sativus)* at harvest in the greenhouse

Organic materials	Fruit length (cm)				Fruit diameter (cm)			
	P1	P2	P3	Mean (V)	P1	P2	P3	Mean (V)
RH	16.36	19.75	22.00	19.37	3.90	5.40	7.00	5.43
MPH	16.04	19.37	21.57	18.99	3.62	5.10	6.99	5.24
MC	16.00	18.31	21.39	18.57	3.60	5.04	6.52	5.05
RH3:PM2	21.30	24.90	26.34	24.18	5.67	7.50	8.60	7.26
MPH3:PM2	20.00	23.79	25.60	23.13	5.40	7.29	8.17	6.95
MC3:PM2	19.87	21.13	24.80	21.93	5.20	7.10	8.00	6.77
*RH	15.71	17.69	20.70	18.03	3.10	5.00	6.00	4.70
*MPH	15.40	17.42	20.02	17.61	2.90	4.38	5.50	4.26
*MC	15.02	17.03	19.69	17.25	2.50	4.25	5.49	4.08
*RH3:PM2	19.50	21.83	23.00	21.44	4.50	6.92	7.65	6.36
*MPH3:PM2	19.00	21.04	22.94	20.99	4.41	6.50	7.50	6.14
*MC3:PM2	17.39	20.00	22.30	19.90	4.00	6.00	7.30	5.77
CONTROL	11.50	13.97	16.00	13.82	2.40	3.00	5.00	3.47
Treatment mean (T)	17.16	19.71	22.03	19.63	3.94	5.65	6.90	5.50

	Fruit length	Fruit diameter
F-LSD $_{(.05)}$ for 2 varieties means (V)	.13	.08
F-LSD $_{(.05)}$ for 2 organic materials means (T)	.26	.16
F-LSD $_{(.05)}$ for 2 V × T means	.46	.29

Table 8. The effects of pelleted and unpelleted composted organic materials on the fruit circumference (cm) and number of fruits per plant of three varieties of cucumber (*Cucumis sativus*) at harvest in the greenhouse

Organic materials	Fruit circumference (cm)				Number of fruits per plant			
	P1	P2	P3	Mean (V)	P1	P2	P3	Mean (V)
RH	12.92	17.48	18.71	16.37	5.67	7.67	11.00	8.11
MPH	12.05	17.25	18.50	15.93	5.67	7.33	10.33	7.78
MC	11.00	17.03	18.20	15.41	5.33	7.33	10.00	7.56
RH3:PM2	15.55	19.79	21.50	18.95	10.67	13.67	15.00	13.11
MPH3:PM2	15.34	19.50	21.30	18.71	10.33	13.33	14.00	12.56
MC3:PM2	15.00	19.30	21.00	18.43	10.00	13.00	13.67	12.22
*RH	10.70	16.59	17.43	14.91	5.00	6.67	9.33	7.00
*MPH	10.20	16.25	17.04	14.50	4.00	6.00	9.00	6.33
*MC	10.00	16.05	17.00	14.35	4.00	6.00	8.00	6.00
*RH3:PM2	14.90	19.00	20.90	18.27	8.67	11.00	12.33	10.67
*MPH3:PM2	14.37	18.57	20.30	17.75	8.00	10.00	12.00	10.00
*MC3:PM2	14.02	18.23	19.00	17.08	7.67	9.33	12.00	9.67
CONTROL	8.92	10.52	14.00	11.15	3.00	4.00	6.33	4.44
Treatment mean (T)	12.69	17.35	18.84	16.29	6.77	8.87	11.00	8.88

	Fruit circumference	Number of fruits per plant
F-LSD $_{(.05)}$ for 2 varieties means (V)	.26	.36
F-LSD $_{(.05)}$ for 2 organic materials means (T)	.54	.74
F-LSD $_{(.05)}$ for 2 V × T means	.94	1.29

Table 9. The effects of pelleted and unpelleted composted organic materials on the total fresh fruit weight (g/plant) of three varieties of cucumber (*Cucumis sativus*) at harvest in the greenhouse

Organic materials	P1	P2	P3	Variety mean (V)
RH	657.00	860.00	900.00	805.67
MPH	640.00	853.00	874.00	789.00
MC	635.00	850.00	865.00	783.33
RH3:PM2	754.00	980.00	1015.00	916.33
MPH3:PM2	740.00	971.00	1013.00	908.00
MC3:PM2	734.00	963.00	1010.00	902.33
*RH	610.00	843.00	855.00	769.33
*MPH	581.00	837.00	842.00	753.33
*MC	576.00	830.00	837.00	747.67
*RH3:PM2	708.00	920.00	1004.00	877.44
*MPH3:PM2	700.00	911.00	996.97	869.22
*MC3:PM2	694.00	905.00	970.00	856.44
CONTROL	358.67	400.33	600.67	452.89
Treatment mean (T)	645.15	855.64	906.36	802.38
F-LSD $_{(.05)}$ for 2 varieties means (V)		2.31		
F-LSD $_{(.05)}$ for 2 organic materials means (T)		4.81		
F-LSD $_{(.05)}$ for 2 V × T means		8.34		

The low nitrogen to carbon ratios in poultry manure help rapid decomposition of the organic materials and subsequent release and uptake of nutrients for plant growth and development according to Adeniyi et al. [13]. The varietal differences obtained in the responses of *Spermarketer, Marketer* and *Poinsett* to the different manure amendments could be attributed to their genotypic differences.

4. CONCLUSION

Generally, the plants grown with unpelleted composts performed better than those grown with pelleted ones in terms of morphological growth characteristics, dry matter accumulation and yield traits. The unpelleted rice husks + poultry manure significantly gave the highest values of morphological growth characteristics, yield and dry matter accumulation, especially in leaves and stems. In all the growth attributes, *Supermarketer* was the best followed by *Marketer*, then *Poinsett.*

COMPETING INTERESTS

Authors have declared that no competing interests exist.

REFERENCES

1. Barker AV, Pilbeam DJ. Handbook of plant nutrition. 5th ed, Taylor and Francis Group; 2007.
2. Schlecht E, Buerkert A, Tielkes E, Batino A. A critical analysis of challenges and opportunities for soil fertility restoration in Sudan-Sahelian West Africa. In: Advances in integrated soil fertility management in Sub-Sahara Africa: Challenges and opportunities, Springer Publishers. Dordrecht, Netherlands; 2007.
3. Powell JM, Unger PW. Alternatives to crop yield residues for sustaining agricultural productivity and natural resources conservation. J. Sust. Agric. 1998;11(2-3):59-84.
4. Powell JM, Fernandez-Rivera S, Hiernanx P, Turner MD. Nutrient cycling in integrated rangeland / crop system of Sahel. Agric Syst.1996;52:143-170.
5. Powell JM, Ikpe FN, Somda ZC. Crop yield and the fate of nitrogen and phosphorous following application of plant materials and faeces to soil. Nutr. Cycl. Agro-ecosyst.1999;54:215-226.
6. Miller JH, Norman J. Organic and compost-based growing media for tree seedling nurseries. World Bank Technical Paper. Number 264 forestry series; 1992.
7. Ekwu LG, Utobo, EB,Oyesola, CA. Vegetative and yield response of cucumber (*Cucumis sativus* L) to staking and nitrogen fertilizer application. J. Appl. Sci. 2007;19(4):7509-7519.
8. Mbagwu JSC. Maize (*Zea mays*) responses to nitrogen fertilizer on an ultisol in southern Nigeria under two tillage and

mulch treatments. J. Sci. Food Agric.1990;52:365-376.
9. Igwe CA, Stahr K. Water stable aggregate of flooded inceptisols from south-eastern Nigeria in relation to mineralogy and chemical properties. Aust. J. Soil Res. 2004;42:171-179.
10. Blanco FF, Folegatti MV. A new method for estimating the leaf area index of cucumber and tomato plants. Hortic. Bras. 2003;2(4):666-669.
11. Obi IU. Statistical methods of detecting differences between treatment means and research methodology issues in laboratory and field experiments. 2nd ed; 2002.
12. Ndubuaku UM, Imegwu CN, Ndubuaku NE. Nutrient compositions of liquid and solid fractions of organic waste fermentation and the influence on growth and yield of okra. Int. J. Cur. Tr. Res. 2013;2(1):415-424.
13. Adeniyi OT, Akanbi WB, Adediran JA. Growth, nutrient uptake and yield of tomato response to different plant residue composts. Food, Agric. and Enivron. 2004;2(1):310-316.

13

Root Morphology and Yield of Wheat (*Triticum aestivum* L.) Crop under Reduced Irrigation in Delhi (India) Weather Conditions

A. K. Mishra[1*], S. S. Parihar[1], V. Visha Kumari[2], T. K. Srivastava[2] and Khan Chand[1]

[1]Water Technology Center, Indian Agricultural Research Institute, New Delhi-110012, India.
[2]Indian Institute of Sugarcane Research, Lucknow, India.

Authors' contributions

This work was carried out in collaboration between all authors. Author AKM designed the study, wrote the protocol and wrote the first draft of the manuscript. Author SSP reviewed the experimental design and all drafts of the manuscript. Authors TKS and VVK managed the analyses of the study. Author VVK took the field measurements of certain parameters, laboratory analysis. Author KC helped in various field operations and maintenance of the experiment, helped the scientists in data collection by identifying the plants. Authors AKM and SSP performed the statistical analysis. All authors read and approved the final manuscript.

Editor(s):
(1) Mao Feng Chai, Forage Improvement Division, The Samuel Roberts Noble Foundation, USA.
Reviewers:
(1) İlkay Yavaş, Crops and Animal Production, Adnan Menderes University, Turkey.
(2) Masmoudi Charfi Chiraz, Olive Tree Institute, Agronomic Institution of Researches and High Education, Tunisia.

ABSTRACT

Aims: Water is one of the most valuable resources for the survival of civilization. Assured supply of water is necessary for sustainable agriculture. Wheat (*Triticum aestivum* L.) is an important staple food crop in India and is cultivated in different agro-ecological regions of northern, eastern and central parts of the country occupying 25.4 million hectare (M ha) and nearly 54% of its area is irrigated. A sizeable 46% area under wheat is still rain dependent which often is faced with reduced irrigation thus decreased crop yield. The aim of this study is to facilitate the farmers of this reason in taking the appropriate decisions as regards providing supplemental irrigation with limited water supply conditions so that comparable yields could be obtained with the water application to critical crop growth stages.

Corresponding author: E-mail: dranilkumarmishra1@gmail.com;

That constitutes about 27% of the irrigated area of India. Changing climate has resulted into recurrence of long rainless spells during winter season (rabi) at places in India known for extensive wheat production which is main staple food of the masses. Wheat crops growth and performance is adversely affected by soil moisture stress so is the growth of roots and consequently the nutrient uptake that reflects into the yield and yield attributes. Hence, the main objectives of present study were to access, the effect of reduced irrigation on yield attributes, root development, dry matter accumulation and partitioning of wheat (*Triticum aestivum* L.) crop under reduced irrigation in Delhi (India) during winter (rabi) weather conditions.

Study Design: Randomized Block Design (RBD) with four replicates.

Place and Duration of Study: Field experiments were conducted at the W-3 Experimental Farm of the Water Technology Centre, IARI, New Delhi during the rabi cropping season of 2011-12. Indian Agriculture Research Institute (IARI) is located at New Delhi (28°38'N, 77°10'E) at an elevation of 248 m above MSL. The crop was sown in the month of Novemebr and harvested at the end of April of the next year. The sandy loam soils of the experimental area, where the field experiments were conducted can be characterized as having low water holding capacity. Delhi falls in the semi arid climatic conditions. The experimental design for the experiment was Randomized Block Design (RBD), replicated four times. The present experiment used a high yielding cultivar of wheat variety HD 2967 having the growth period of 120 days.

Methodology: For root analysis, the roots were taken from a depth of 0-90 cm and were analyzed for different properties like root length, surface area and volume using root scanner (EPSON expression 1640XL, Japan). The same roots were then used for measuring wet mass and kept in an oven at 70°C for three days before taking dry mass (Figs. 3 and 4). WinRhizo™ software was used with an approved scanner, Yield and yield attributes were measured using standard experimental procedures. Observations on plant growth (crop morphometry, photosynthetic activity (LAI), dry mass partitioning), soil moisture dynamics and yield of wheat *(Triticum aestivum* L.*)* crop were taken under a well structured schedule.

Results: There was no significant difference in the plant height in the early stages (30 days after sowing (DAS)) in wheat during the winter (rabi) season of 2011-12. However, in the later stages i.e., during late vegetative and reproductive phases, various treatments exhibited significant differences in yield and yield attributes. Also, in number of tillers/m^2 and leaf area index (LAI) a highly significant differences were observed among treatments. It was worth noticing that skipping irrigation in CRI and milking stages has significant decrease in the dry matter accumulation even though the treatment with no water deficiency (T8) recorded highest dry matter accumulation. Significant differences among the treatments were also observed especially in root length. The treatment T1 with no irrigation (Control) recorded in significant reduction with T8 treatment (no water deficiency), which clearly indicated that root of the crop elongated in stress condition to fulfill its metabolic activities. crop tries to utilize the available soil moisture in this case. Root length has also shown a direct correlation with soil moisture availability.

Conclusion: Skipping of irrigation during tillering stage has drastically reduced the elongation of root among all the treatment. T4 treatment (100% soil moisture deficient at booting stage) has resulted in highest root length which indicates that irrigation is critical during the flowering and milk stages of wheat. Root surface has been drastically reduced in treatment T1 (No irrigation in all growth stages).

Keywords: Rainfall; reduced irrigation; leaf area index; irrigation; root morphology; root biomass; wheat (Triticum aestivum L.) crop; water Management.

1. INTRODUCTION

Water is one of the most valuable resources for the survival of civilization. Assured supply of water is necessary for sustainable agriculture. Wheat (*Triticum aestivum* L.) is an important staple food crop in India and is cultivated in different agro-ecological regions of northern, eastern and central parts of the country occupying 25.4 million hectare (M ha) and nearly 54% of its area is irrigated. That constitutes about 27% of the irrigated area of India [1]. The agriculture sector is the largest consumer of water resources in Indo-Gangetic plains which is the main cereal producing area in India. Well-drained clay loams, loams and sandy loams are suitable for this crop. The majority of farmers of

Indo-Gangetic plains are using water irrationally [2,3]. Wheat production in India generally happens in cooler month (winter season) where there is limited water loss from the soil surface. Due to increased cropping intensity water is becoming a limiting factor in wheat production areas in India [4,5]. Hence, the yield and quality of wheat (*Triticum aestivum* L.) crop often suffers due to insufficient water supply and improper scheduling of irrigation [6-8]. Again, excessive irrigation under shallow water table condition especially in lower portions of the Indo-Gangetic belt, will not only aggravate the problem of water logging but will also reduce the irrigated area and total yield [9-11]. The appropriate crop water requirement (CWR), critical crop growth stages and irrigation schedules are highly essential for realizing targeted level of crop production. The appropriate water amount, its time of application (schedules) are vital for optimal water productivity of any crop [12].

Crop water use varies substantially during the growing period due to variation in crop canopy and climatic conditions [13]. Total crop water use is nothing but the sum of root water uptake plus soil evaporation plus interception, and the spatial and temporal pattern of soil water use by crops can best be obtained from the accurate profile description of hydraulic conductivity, soil water flux and rooting pattern [14]. Field water balance is commonly used to measure total water use or actual crop evapotranspiration (ET_a) when lysimeter facilities are not available and it is found that ET_a increases with the increase in number of irrigation from one to adequate [15]. The prediction of ET_a and crop coefficients (K_c) as a function of growth period is very much important for determining crop water use and scheduling irrigation at a regional level [13]. Under stress conditions, ET_a is calculated by the combining effect of K_c and soil water stress coefficient (K_s).

In India the location specific crop coefficients (K_c's) have not been worked out for many crops. Hence, for irrigation scheduling based on the estimation of crop water requirements all the researchers invariably use the FAO CROPWAT-56 data for K_c [13]. The experimental verification being tedious, time consuming and difficult; many attempts have not been made in the past to experimentally verify the results of such arbitrary assumption [1,16].

Much is known about the CWR of wheat using field water balance and/or lysimeter study in field experimental plots at various agro-ecological conditions of India [2,17,18]. Therefore, the thorough knowledge of optimum time and amount of limited water to be applied to obtain higher productivity is essential. Under normal conditions, four-to-six irrigations were recommended for optimum wheat production in India [19-24].

Much is known about the CWR of wheat using field water balance and/or lysimeter study in field experimental plots at various agro-ecological conditions of India [2,17,18]. Therefore, the thorough knowledge of optimum irrigation schedules for realizing higher crop and water productivities are a must. For obtaining optimum water productivity in wheat crop the researchers have advised a minimum of four and a maximum of six irrigations [19-24]. Several researchers have already made crucial attempts to identify the critical crop growth stages corresponding to Indian weather conditions in wheat crop [25-27,87] that has resulted into a clear understanding of water sensitive growth stages for applying irrigation.

When water becomes scarce, demand management becomes the key to the overall strategy for managing water [28]. All the available residual soil moisture can be used by a crop since dry soil conditions promote root elongation and branching [29]. Hence, most of the residual soil water in the soil profile can be utilized. Water deficit has become a leading environmental constraint that limits crop photosynthesis, productivity and yield [6,30,31]. A decline of photosynthesis in water stressed plants can be caused by stomata closure and impairments in photochemical and/or biochemical reactions [32,33]. Water deficit also results in increase dry matter allocation to roots [34,35] which can enhance water uptake [36]. Roots, in wheat crop comprise close to half total plant biomass [37] and are critically involved in water up take and nutrient supply. Deeper roots can extract more water from depth thus avoiding water deficits at critical growth stages resulting in higher harvest indices and reduced water loss by deep drainage [38,39] studied the effective management of irrigation water for wheat crop under stressed conditions using simulation modeling. It was concluded that under water scarcity condition, when soil water stress is imposed during non-critical stages of growth, irrigation is to be scheduled at 45% maximum allowable depletion of available soil water for wheat crop grown in sandy loam soils in sub-humid regions in order to obtain maximum grain yield and above ground dry matter. Also, in order to obtain the highest water use efficiency (WUE), irrigation is to be scheduled at 45% Maximum Allowable Depletion (MAD) of available soil water for wheat crop in sandy loam soil. However, increased early vigour leads to faster, deeper root growth and more adventitious roots in the top soil thus improving water and nutrient use and reducing evaporative losses from the top soil [40,41]. Generally, 70% of the total root volume is found in the top 0-30 cm soil layer, where most nutrients are present in the majority of the agricultural soils [42]. The existing root length density of wheat is not sufficient to extract all the available water deep in the soil profile [43,44].

Further, the grain yield in *rabi* (winter) wheat (*Triticum aestivum* L.) is influenced not only by shoot characteristics, morphological traits such as plant height and number of tillers, and phonological periods such as time to anthesis and maturity. Information on the relationship between grain yield but also the root characteristics, such as root morphology, root biomass and shoot biomass in wheat [12,45-48]. It was reported by [49] that water stress during tillering until physiological maturity causes significant reduction of wheat grain yield cultivars. Also, this reduction results from both grain weight reduction and number of grain per spike. In a controlled research, [50] reported the 25 and 50% reduction of water consumption may decrease grain wheat yield 21.8% and 40.7% respectively. In another experiment, [51] also reported that different irrigation treatments have significant effect over wheat protein index; therefore, the increase in water stress would lead to grain protein percentage rise in all cultivars. They report that the decrease in moderate and intense stresses would bring a drop in grain yield 23% and 46% respectively. The increase in water use leaded to increase of 1000- grain weight in cultivars was reported by [52]. It should be also taken into consideration that different genotype behaves differently in water stress condition. Genotypic variations related to photosynthetic activity, productivity and grain yield under water deficit has also been extensively reported [53]. Therefore, proper management of important inputs particularly irrigation water using modern technology is essential for increasing production and for giving high return to the farmers. Crop growth simulation models could be used as useful tools for determining crop growth, development and to formulate irrigation management strategies for efficient use of inputs [54,55].

Despite their crucial role of roots in realizing higher production; very little attention has been paid to the study of root morphology, traits and root biomass as compared to more easily assessable above ground (shoot traits). Drought is the most common crop stress globally and characters that improve water-use-efficiency (WUE) such as sub-soil water extraction by roots can be enhanced through agronomic management or plant breeding to increase yield. However, the benefits depend on the seasonal pattern of water availability as influenced by rainfall distribution, soil type and management [56]. Under soil water deficit conditions therefore, the crop water extraction depends on root distribution and depth [57]. As a result to enhance the food security and taking into the local and global water scarcity conditions into consideration, the present study was conducted with wheat variety HD 2967 for studying the effect of soil moisture deficit during different crop growth stages on root and shoot growth (root morphometry and root dynamics and photosynthetic activity (LAI)), soil moisture dynamics, and grain yield of wheat (*Triticum aestivum* L.) crop. Apart from these an analysis of the water dynamics would also enabled to apply the precise amount of water to the crop root zone to avoid the deep percolation losses which accounts for the maximum wastage during the irrigation.

The deep percolation losses accounts for the maximum wastage during the irrigation which can also be attempted by analysis of the water dynamics apart from all the above mentioned parameters which can result into an application of the precise amount of water to the crop root zone to avoid the losses. The results of the present study would not only help the farming community of Delhi NCR region in significant amount of water saving during the entire crop growth season of *rabi* (winter season crop) but also help in reallocating the balance amount of water for bringing more areas under irrigation.

2. MATERIALS AND METHODS

2.1 Experimental Site

The Indian Agriculture Research Institute (IARI) is located at New Delhi (28°38'N, 77°10'E) at an elevation of 248 m above MSL. For accessing the effects of reduced irrigation on soil moisture, root dynamics, yield and water productivity of wheat crop *(Triticum aestivum* L.) crop the field experiments were conducted at WTC farm No.3. During the winter season of the year 2011-12, the field investigations were carried out with wheat variety HD2967 (duration 120 days) in sandy loam soil in Randomized Block Design (RBD), replicated four times and eight different treatments of with an objective to identify the most critical crop growth period corresponding to the ambient weather conditions most adversely affecting the crop during the limited irrigation.

The eight treatments were decided in such a manner that moisture limiting stages can be easily identified against the control (i.e complete soil moisture deficit meaning absolutely no post sowing irrigation). Other treatments were; complete soil deficit during crown root initiation (CRI) stage; complete soil moisture deficit during the tillering stage (TS); complete soil moisture deficit during booting stage (BS); complete soil moisture deficit during flowering stage (FS); complete soil moisture deficit during milking/dough stage (DS); complete soil moisture deficit during grain filling stage (GFS); and fully irrigated in all growth stages (providing irrigation in all the crop growth stages) i.e. absolutely no deficit in any stage having ample soil moisture in all growth stages).

Further bifurcation of the growth stages were avoided due to the limitations of the field, facilities and manpower. The seeds of wheat variety HD 2967 were machine sown at a distance of 22 cm row to row spacing. The basal dose of the fertilizers (NPK) was applied at the rate of 140:60:40 kg/ha [58] based on the previous results of soil analysis. The area of the each experiment plot was 36 m^2 (6mX6m) leaving a buffer strip of width of 1.5 m between two adjacent plots which provided the much required buffer to avoid the seepage of irrigated water across different plots. The crop was irrigated based on the soil water availability. During the entire experimental period, the crop could receive rainfall only once (recorded to be 5 cm). The structured details of all the treatments are given below:

2.2 Treatment Details

The field experiment in winter (Rabi) season 2011-12 was conducted with following eight treatments T1 - Control (Pre-sowing+No irrigation in all growth stages) T2 – No irrigation during CRI stage; T3 – No irrigation during tillering stage; T4 –No irrigation during Booting stage; T5 – No irrigation during Flowering stage; T6 – No irrigation Milk/Dough stage; T7 –No irrigation during grain filling stage; T8 – Full Irrigation in all the crop growth stages There were three replicates in 36 sq. m. plot size with row to row spacing of 22.5 cm. The crop was sown using a seed drill machine in RBD experimental design with full doses of P & K and half dose of N as basal @ N:P:K:: 120:60:60 on 11.11. 2011, with wheat variety Pusa HD 2967 and harvested on 29th April, 2012.

2.3 Field Sampling of the Roots and Image Analysis

2.3.1 Root auger and root scanner

Root sampling was done using soil core method with the help of soil auger at different vegetation stages and soil depths (15-90 cm). Roots were washed properly and collected on a sieve with 0.5 mm mesh screen. Debris and dead roots were manually removed from vital roots by considering their color and flexibility. The auger body (Fig. 1) consists of a steel cylinder holding an exchangeable, toothed drilling-crown. The drilling crown is made from hardened steel and its toothing allows cutting through the rooted soil. The top end of the cylinder is closed and has a breather hole. The auger body is sturdy and as such suitable for use in hard soils. The cylinder's diameter is 8 cm, its operational depth is 15 cm and its volume is 753.60 cm^3. Roots are floated in water in acrylic trays on the scanner (Fig. 2). This allows the roots to be arranged to reduce overlap and crossing of roots. For root analysis, the roots were taken from a depth of 90 cm and were analyzed for different properties like root length, surface area and volume using root scanner (EPSON expression 1640XL, Japan). The same roots were then used for measuring wet mass and kept in an oven at 70°C for three days before taking dry mass (Figs. 3 and 4). WinRhizo TM software was used with an approved scanner, which allows the roots to be lit from above and below while being scanned (Fig. 3). Measurements involved total root length, average root diameter, projected and surface area, plus length and area measurements as a function of different root diameter classes. The images were saved as gray scale image as given below. Analysis of the image is a very important step. Even though WinRhizoTM can automatically set these, but we need manually tweak them from time to time. If the roots are voluminous the smaller region of the image should be analyzed for clarity. The color traces on the root indicate where roots have been detected. The output results are given ready for analysis [58].

2.4 Leaf Area Index (LAI) and Dry Matter Partitioning (DMP)

The gravimetric method [50] was employed for field measurement of soil moisture and its dynamics. Soil samples from four different depths (0-15, 15-30, 30-45 and 45-60 cm) were collected from the field and the soil moisture content was estimated using gravimetric method. A Laser leaf area meter (model: CI-203) was used for estimating the leaf area index. Since this instrument requires more time to estimate the leaf area; only representative samples were analyzed with the same. LAI for the remaining stages of the crop were computed by length and breadth method (using crown leafs of five plants and taking their average length and widths). A correction factor of 0.65 was multiplied as suggested by [59] in the average leaf area estimated by initial measurements. Five (5) representative plants were taken to study the dry matter production in each stage for estimating the dry matter partitioning.

2.5 Field Scout Digital Moisture Sensor

The Field Scout Digital Moisture Sensor allowed us to monitor and record soil moisture quickly and accurately. The Field Scout Digital Moisture Sensor having two volumetric water content modes; one for standard soils and one for higher clay soils with TDR technology (time-domain measurement technology was used for measuring the soil moisture and validated with gravimetric method by constructing a calibration chart. The meter converts a measured electrical signal into % soil moisture content using an equation valid over a wide range of mineral soils in volumetric water content (VWC) mode. However, In irrigation mode, the meter displays relative water content (RWC) on a scale of 0 to 100 corresponding to a user-defined upper and lower soil moisture reference level. It is fitted with two 3.8 cm rod, two 7.5 cm rod, two 12 cm rod, and two 20 cm rod [60].

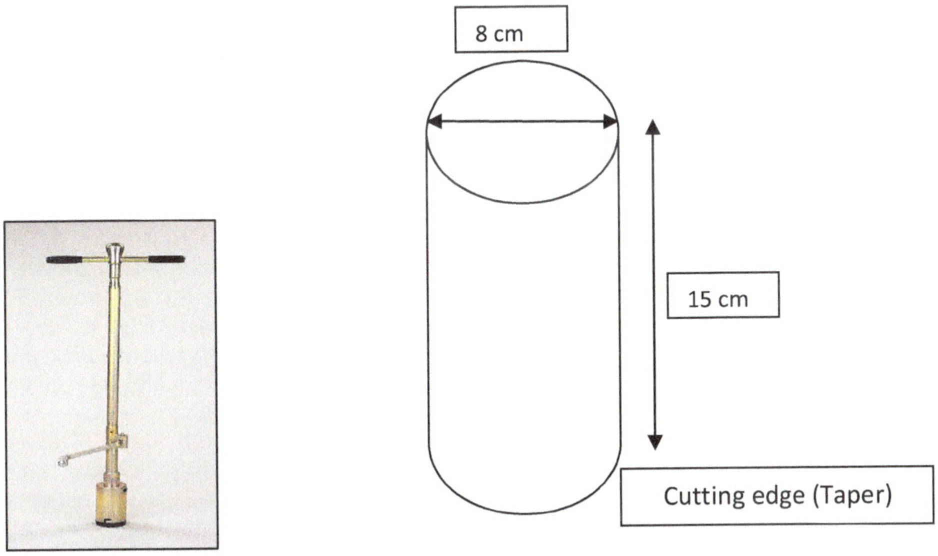

a. Root Augur assembly with core sampling mechanism

b. Root Augur core sampling mechanism showing the dimensions

Fig. 1. Root auger

Source: http://pkd.eijkelkamp.com

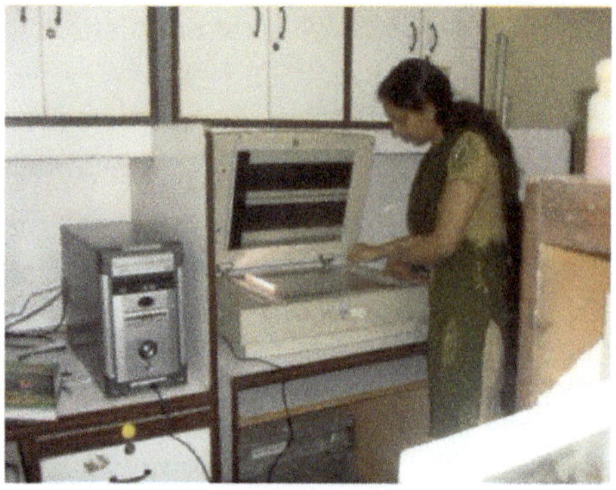

Fig. 2. The root morphology analysis using Root Scanner (EPSON expression 1640XL, Japan)

Fig. 3. A sample grey scale image (*.TIF image) of wheat crop roots

2.6 Field Measurement of Yield Parameters

For estimating yield parameters, twenty ear heads selected at random from each plot at harvest were used for recording the grain weight per ear head in g. The net plot was marked, harvested separately and total biomass yield from each net plot was recorded. After threshing, grains were separated, cleaned and weighed. Straw yield per net plot was worked out by subtracting respective total grain weight from the total biomass yield. Later the grain and straw yield per net plot was computed on hectare basis and expressed in q per ha. Harvest index was calculated [7] by using formula of as: HI = EY/BY and HI = GY/TBY; where, HI = Harvest Index (q/ha); EY= Economic yield (q/ha); BY = Biological yield; (q/ha) GY=Grain yield (q/ha) and TBY= Total biomass yield (q/ha). The analysis of the above ground biomass and the average grain weight per spikelet (gm) (average of 20 spiklets) has been presented and conclusions have been drawn on the basis of information so obtained.

3. RESULTS AND DISCUSSION

3.1 Analysis of Root Morphology

Several researchers [9,39] have hypothesized that the ability of a plant to change its root distribution in the soil is an important mechanism for drought avoidance. Different patterns of root growth were observed in treatments with different soil water regimes [10,40].

3.2 Effect of Irrigation Treatments on Root Length (cm) of Wheat (*Triticum aestivum* L.) Crop under Reduced Irrigation at Delhi Weather Conditions

The treatment and replication wise root lengths of wheat (*Triticum aestivum* L.) crop under reduced irrigation after 70 DAS and 100 DAS are presented in Table 1 wherein the mean root lengths have also been worked out. It was observed that after 70 DAS the minimum root growth of only 931.06 cm was recorded in T5 (i.e – No irrigation during Flowering stage) followed with T3 (i.e. – no irrigation during Tillering stage); in which the total root length was 1001.45 cm. The maximum root development and growth (1920.06 cm) was observed in treatment T4 (i.e–no irrigation during booting stage); preceded by T2 (1512.79 cm) (i.e. no irrigation during CRI stage). Whereas, in case of 100 DAS, the highest root growth of 2715.32 cm was recorded in the treatment T6 (no irrigation milk/dough stage); followed with 2456.24 cm in T7 which is no irrigation during grain filling stage. The Post-hoc analysis was done as suggested by [60] and is presented in Table 2. On persuasion of both these (Tables 1 and 2) it is clear that the root lengths were significantly affected by the reduced levels of soil moistures in different treatments that coincided with the crop sensitive stages in varying degrees. In Table 2, where means with the same letter are depicted as being not significantly different; one can easily notice that soil moisture deficit has not very clear cut distinction among the treatments but is indicative only. While the soil moisture deficit in certain stages have nominal effect on crop root growth inferred by the way of non significant mean differences after 70 DAS. On the contrary, in the later stages of the crop development when the crop completed its full growth potential and neared maturity the effective soil moistures deficits have altogether changed scenarios. After 100 DAS the minimum root growth was observed in treatment T3 (1275.57 cm) (i.e. no irrigation during Tillering stage) followed with T5 once again with 1648.39 cm. This clearly confirms that T5; i.e. no irrigation during flowering stage can seriously hamper the root growth pattern. However, various treatments showed significant difference in mean root lengths after 70 and 100 DAS with non-uniformity and a few showed no significant difference (Table 2). A closer look of Fig. 5 and Fig. 6 will support this fact because while the soil moisture has been imposed among various treatments there has been a few good showers during the winter season that might have replenished the root zone soil moisture which is evident from Fig. 2 too. These results are in strict agreement with the findings of some of the past researchers [40,76-78]. The functional relationship developed between the number of days after showing and root length (cm) for all treatments at two stages (70 DAS and 100 DAS) have been worked after fitting the most appropriate trend lines and are given below with their respective R^2 (eqn. 1 & 2). Functional relationship between number of days after sowing and the root length for:

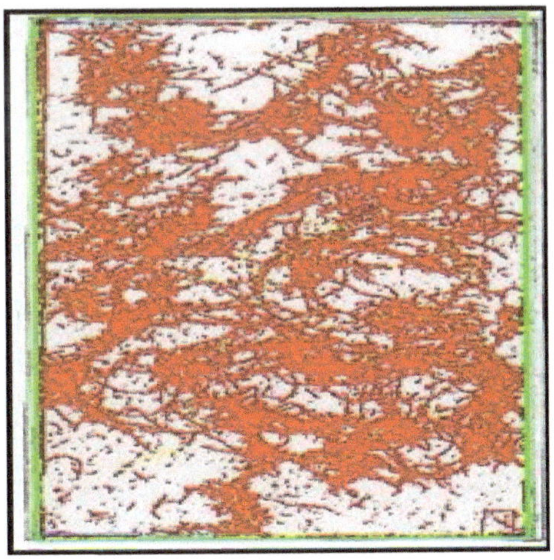

Fig. 4. A sample Win Rhizo™ analyzed image of wheat (*Triticum aestivum* L.) crop roots

3.2.1 Functional relationship between number of days after sowing and the root length 70 das

A sixth order polynomial between Number of Days after Sowing and the Root Length (70 DAS) of the type below; was fitted and the value of R^2 was worked out:

$$y = -2.94x^6 + 79.68x^5 - 844.6x^4 + 4430.x^3 - 11975x^2 + 15591x - 6147 \quad (1)$$

$R^2 = 0.255$

3.2.2 Functional relationship between number of days after sowing and the root length 100 das

A sixth order polynomial between Number of Days after Sowing and the Root Length (100 DAS) of the type below; was fitted and the value of R^2 was worked out to be:

$$y = -2.390x^6 + 68.73x^5 - 783.6x^4 + 4457.x^3 - 12964x^2 + 17508x - 6077 \quad (2)$$

$R^2 = 0.465$

R^2 values in both the functional relationships (eq. 1 & 2) are not very encouraging that supports the statement made as above regarding the root growth patterns after 70 and 100 DAS. It can be seen from these values that after 70 DAS till the 100 DAS the plant responded in a much balanced manner as regards to plant root length than 70 DAS which showed an indefinable/ zig-zag pattern.

3.3 Effect of Irrigation Treatments on Root Surface Area (Cm^2) of Wheat (*Triticum aestivum* L.) Crop

The treatment and replication wise root surface area of wheat (*Triticum aestivum* L.) crop under reduced irrigation after 70 DAS and 100 DAS are presented in Table 3 wherein the mean root surface area have also been worked out (Table 3). It was observed that after 70 DAS the minimum root surface area of only 174.61 cm^2 was recorded in T7 (i.e 100% soil moisture deficit during grain filling stage) followed with 265.71 cm^2 in T2 (i.e. – 100% soil deficit during CRI stage). The maximum root surface area (400.71 cm^2) was observed in treatment T4 (i.e–no irrigation during booting stage); preceded by T8 (331.79 cm^2) (i.e. no deficit (providing irrigation in all the stages) which is quite obvious too.

Whereas, in case of 100 DAS, the highest root surface area of 411.25 cm^2 was recorded in the treatment T7 (100% soil moisture deficit during grain filling stage); followed with 401.48 cm^2 in T6 which is 100% soil moisture deficit during milk/dough stage. The Post-hoc analysis is presented in Table 4. On persuasion of Tables 3 and 4 it is quite clear that the root surface area were significantly affected by the levels of treatments though a definite trend as per the treatment details could not be established from the data of root surface area of various treatments. However, a clear cult idea about root surface area emerged as the moisture sensitive stages are also equally sensitive to root development and root surface area. In Table 4, where means with the same letter are depicted as being not significantly different one can easily notice that soil moisture deficit has not very clear cut distinction among the treatments but is indicative (Fig. 7). As explained in section 3.1.1; the root length as well as root surface area might have experienced the similar stress conditions and the trend (Fig. 7) is almost similar. While the soil moisture deficit in certain stages have nominal effect on crop root surface area inferred by the way of non-significant mean differences after 70 DAS (Tables 3 and 4). Similarly, at a later stage of the crop development when the crop completed its full growth potential and neared maturity the effective soil moistures deficits have altogether changed scenarios. After 100 DAS the root surface area was observed to be at a minimum in treatment T1 (complete soil moisture deficit in all growth stages) and the surface area went on increasing with the corresponding reduction in soil moisture deficit (Fig. 8). The functional relationship developed between the number of days after showing and root surface area for all treatments at two stages (70 DAS and 100 DAS) have been worked after fitting the most appropriate trend lines and are given below with their respective R^2. Functional relationship between number of days after sowing and the root surface area are given in eqns. 3 and 4:

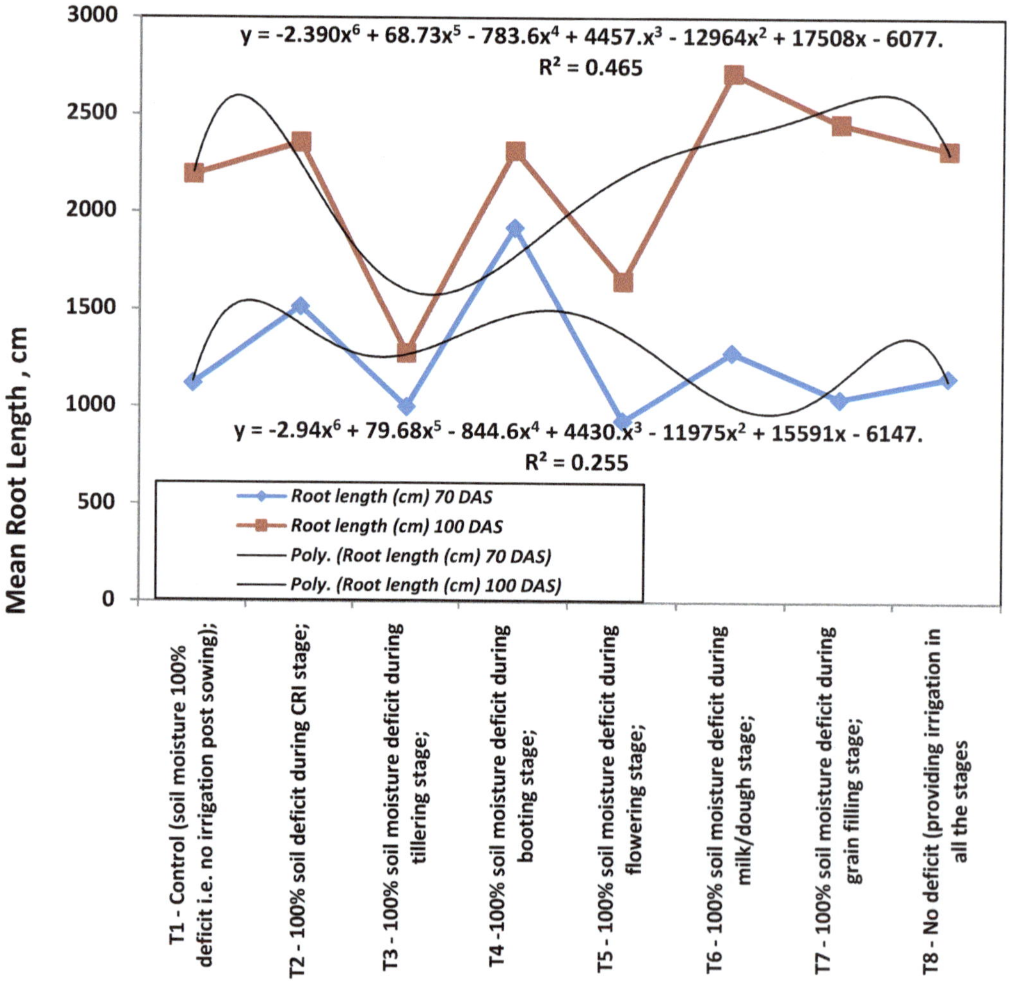

Fig. 5. Effect of soil moisture treatments on mean root length (cm) of wheat (*Triticum aestivum* L.) crop

Table 1. Effect of irrigation treatments on root length (cm) of wheat (*Triticum aestivum* L.) crop under reduced irrigation at Delhi weather conditions

Treatments	70 DAS					100 DAS				
	R1	R2	R3	R4	Mean	R1	R2	R3	R4	Mean
T1	1088.26	1015.22	1245.32	1124.24	1118.26	2166.69	2155.26	2256.35	2184.26	2190.64
T2	1520.47	1458.26	1548.21	1524.21	1512.79	2568.65	2456.32	2154.26	2243.68	2355.73
T3	985.39	1021.25	995.26	1002.30	1001.05	1246.10	1256.35	1325.26	1274.56	1275.57
T4	1938.27	1854.26	1942.50	1945.20	1920.06	2492.00	2153.65	2245.68	2365.26	2314.15
T5	845.55	958.24	1024.21	896.25	931.06	1647.45	1745.26	1548.56	1652.30	1648.39
T6	1273.19	1321.20	1245.26	1284.56	1281.05	2805.95	2463.50	2845.26	2746.56	2715.32
T7	1030.43	1025.23	985.62	1140.30	1045.40	2564.33	2456.36	2449.65	2354.62	2456.24
T8	1188.79	1125.26	1124.20	1184.65	1155.73	2311.45	2263.54	2453.35	2259.23	2321.89
CD at 5% 82.523, C.V-4.5%						*CD at 5% 171.64, C.V- 5.1%*				

T1 - Control (Pre sowing +No irrigation in all growth stages), T2 – No irrigation during CRI stage; T3 – No irrigation during Tillering stage; T4 –No irrigation during Booting stage; T5 – No irrigation during Flowering stage; T6 – No irrigation Milk/Dough stage; T7 –No irrigation during grain filling stage; T8 – Full Irrigation in all the crop growth stages

Table 2. Post hoc analysis for root length as affected by reduced irrigation treatments of wheat (*Triticum aestivum* L.) crop at Delhi weather conditions

Treatments	Root length (cm)*			
	70 DAS	% of growth during 70 DAS	100 DAS	% of growth during 70 -100 DAS
T1	1118.26 [de]	33.8	2190.64 [c]	66.2
T2	1512.79 [b]	39.1	2355.73 [bc]	60.9
T3	1001.05 [fg]	44.0	1275.57 [e]	56.0
T4	1920.06 [a]	45.3	2314.15 [bc]	54.7
T5	931.06 [g]	36.1	1648.39 [d]	63.9
T6	1281.05 [c]	32.1	2715.32 [a]	67.9
T7	1045.40 [ef]	29.9	2456.24 [b]	70.1
T8	1155.73 [d]	33.2	2321.89 [bc]	66.8

*T1 - Control (Pre sowing +No irrigation in all growth stages), T2 – No irrigation during CRI stage; T3 – No irrigation during Tillering stage; T4 –No irrigation during Booting stage; T5 – No irrigation during Flowering stage; T6 – No irrigation Milk/Dough stage; T7 –No irrigation during grain filling stage; T8 – Full Irrigation in all the crop growth stages;*Means with the same letter are not significantly different*

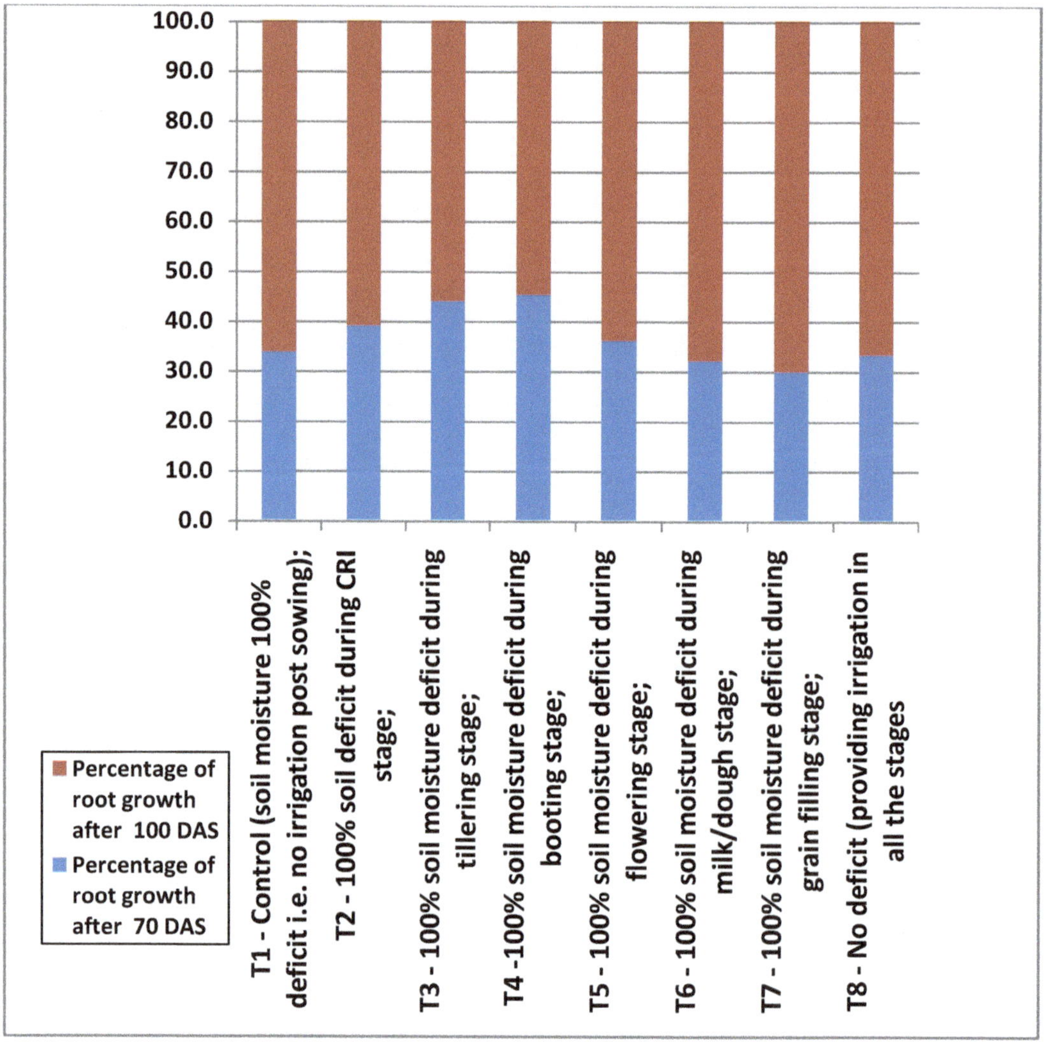

Fig. 6. Relative percentage growth of root length of wheat (*Triticum aestivum* L.) crop after 70 and 100 DAS under different treatment combinations

3.3.1 Functional relationship between number of days after sowing and the root surface area

3.3.1.1 Functional relationship between number of days after sowing and the root surface area 70 das

A fifth order polynomial between Number of Days after Sowing and the Surface Area (70 DAS) of the type below; was fitted and the value of R^2 was worked out to be:

$$y = 0.384x^5 - 5.241x^4 + 12.58x^3 + 70.66x^2 - 283.7x + 512.7 \quad (3)$$

$R^2 = 0.893$

3.3.1.2 Functional Relationship between Number of Days after Sowing and the Root Surface Area 100 DAS

A fourth order polynomial between Number of Days after Sowing and the Surface Area (100 DAS) of the type below; was fitted and the value of R^2 was worked out to be:

$$y = -3.639x^4 + 66.57x^3 - 413.1x^2 + 1005.x - 464.0 \quad ....(4)$$

$R^2 = 0.906$

Based on the R^2 values in both the functional relationships (eq. 3 & 4) it is quite clear that the root surface area showed a well defined trend in the growth patterns after 70 and 100 DAS as expecetd. In this case too, the wheat plant responded in a much balanced manner as regards to plant root surface area increased. This means that although the roots have not elongated much in length they have gained horizontal thickness and became stronger to cope up with the soil moisture stress induced mechanical impedance. The pattern was much more matured and explicit after 70 DAS to 100 DAS (R^2 = 9.06).

3.4 Effect of Irrigation Treatments on Root Surface Volume (X100 Cm^3) of Wheat (*Triticum aestivum* L.) Crop

The treatment and replication wise root volume of wheat (*Triticum aestivum* L.) crop under reduced irrigation after 70 DAS and 100 DAS are presented in Table 5 wherein the mean root volume have also been worked out (Table 5). It was observed that after 70 DAS the minimum root volume of only 365 cm^3 was recorded in T1 (i.e. pre sowing but no irrigation in all growth stages) followed with 426 cm^3 in T2 (i.e. – 100% soil deficit during CRI stage). The maximum root volume (524 cm^3) was observed in treatment T5 (i.e. no irrigation during flowering stage); preceded by T7 (517 cm^3) (i.e. No irrigation during grain filling stage) which is natural. Whereas, in case of 100 DAS, the highest root volume of 816 cm^3 was recorded in the treatment T2 (No irrigation during CRI stage); followed with 672 cm^3 in T2 which is No irrigation during CRI stage. The Post-hoc analysis is presented in Table 6. On persuasion of Tables 5 and 6 it is quite clear that the root volume were significantly affected by the levels of treatments. In case of route volume a definite trend as per the treatment details could be established from the data of root volume of various treatments for 70 DAS as well as 100 DAS (eqn. 5 and 6). However, in this case also it is quite difficult to draw a final conclusion and a clear cult idea about root volume fluctuations did not emerged. However, it could be inferred from the trends that the root volume is not affected in the same intensity as the root length and root surface area (Fig. 9) which was very sensitive and root surface area which showed a relatively lower sensitivity to the moisture sensitive stages. Nevertheless, from this exercise at least it has emerged that root and shoot growth are related and they exhibit almost a similar behavior to the input management with same or similar growth patterns. Majority of the root development seems to have concentrated in the period after 70 DAS (Fig. 10). More precise studies would therefore, be desirable to clearly establish the relationships by doing precise and controlled experiments in Phytotrones, therefore. The functional relationship developed between the number of days after showing and root volume (cm^3) for all treatments at two stages (70 DAS and 100 DAS) have been worked after fitting the most appropriate trend lines and are given below with their respective R^2 (eqn. 5 & 6). Functional relationship between number of days after sowing and the root volume (cm^3) for 70 and 100 DAS are discussed below:

<u>3.4.1 Functional relationship between number of days after sowing and the root volume</u>

3.4.1.1 Functional relationship between number of days after sowing and the root volume (x100) 70 DAS

A sixth order polynomial between Number of Days after Sowing and the Root Volume (X100) (700 DAS) of the type below; was fitted and the value of R^2 was worked out to be:

$$y = -2.94x^6 + 79.68x^5 - 844.6x^4 + 4430.x^3 - 11975x^2 + 15591x - 6147 \quad (5)$$

R^2 = 0.255

3.4.1.2 Functional relationship between number of days after sowing and the root volume (x100) 100 DAS

A sixth order polynomial between Number of Days after Sowing and the Root Volume (X100) (100 DAS) of the type below; was fitted and the value of R^2 was worked out to be:

$$y = -2.390x^6 + 68.73x^5 - 783.6x^4 + 4457.x^3 - 12964x^2 + 17508x - 6077 \quad (6)$$

R^2 = 0.465

Like the root length in the case of root volume too the R^2 values in both the functional relationships (eq. 5 & 6) are not very encouraging that supports the statement made as above regarding the root growth patterns after 70 and 100 DAS. It can be seen from these values that after 70 DAS till the 100 DAS the plant responded in a much balanced manner as regards to plant root volume than 70 DAS which showed an indefinable/ zig-zag pattern.

Table 3. Effect of irrigation treatments on root surface area (cm^2) of wheat (*Triticum aestivum* L.) crop under reduced irrigation at Delhi weather conditions

Treatments	70 DAS					100 DAS				
	R1	R2	R3	R4	Mean	R1	R2	R3	R4	Mean
T1	292.45	325.26	295.63	311.25	306.15	183.09	193.56	203.65	189.65	192.49
T2	261.76	248.56	248.26	304.25	265.71	373.87	384.65	345.21	356.24	364.99
T3	251.15	325.26	265.63	289.65	282.92	347.86	325.36	365.58	324.26	340.77
T4	450.77	355.26	412.25	384.56	400.71	300.13	285.36	288.59	299.65	293.43
T5	355.01	322.12	322.15	311.25	327.63	225.40	255.64	254.29	245.25	245.14
T6	295.34	256.23	286.36	298.56	284.12	419.08	400.12	374.48	412.25	401.48
T7	153.71	165.23	211.25	168.26	174.61	441.51	402.26	385.99	415.26	411.25
T8	360.11	340.26	302.25	324.56	331.79	332.59	351.26	298.57	320.26	325.67
CD at 5% 40.86	*C.V-9.36 %*					*CD at 5% 25.97, C.V-5.48%*				

T1 - Control (soil moisture 100% deficit i.e. no irrigation post sowing); T2 - 100% soil deficit during CRI stage; T3 - 100% soil moisture deficit during tillering stage; T4 -100% soil moisture deficit during booting stage; T5 - 100% soil moisture deficit during flowering stage; T6 - 100% soil moisture deficit during milk/dough stage; T7 -100% soil moisture deficit during grain filling stage; T8 - No deficit (providing irrigation in all the stages)

Table 4. Post hoc analysis for root surface area (cm^2) as affected by reduced irrigation treatments of wheat (*Triticum aestivum L.*) crop at Delhi weather conditions

Treatments	Root surface area (cm^2)			
	70 DAS	% of growth during 70 DAS	100 DAS	% of growth during 70 -100 DAS
T1	306.15^{bc}	61.4	192.49^{f}	38.6
T2	265.71^{c}	42.1	364.99^{b}	57.9
T3	282.92^{c}	45.4	340.77^{bc}	54.6
T4	400.71^{a}	57.7	293.43^{d}	42.3
T5	327.63^{b}	57.2	245.14^{e}	42.8
T6	284.12^{c}	41.4	401.48^{a}	58.6
T7	174.61^{d}	29.8	411.25^{a}	70.2
T8	331.79^{b}	50.5	325.67^{c}	49.5

**Means with the same letter are not significantly different; T1 - Control (soil moisture 100% deficit i.e. no irrigation post sowing); T2 - 100% soil deficit during CRI stage; T3 - 100% soil moisture deficit during tillering stage; T4 -100% soil moisture deficit during booting stage; T5 - 100% soil moisture deficit during flowering stage; T6 - 100% soil moisture deficit during milk/dough stage; T7 - 100% soil moisture deficit during grain filling stage; T8 - No deficit (providing irrigation in all the stages)*

3.5 Effect of Irrigation Treatments on Root Biomass (X100 g) of Wheat (*Triticum aestivum* L.) Crop

The treatment and replication wise root biomass (x100 g) of wheat (*Triticum aestivum* L.) crop under reduced irrigation after 70 DAS and 100 DAS are presented in Table 7 wherein the mean root biomass have also been worked out and subjected to statistical analysis (Table 8). It was observed that after 70 DAS the minimum root biomass of only 1400 g recorded in T1 (i.e pre sowing but no irrigation in all growth stages) followed with 1950 g in T2 (i.e. – 100% soil deficit during CRI stage). The maximum root biomass (2910 g) was observed in treatment T7 (i.e.– No irrigation during grain filling stage); preceded by T3 (2680 g) (i.e. – No irrigation during Tillering stage;) which is quite unnatural. Whereas, in case of 100 DAS, the highest root biomass of 7350 g was recorded in the treatment T8 which is Full Irrigation in all the crop growth stages followed with 7352 g in T1 (i.e. Pre sowing +No irrigation in all growth stages) (Fig. 11). The functional relationship between root biomass after 70 and 100 DAS have also been worked out from the trendline that was plotted (Fig. 11). The Post-hoc analysis is presented in Table 8. On persuasion of Tables 7 and 8 it is quite clear that the root biomass were significantly affected by the levels of treatments (Fig. 12). In case of route biomass a definite trend as per the treatment details could be established from the data of root biomass of various treatments for 70 DAS as well as 100 DAS (Eqn. 7 and 8). Nevertheless, it is noticeable that the relative percentage of root biomass of wheat (*Triticum aestivum* L.) crop after 70 DAS was significantly lower (nearly 1/3rd) as compared to the root biomass accumulation in next 30 days as recorded after 100 DAS (Fig. 12) under different treatment combinations. However, in this case also it is quite difficult to draw a final conclusion and a clear cult idea about root biomass fluctuations did not emerged. However, it could be inferred from the trends that the root biomass is not affected in the same capacity as the root length which was very sensitive and root surface area and root volume which showed a relatively lower sensitivity to the moisture sensitive stages. Nevertheless, from this exercise at least it has emerged that root and shoot growth are related and they exhibit almost a similar behavior to the input management with same or similar growth patterns. More precise studies would therefore, be desirable to clearly establish the relationships by doing precise and controlled experiments. Functional relationship between number of days after sowing and the root biomass was developed (eqn. 7 & 8).

Table 5. Effect of irrigation treatments on root volume (x100 cm^3) of wheat (*Triticum aestivum* L.) crop under reduced irrigation at Delhi weather conditions

Treatments	70 DAS					100 DAS				
	R1	R2	R3	R4	Mean	R1	R2	R3	R4	Mean
T1	3.21	4.25	4.26	3.65	3.84	8.63	8.56	8.15	8.65	8.50
T2	4.55	5.24	4.26	4.26	4.58	6.59	6.57	6.45	7.26	6.72
T3	5.66	4.26	5.26	5.24	5.11	7.32	7.15	7.26	7.48	7.30
T4	4.26	4.56	4.59	4.59	4.50	8.34	8.54	8.45	8.59	8.48
T5	4.22	4.85	4.59	4.69	4.59	7.86	7.56	7.58	8.05	7.76
T6	5.62	4.56	5.24	5.12	5.14	5.45	6.25	5.86	6.12	5.92
T7	5.58	6.25	5.48	5.17	5.62	8.83	8.21	8.59	8.12	8.44
T8	4.25	5.85	4.26	4.59	4.74	8.68	8.54	8.95	8.45	8.66
	CD at 5% 0.73 C.V- 10.42 %					*CD at 5% 0.39, C.V- 3.47%*				

T1 - Control (Pre sowing +No irrigation in all growth stages), T2 – No irrigation during CRI stage; T3 – No irrigation during Tillering stage; T4 –No irrigation during Booting stage; T5 – No irrigation during Flowering stage; T6 – No irrigation Milk/Dough stage; T7 –No irrigation during grain filling stage; T8 – Full Irrigation in all the crop growth stages

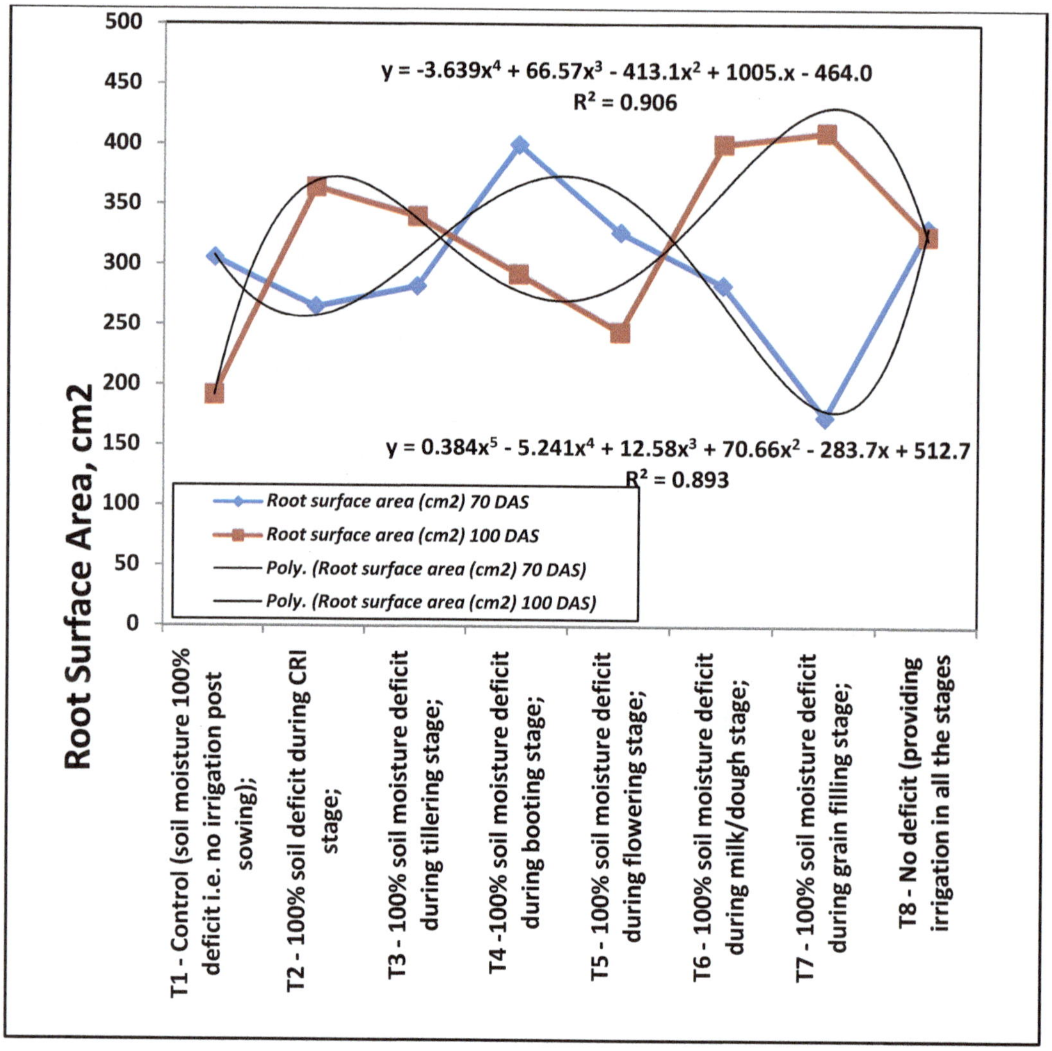

Fig. 7. Effect of soil moisture treatments on mean root surface area (cm2) of wheat (*Triticum aestivum* L.) crop

3.5.1 Functional relationship between number of days after sowing and the root biomass

3.5.1.1 Functional relationship between number of days after sowing and the root biomass (x100 g) 70 DAS

A sixth order polynomial between Number of Days after Sowing and the root biomass (x100 g) (70 DAS) of the type below; was fitted and the value of R^2 was worked out to be:

$$y = 0.041x^6 - 1.173x^5 + 12.87x^4 - 68.90x^3 + 185.1x^2 - 226.7x + 112.7 \quad (7)$$

$R^2 = 0.982$

3.5.1.2 Functional relationship between number of days after sowing and the Root biomass (x100 g) 100 DAS

A sixth order polynomial between Number of Days after Sowing and the Root Biomass (x100 g)) of the type below; was fitted and the value of R^2 was worked out to be:

$$y = -0.207x^6 + 5.308x^5 - 52.17x^4 + 247.2x^3 - 579.8x^2 + 615.7x - 162.6 \quad (8)$$

$R^2 = 0.994$

Based on the R^2 values in both the functional relationships (eq. 7 & 8) it is quite clear that the root biomass which is a combined indicator of all the parameters listed above viz., root length, root surface area and root volume, showed a very well defined trend in the growth patterns after 70 and 100 DAS. This clearly established the impact of soil moisture deficit during initial to developmental to maturity stages that soil moisture has a very strong influence on the total root biomass. In this case, during both periods i.e. 70 DAS as well as 100 DAS the trend of the root biomass accumulation curves could be defined with a very high degree of accuracy (R^2 = 0.982 for 70 DSA and = 0.994 for 100 DAS). Hence, wheat crop responded in a much balanced manner as regards to plant root biomass and the influence of the soil moisture deficit has been pronounced except in the treatments which go slightly disturbed due to intermittent rains (Fig. 13).

3.6 Effect of Reduced Irrigation on Plant Height of Wheat Crop

There was no significant difference in plant heights amongst various treatments in early vegetative stages of wheat crop (40 DAS). However, in the later stages i.e., (70 and 100 DAS) during late vegetative and reproductive phases; various treatments exhibited significant differences in the plant heights as compared to the control (T1). T1, T3 and T6 were amongst the treatments which were not significantly different with respect to plant height during late vegetative stages. Similarly, the treatments T1, T3, T5, T6 and T7 did not exhibit significant differences in plant heights although; all of them had significantly different heights during reproductive stage those were significantly different with the control.

Table 6. Post hoc analysis for root volume (x 100 cm^3) as affected by reduced irrigation treatments of wheat (*Triticum aestivum* L.) crop at Delhi weather conditions

Treatments	Root volume (x 100 cm^3) *			
	70 DAS	% of growth during 70 DAS	100 DAS	% of growth during 70 -100 DAS
T1	3.84 ^{c}	30.0	8.50 ^{a}	70.0
T2	4.58 ^{bc}	38.8	6.72 ^{d}	61.2
T3	5.11 ^{ab}	41.8	7.30 ^{c}	58.2
T4	4.50	35.1	8.48 ^{a}	64.9
T5	4.59 ^{b}	37.7	7.76 ^{b}	62.3
T6	5.14 ^{ab}	46.4	5.92 ^{e}	53.6
T7	5.52 ^{a}	38.0	8.44 ^{a}	62.0
T8	4.74 ^{b}	34.6	8.66 ^{a}	65.4

*T1 - Control (Pre sowing +No irrigation in all growth stages), T2 – No irrigation during CRI stage; T3 – No irrigation during Tillering stage; T4 –No irrigation during Booting stage; T5 – No irrigation during Flowering stage; T6 – No irrigation Milk/Dough stage; T7 –No irrigation during grain filling stage; T8 – Full Irrigation in all the crop growth stages; *Means with the same letter are not significantly different*

Table 7. Effect of irrigation treatments on root biomass (x100 g) of wheat (*Triticum aestivum L.*) crop under reduced irrigation at Delhi weather conditions

Treatments	70 DAS					100 DAS				
	R1	R2	R3	R4	Mean	R1	R2	R3	R4	Mean
T1	4.1	13.6	18.1	15.5	14.0	33.1	73.9	69.8	70.5	73.5
T2	5.8	23.8	22.3	18.1	19.5	30.2	43.3	42.4	46.8	48.8
T3	7.2	24.1	22.4	27.6	26.8	37.4	52.3	51.9	54.3	54.6
T4	5.5	19.4	20.9	21.1	20.7	37.5	71.2	72.2	72.6	72.8
T5	5.4	20.5	22.3	21.5	21.5	36.1	59.4	57.3	61.0	62.5
T6	7.2	25.6	23.9	26.8	26.3	28.0	34.1	36.6	35.9	36.2
T7	7.1	34.9	34.3	28.3	29.1	49.6	72.5	70.5	69.8	68.5
T8	5.4	24.9	24.9	19.6	21.8	41.1	74.1	76.4	75.6	73.2
	CD at 5% 0.974 C.V- 10.42 %					*CD at 5% 0.499, C.V- 3.47%*				

T1 - Control (Pre sowing +No irrigation in all growth stages), T2 – No irrigation during CRI stage; T3 – No irrigation during Tillering stage; T4 –No irrigation during Booting stage; T5 – No irrigation during Flowering stage; T6 – No irrigation Milk/Dough stage; T7 –No irrigation during grain filling stage; T8 – Full Irrigation in all the crop growth stages

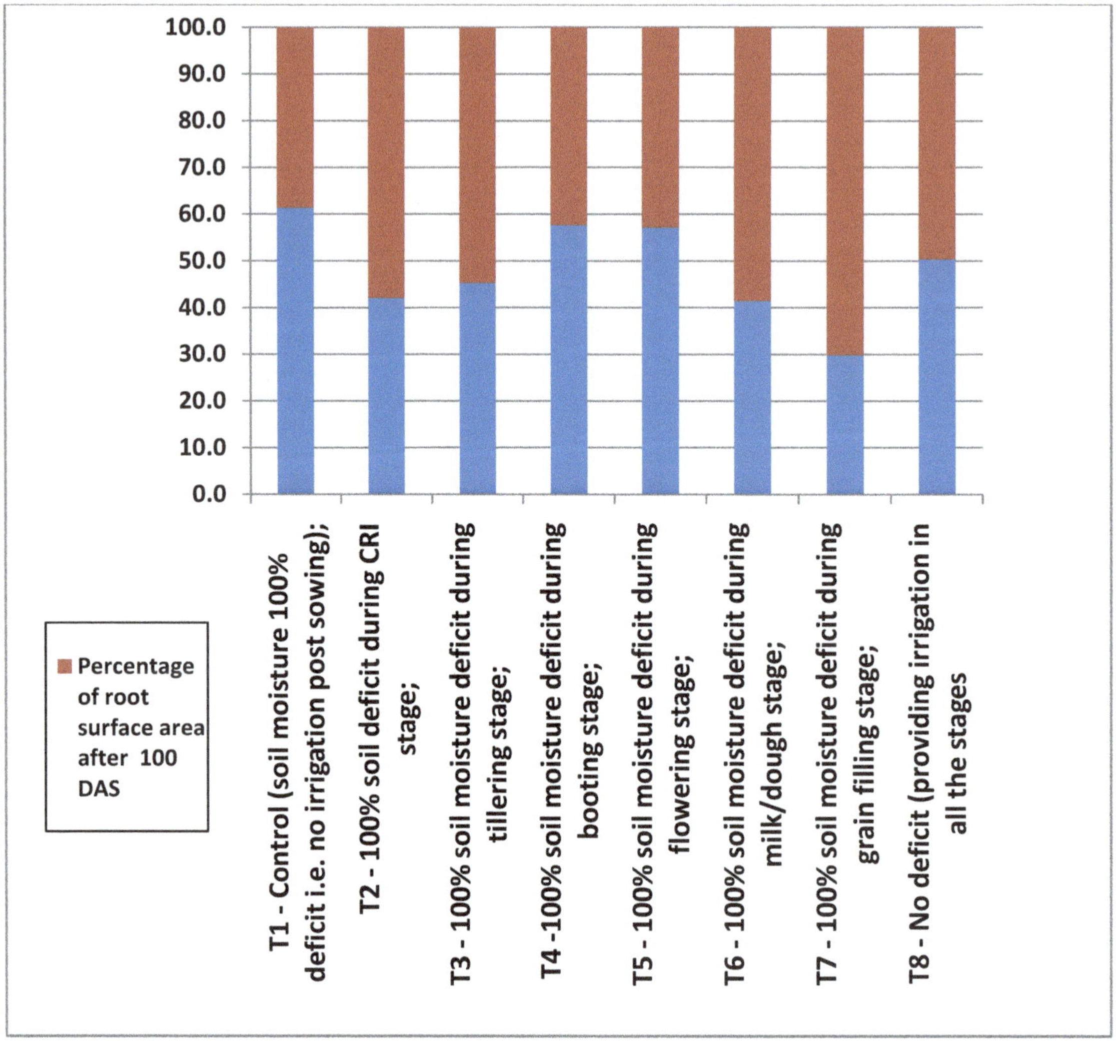

Fig. 8. Relative percentage growth of root surface area of wheat (*Triticum aestivum* L.) crop after 70 and 100 DAS under different treatment combinations

Table 8. Post hoc analysis for root biomass (x 100 g) as affected by reduced irrigation treatments of wheat (*Triticum aestivum* L.) crop at Delhi weather conditions

Treatments	Root biomass (x100 g) *			
	70 DAS	% of growth during 70 DAS	100 DAS	% of growth during 70-100 DAS
T1	14.0[c]	30.0	73.5[a]	70.0
T2	19.5[bc]	38.8	48.8[d]	61.2
T3	26.8[ab]	41.8	54.6[c]	58.2
T4	20.7	35.1	72.8[a]	64.9
T5	21.5[b]	37.7	62.5[b]	62.3
T6	26.3[ab]	46.4	36.2[e]	53.6
T7	29.1[a]	38.0	68.5[a]	62.0
T8	21.8[b]	34.6	73.2[a]	65.4

*T1 - Control (Pre sowing +No irrigation in all growth stages), T2 – No irrigation during CRI stage; T3 – No irrigation during Tillering stage; T4 –No irrigation during Booting stage; T5 – No irrigation during Flowering stage; T6 – No irrigation Milk/Dough stage; T7 –No irrigation during grain filling stage; T8 – Full Irrigation in all the crop growth stages *Means with the same letter are not significantly different*

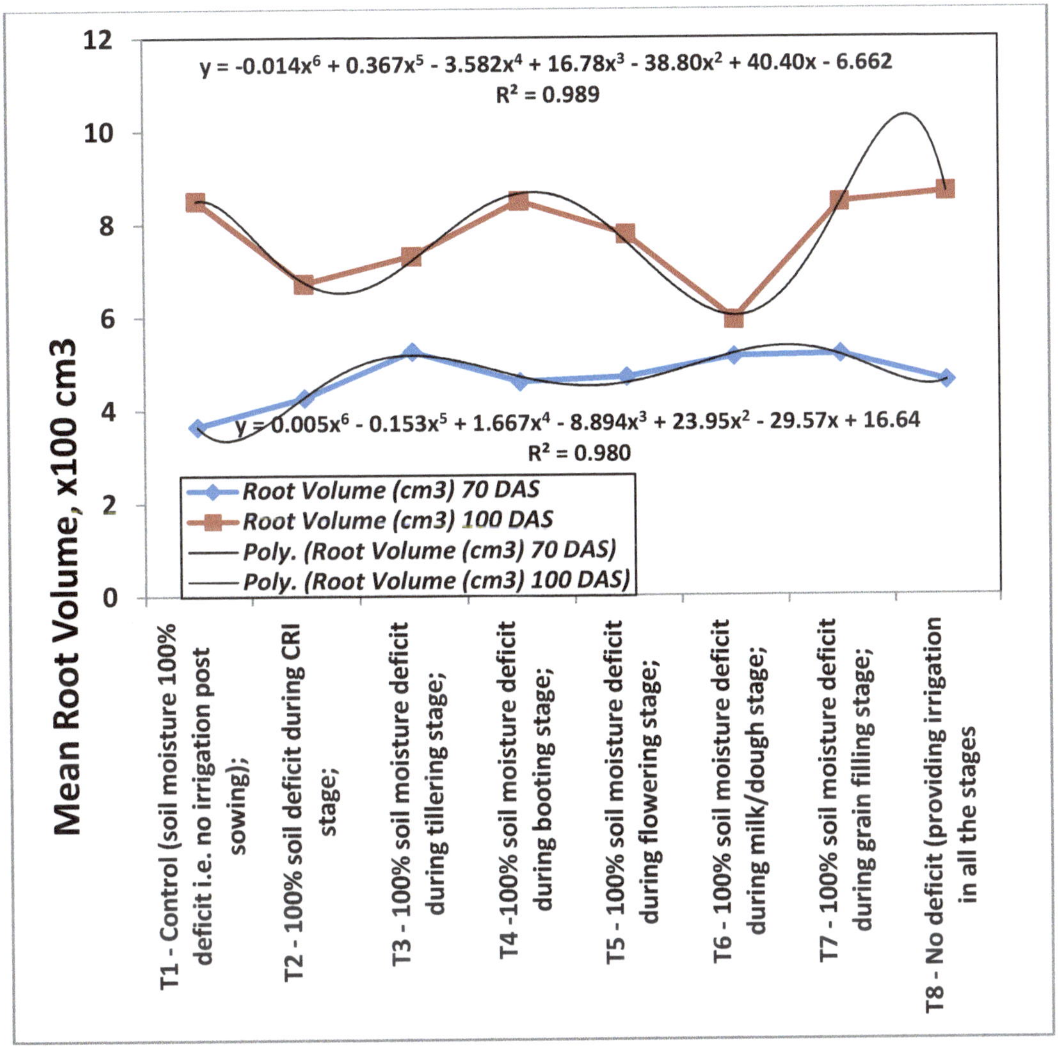

Fig. 9. Effect of soil moisture treatments on mean root volume (x100 cm3) of wheat (*Triticum aestivum* L.) crop

3.7 Effect of Reduced Irrigation on Plant Growth, Development of Number of Tillers

Similarly, significant differences were observed in number of tillers/m2. Persistence of the higher assimilatory surface leaf area is a pre-requisite for prolonged photosynthate activity vis-à-vis higher dry matter accumulation and ultimately crop productivity. Since plant height, number of tillers plays and leaf area plays an important role in photosynthesis [39,40,61], it may be understood that these growth parameters contributes significantly in the development of "source". Since there are significant differences in source development, it may be interpreted that it will also affect the yield of the crop. Yield results reveal that there is a positive correlation between LAI and grain weight (r = 0.728).

3.8 Effect of Reduced Irrigation on Leaf Area Index (LAI)

An indicator of the overall health conditions of the crop is the Leaf Area Index (LAI) of the plant which eventually is a combined result of soil moisture status, soil fertility and disease free atmosphere. Consequently, the LAI has been consistently high in almost all growth stages in all the treatments but the trend has remained the same. This has been directly related with the soil moisture status and can rightly be attributed to the main contributor. The leaf area index (LAI) in different plant growth stages, were found to be statistically significant differences as a result of reduced irrigation. Similar results were obtained by other workers too [9,39]. As our main emphasis here is the root architecture hence, this factor (Leaf Area Index (LAI)) is not being subjected to elaborate discussions.

3.9 Dry Matter Production

Leaf Area Index (LAI) is closely related with the dry matter production [11]. Dry matter accumulation was estimated in three main crop stages i.e., tillering, flowering and milking stages respectively. In order to estimate the dry matter partitioning it was accounted for root, stem, leaves and spikelets separately. It is worth noticing that skipping irrigation in CRI and milking stages had significant decrease in the dry matter accumulation [40,30] even though the treatment with no water deficiency (T8) recorded highest dry matter accumulation. To calculate the amount of irrigation water, differences among treatment in allocating the sources from the sink were observed.

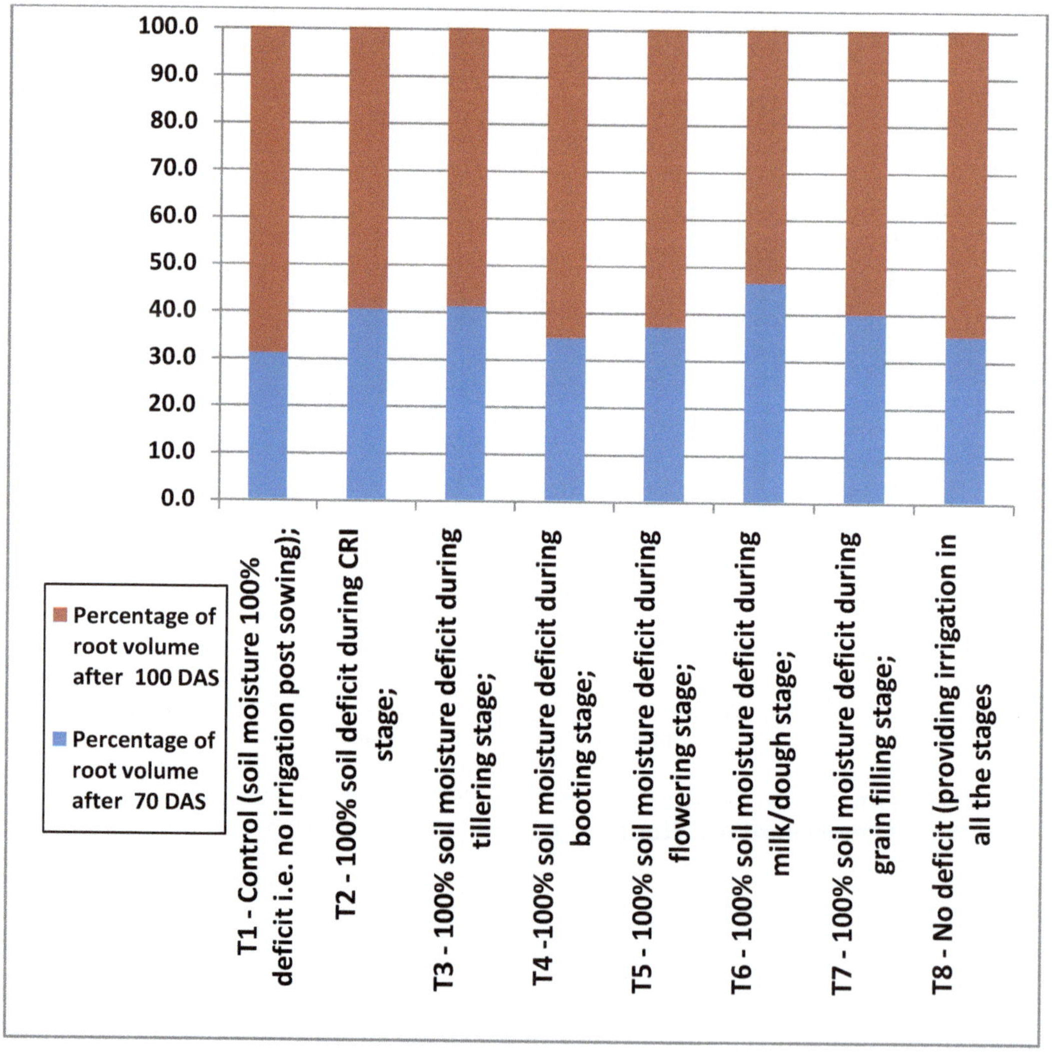

Fig. 10. Relative percentage growth of root volume of wheat (*Triticum aestivum* L.) crop after 70 and 100 DAS under different treatment combinations

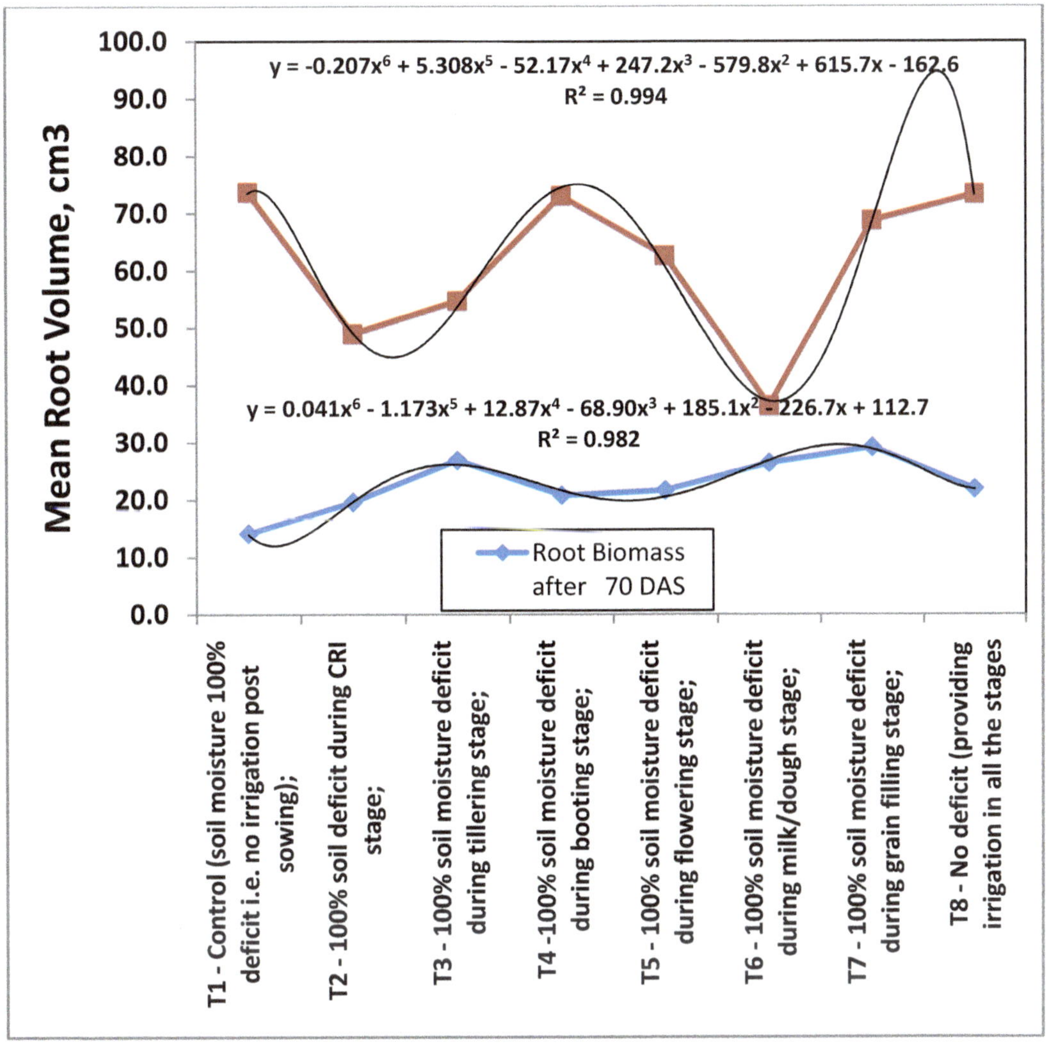

Fig. 11. Effect of soil moisture treatments on mean root biomass (x100 g) of wheat (*Triticum aestivum* L.) crop after 70 and 100 DAS

3.10 Soil Moisture Dynamics

Soil moisture content was estimated. The soil moisture changes during the crop period were also recorded by book keeping method. After the pre-sowing irrigation which coincides with the onset of winters in Northern India, the temperature becomes low to extremely low the evapotranspiration losses are limited to minimum. Hence, the second and third stages of crop keep on surviving on the root zone soil moisture. Estimating the soil moisture content before irrigation can be an important criteria to schedule irrigation and also irrigating with the required moisture content [29,62]. The effective rainfall due to rainfall as estimated using standard procedures which has been duly accounted in the irrigation scheduling procedures (Fig. 13).

3.11 Yield and Yield Attributes

Although, there was no significant difference in the grain weight among T3, T4 and T8 the treatment with no irrigation (T1) has shown drastic reduction in grain weight (Table 9). It is also observed that treatment T2 has adversely affected the grain weight. The results are presented in Fig. 14. The relationship between grain weight and LAI is also illustrated through regression analysis (Fig. 14). Due incessant rainfall in the field in the preceding week of harvesting that resulted into inability in threshing of the produce having higher moisture contends,

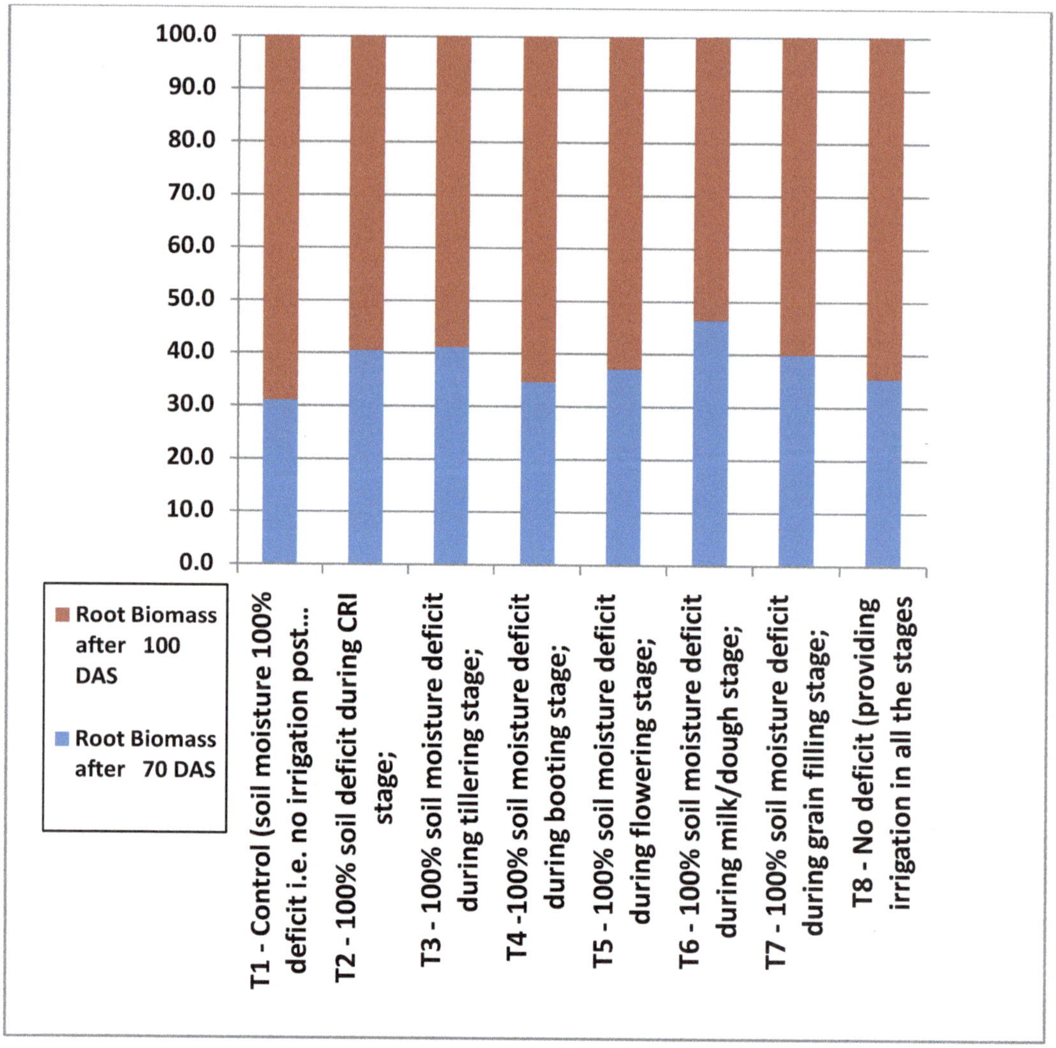

Fig. 12. Relative percentage of root biomass of wheat (*Triticum aestivum* L.) crop after 70 and 100 DAS under different treatment combinations

the results of grain yields could not be reported. An equation between LAI and plant dry matter accumulation has been developed based on the research results.

Grain weight (y) = 0.682. Leaf Area Index (x) + 0.8083 (9)

$R^2 = 0.538$

A general variation of the major climatic parameters at the study area (Fig. 13) indicated that though the maximum and minimum temperatures have behaved quite normally, the rainfall at the station was quite consistent. There were several rainstorms (seven) of the order of 5mm to more than 20 mm which has affected the crop yield quite favourably by adding moisture to the soil during this year. During the third week of November (sowing period) to second week of February the minimum temperature reduced drastically which in turn had resulted into a substantial reduction in the evaporative demand of the atmosphere. This might have been the major reason for non-significant changes in the crop parameters in the initial two crop growth stages followed with a very heavy pre-sowing irrigation which has resulted into non-significant variation among the first two stages. In the later parts of the crop growing season however, the significant changes in different crop growth parameters have been observed and recorded due to improved climatic conditions. This is a clear cut reason for keeping the crop without water for so long by some farmers after sowing it with pre-sowing irrigation of more than 50 mm or fully saturating the soil profile

3.12 Effect of Soil Moisture Stress on Root Morphometry, Architecture and Physiology

Plant roots are strongly affected by soil water potential and physical characteristics. A plant responds to a lack of water by halting growth and reducing photosynthesis and other plant processes in order to reduce water use [63-67]. As water loss progresses, leaves of some species may appear to change color—usually to blue-green. Foliage begins to wilt and, if the plant is not irrigated, leaves will fall off and the plant will eventually die. Similar changes have been found to be occurring in the plat root system as well [68-71]. Soil Moisture stress lowers the water potential of a plant's root and upon extended exposure, abscisic acid is accumulated and eventually stomata closure occurs [72-75]. This reduces a plant's leaf relative water content. The time required for drought stress to occur depends on the water-holding capacity of the soil, environmental conditions, stage of plant growth, and plant species. Plants growing in sandy soils with low water-holding capacity are more susceptible to drought stress than plants growing in clay soils. A limited root system will accelerate the rate at

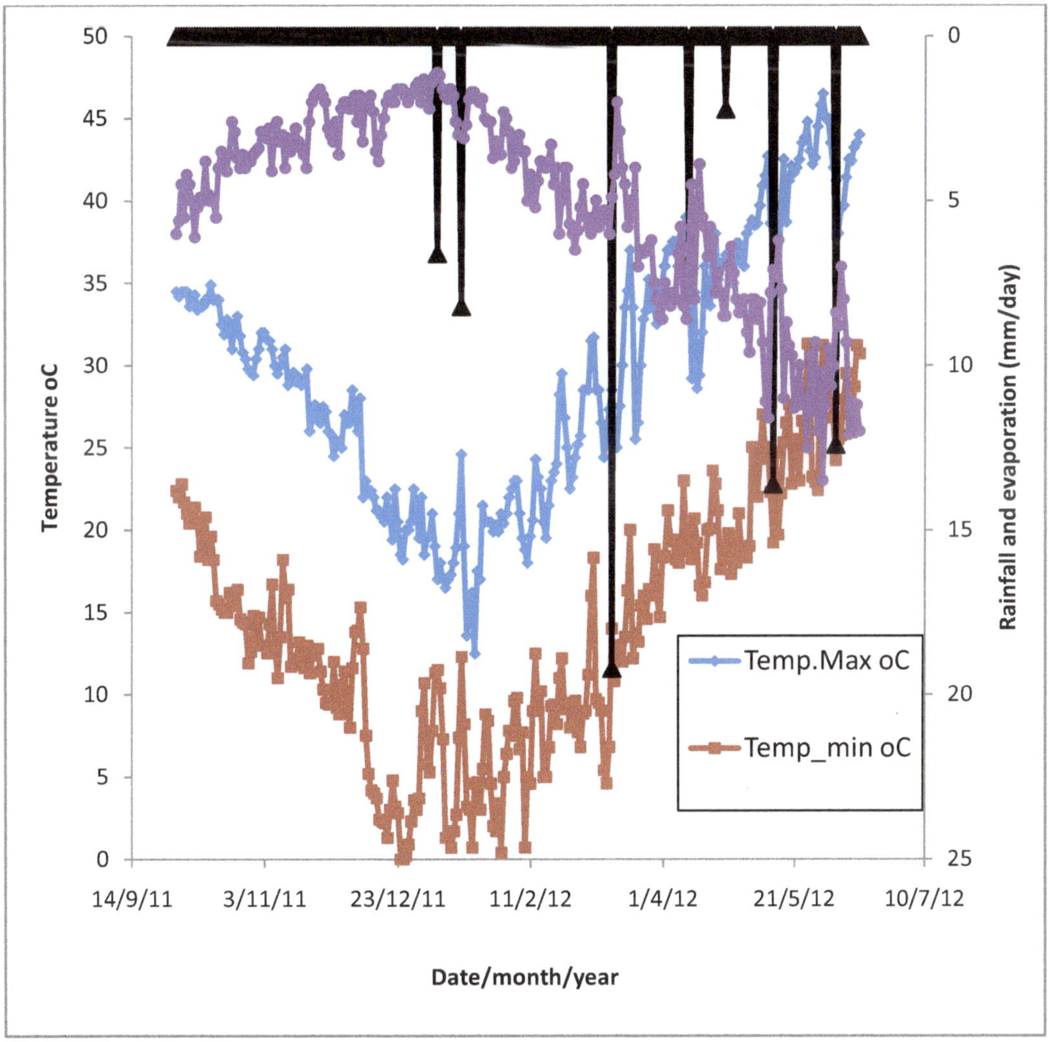

Fig. 13. General variations in the major climatic parameters during the wheat (*Triticum aestivum* L.) crop growing season (rabi) of the year 2011-12 at IARI Pusa observatory, New Delhi, India

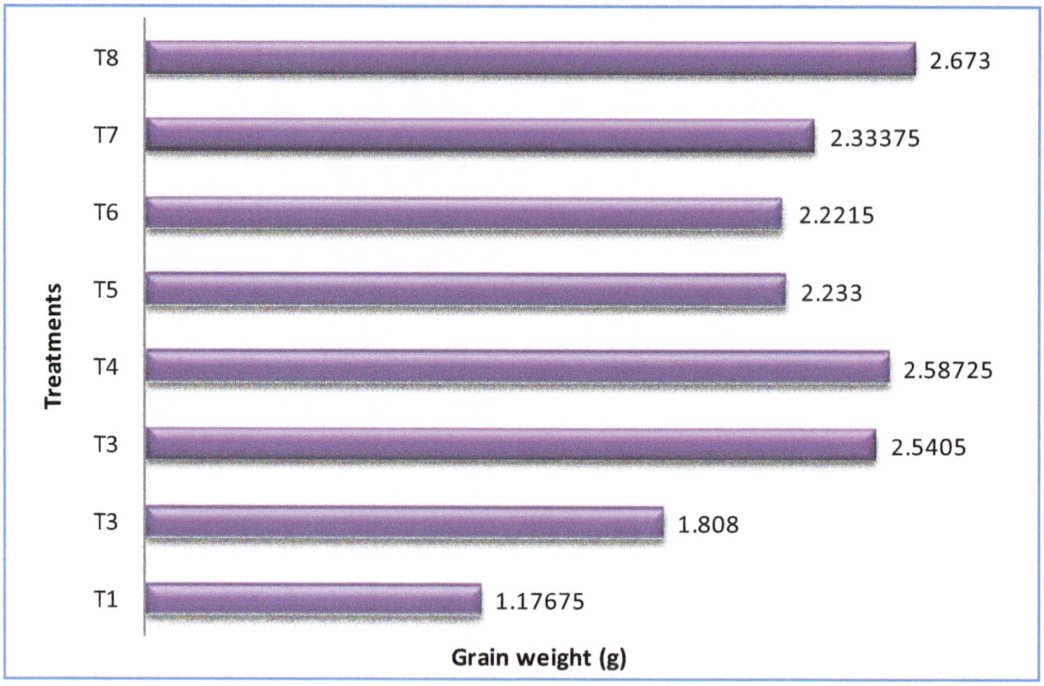

Fig. 14. Grain weight (gram/spikelet) in wheat (*Triticum aestivum)* crop under reduced irrigation at different stages of crop growth (Var. Pusa Hybrid HD-2967)

T1 - Control (Pre sowing +No irrigation in all growth stages), T2 – No irrigation during CRI stage; T3 – No irrigation during Tillering stage; T4 –No irrigation during Booting stage; T5 – No irrigation during Flowering stage; T6 – No irrigation Milk/Dough stage; T7 –No irrigation during grain filling stage; T8 – Full Irrigation in all the crop growth stages

which drought stress develops. A root system may be limited by the presence of competing root systems, by site conditions such as compacted soils or high water tables, or by container size (if growing in a container). A plant with a large mass of leaves in relation to the root system is prone to drought stress because the leaves may lose water faster than the roots can supply it. Newly installed plants and poorly established plants may be especially susceptible to drought stress because of the limited root system or the large mass of stems and leaves in comparison to roots [76,77]. Soil mechanical impedance against root growth and development is caused mainly by natural processes and by the use of heavy machinery for soil cultivation [78,79]. The root system of an individual plant consists of several component roots of different nature. Those components differ in external morphology, physio-logical function and genetic control. According to [80], root system structure of cereal plant consists of seminal, seminal adventitious, nodal and lateral roots. The seminal and nodal roots build up the framework, while lateral roots of different orders build network of the roots in soil. The cereal species develop two types of root system, depending on the angle of growth of branches (lateral roots) and their distribution in a soil profile [81]. Highlighting on water relation of plants in soil stresses, many studies indicate that leaf water status is influenced through several mechanisms [82,83]. According to [84] root borne signals affect the rate of development in the apical meristem, cell division and cell expansion in the expanding leaves and they induce stomata behavior. Root signals are expected to be electrical and hormonal (ethylene, ABA, auxin and likely cytokinin signaling cascades) and are involved in mediating physiological effects. Understanding of processes in which the photosynthesis and gas exchange rate are depressed by soil compaction requires more physiological studies on roots and shoots. Since development of a whole root system consisting of root components was closely related to productivity of wheat; the crop yield also was affected due to soil moisture stress in different growth stages (Tables 2, 9). However, there are huge genetic variations among the cultivars with regard to shoot and root characteristics [85,86] as well as ambient climatic and edaphic characteristics of the place of experiment [87,88].

Table 9. Grain weight and above ground biomass as affected by different irrigation treatments (different stages) of wheat (*Triticum aestivum* L.) crop under reduced irrigation at Delhi weather conditions

Treatments	Yield parameters	
	Grain weight/ spikelet (g)	Above ground biomass (t/ha)
T1	1.18	8.58
T2	1.81	13.76
T3	2.54	19.79
T4	2.59	19.19
T5	2.23	18.07
T6	2.22	15.79
T7	2.33	22.56
T8	2.67	21.14
S.E.(d)	*0.134*	*1.764*
C.D. (5 %)	*0.29*	*3.692*

T1 - Control (Pre sowing +No irrigation in all growth stages), T2 – No irrigation during CRI stage; T3 – No irrigation during Tillering stage; T4 –No irrigation during Booting stage; T5 – No irrigation during Flowering stage; T6 – No irrigation Milk/Dough stage; T7 –No irrigation during grain filling stage; T8 – Full Irrigation in all the crop growth stages

4. CONCLUSION

In this experiment; the water requirement in the initial stages of crop growth and development was low as compared to other stages due to extreme low temperature, high humidity, and low sunshine hours and near zero wind that brought down the evaporative demand of the atmosphere. However, during the later part of the year, the evaporative demand has eventually increased resulting into higher atmospheric/ evaporative demand and thus the effect of reduced irrigation was pronounced. Control treatment T1 (with no irrigation) recorded a significant difference with T8 treatment (no water deficiency) in all traits studies (the root morphology, leaf area index, soil moisture dynamics and crop yields), that were closely associated but these traits showed significant differences among the treatments. It was established that under different critical crop growth stages plants response to the stress conditions varied differently to fulfill their metabolic activities. However, the successive critical growth stages were not completely different to each other based on statistical significance tests.

Results of the present study confirmed that root volumes as well as biomass were significantly affected due to skipping irrigation in crop sensitive stages eg. CRI, flowering and grain filling crop growth stages. Root growth and development are important parameter in managing the crop in reduced irrigation conditions. The results have clearly demonstrated that that each root related parameter varies with each treatment (Tables 2, 4, 6, 8). Root dynamics study is important to understand the nutrient and water uptake since it is directly related with yield. Functional relationships developed in this study may be used for understanding the root growth and development under soil moisture stress conditions in various crop growth stages. The study needs to be conducted under rain shelter or phytotrones for better control of the parameters.

COMPETING INTERESTS

Authors have declared that no competing interests exist.

REFERENCES

1. Singh RD, Sinha HN. Water management practices for wheat. Indian J. Soil Conserv. 1987;15:101–106.
2. Singh P. Water stress and plant parameters for wheat. Indian Soc. Agric. Engrgs. 1989;85:34–40.
3. Singh V, van Oosterom EJ, Jordan DR, Messina CD, Cooper M, Hammer GL. Morphological and architectural development of root systems in sorghum and maize. Plant and Soil. 2010;333:287-299.
4. Nagesh AKB. The physiological and genetic bases of water use efficiency in winter wheat. Unpub. PhD thesis, University of Nottingham, UK. 2006;272.
5. Chopra RK, Shukla S, Singh K, Kadam SB, Singh NK. Characterization of High Yielding and Drought Tolerant RILs Identified from Wheat Cross WL711 x C306 RIL Mapping Population using Drought Susceptibility Index (DSI) as Selection Criteria. Indian J. Plant Genet. Resour. 2013 26(1):25–31.
6. Shao HB, Liang ZS, Shao MA, Wang BC. Changes of some physiological and biochemical indices for soil water deficits among 10 wheat genotypes at seedling

stage, Colloids Surf. B: Biointerfaces. 2005;42(2):107-211.
7. Donald CM. In search of yield. J Aus. Inst. Agric. Sci., 1962;28:171-178.
8. Mirbahar AA, Markhandl GS, Mahar AR, Abro SA and Kanhar NA. Effect of water stress on yield and yield components of wheat (*Triticum aestivum* L.) varieties. Pak. J. Bot. 2009;41(3):1303-1310.
9. Pandey RK, Herrera WAT and Pendleton JW. Drought response of grain legumes under irrigation gradient. III. Plant growth. Agron. J. 1984;76:557–560.
10. Gajri PR, Arora VK and Parihar SS. Tillage management for efficient water and nitrogen use in wheat following rice Soil Tillage Res. 1992;24:167–182.
11. Bhan VM. Analysis of growth in upland rice. Risco. 1967;16:113-118.
12. Yitaew M, Brown P. Predicting daily evapotranspiration from short-term values. J. Irrig. Drain. Eng. 1990;115:387–398.
13. Allen RG, Pereira LS, Raes D, Smith M. Crop evapotranspiration-guidelines for computing crop water requirements, irrigation and drain, Paper No. 56. FAO, Rome, Italy. 1998;300.
14. Proffitt APB, Berliner PR, Oosterhuis DM. A comparative efficiency of wheat grown under high and low frequency irrigation. Agron. J. 1985;77:655–662.
15. Prihar SS, Sandhu BS. Irrigation of field crops-principles and practices. I.C.A.R., New Delhi, India. 1987;326
16. Prihar SS, Cheri KL, Sandhu KS, Sandhu BS. Comparison of irrigation schedules based on pan evaporation and growth stages of winter wheat. Agron. J. 1976;60: 650–653.
17. Sharma DK, Kumar A, Singh KN. Effect of irrigation scheduling on growth, yield and evapotranspiration of wheat in sodic soils. Agric. Water Manage. 1990;18:267–276.
18. Tyagi NK, Sharma DK, Luthra SK. Evapotranspiration and crop coefficients of wheat and sorghum. J. Irrig. Drain. Eng. 2000;126:215–222.
19. Singh P. Evapotranspiration, evaporation and crop yield relationship for wheat in humid climate. J. Inst. Eng.1991;77:65–71.
20. Bandyopadhyay PK, Mallick S. Actual evapotranspiration and crop coefficients of wheat (*Triticum aestivum*) under varying moisture levels of humid tropical canal command area. Agric. Water Manage. 2003;59:33-47.
21. Blum A. Drought resistance, water-use efficiency, and yield potential-are they compatible, dissonant, or mutually exclusive? Aust. J. Agric. Res. 2005; 56:1159-1168.
22. Blum A. Effective use of water (EUW) and not water-use efficiency (WUE) is the target of crop yield improvement under drought stress. Field Crops Res. 2009; 112:119-123.
23. Cattivelli L, Rizza F, Badeck F, Mazzucotelli E, Mastrangelo A, Francia E, Mare C, Tondelli A, Stanca A. Drought tolerance improvement in crop plants: an integrated view from breeding to genomics. Field Crops Res. 2008;105:1-14.
24. Abd El-Ghany HM, Abd El-Salam, MS Hozyen M, Afifi, MHM Effect of reduced irrigation on some growth stages of wheat. J. Applied Sci. Res. 2012;8(5):2776-2784.
25. Laaboudi A, Mouhouche B. Water requirement modeling for wheat under arid climatic conditions. Hydrol. Curr. Res. 2012;3:130. DOI:10.4172/2157-7587.1000130.
26. Savicka M and N Škute. Some morphological, physiological and biochemical characteristics of wheat seedling *Triticum aestivum* L. organs after high-temperature treatment. EKOLOGIJA. 2012;58(1):9–21.
27. Zarkti H, Ouabbou H, Udupa SM and Hilali A. Agro-morphological variability in durum wheat landraces of Morocco. Aus. J. Crop Sci. 2012;6(7):1172-1178.
28. Molden D, Sakthivadivel R and Zaigham H. Basin-level use and productivity of water: Examples from South Asia, IWMI Research Report 49, Colombo: International Water Management Institute. EKOLOGIJA. 2001;58(1):9–21.
29. Sharma RB and Ghildyal BP. Soil water-root relations in wheat: water extraction rate of wheat roots that developed under dry and moist conditions. Agron. J. 1977;69(2):231–233. DOI:10.2134/agronj1977.00021962006900020009x.
30. Huang X, Xiao X, Zhang S, Korpelainen H, Li C. Leaf morphological and physiological responses to drought and shade in two Populus cathayana populations. Biol. Plant. 2009;53(3):588-592.
31. Zhang YX, Wu JC, Cao FL, Zhang YP. Effects of water stress on photosynthetic activity, dry mass partitioning and some

associated metabolic changes in four provenances of neem (*Azadirachta indica* A. Juss). Photosynthetica. 2010;48(3): 361-369.

32. Tang AC, Kawamitsu Y, Kanechi M, Boyer JS. Photosynthetic oxygen evolution at low water potential in leaf discs lacking an epidermis. Ann. Bot. 2002;89:861-870.
33. Yu DJ, Kim SJ, Lee HJ. Stomatal and non-stomatal limitations to photosynthesis in field-grown grapevine cultivars. Biol. Plant. 2009;53:133-137.
34. Sponchiado BN, White JW, Castillo JA, Jones PG. Root growth of four common bean cultivars in relation to drought tolerance in environments with contrasting soil types. Exp. Agric. 1980;25:249–257.
35. Merrill SD, Tanaka DL Hanson JD. Root length growth of eight crop species in Haplustoll soils. Soil Sci. Soc. Am. J. 2002; 66:913–923.
36. Bryla DR, Bouma TJ, Hartmond U, Eissenstat DM. Influence of temperature and soil drying on respiration of individual roots in citrus: Integrating greenhouse observations into a predictive model for the field. Plant Cell Environ. 2001;24:781-790.
37. Atta B, Mahmood MT, Trethowan RM. Relationship between root morphology and grain yield of wheat in north-western NSW, Australia. AJCS. 2013;7(13):2108-2115.
38. Ludlow MM, Muchow RC. A critical evaluation of traits for improving crop yields in water-limited environments. Adv. Agron. 1990;43:107-153.
39. Panda RK, Behera SK and Kashyap PS. Effective Management of Irrigation Water for Wheat Crop Under Stressed Conditions Using Simulation Modeling Seventh International Water Technology Conference Egypt . 2003;35-62.
40. Richards RA. Crop improvement for temperate Australia: Future opportunities. Field Crops Res. 1991;26:141-169.
41. Richards RA, Condon AG, Rebetzke GJ. Traits to improve yield in dry environments. In: Reynolds MP, Ortiz-Monasterio JI, McNab A (eds.) Application of physiology in wheat breeding, Mexico, D.F., CIMMYT, 2001;88-100.
42. Richards RA, Rebetzke GJ, Condon AG, Watt M, Spielmeyer W, Ellis MH, Bonnett DG, Dolferus R. Genetic improvement of wheat for dry environments - A trait based approach. In: Appels R, Eastwood R, Lagudah E, Langridge P, Mackay M, McIntyre M, Sharp P (eds) Proceedings of the 11th International Wheat Genetics Symposium, 24-29 August, Brisbane, Australia. 2008;1-6.
43. Manske GGB, Vlek PLG. Root architecture - wheat as a model plant. In: Waisel Y, Eshel A, Kafkafi U (eds.) Plant Roots: The Hidden Half', 3rd edition, Marcel Dekker, New York USA. 2002;149-259.
44. Clarke JM, Townley-Smith TF. Screening and selection techniques for improving drought resistance. In: Vose PB, Blixt SG (eds) Crop Beeding: A Contemporary Basis, Pergamon, Oxford UK. 1984;137-162.
45. Lazarow PB and Fujiki Y. Biogenesis of peroxisomes. Annu. Rev. Cell Biol. 1985;1:489-530.
46. Yan F, Bing-cheng X, Turner NC, Li FM. Grain yield, dry matter accumulation and remobilization, and root respiration in winter wheat as affected by seeding rate and root pruning. Europ. J. Agronomy. 2010;33:257–266.
47. Wasson AP, Richards RA, Chatrath R, Misra SC, Prasad SVS, Rebetzke GJ, Kirkegaard JA, Christopher J, Watt M. Traits and selection strategies to improve root systems and water uptake in water-limited wheat crops. J. Exp. Bot. 2012;63: 3485-3498.
48. Waines, JG. Determination of optimum root and shoot size in bread wheat for increased waterand nutrient-use efficiency and grain yield. Report to California Wheat Commission: GH 2011-2012 Experiments. 2013; UC Riverside.
49. Guttieri MJ, Stark JC, O'Brien K, Souza E. Relative sensitivity of spring wheat grain yield and quality parameters to moisture deficit. Crop Sci. 2001;41:327-335.
50. Ramezanpoor M, Dastfal M. Evaluation bread and durum wheat cultivars tolerance to water stress. Paper presented at the 8rd Iran agronomy and plant breeding conference, University of Rasht, Gilan, Iran; 2004
51. Guttieri MJ, Ahmad R, Stark JC, Suza E. End-use quality of six hard spring wheat cultivars at different irrigation levels. Crop Sci. 2000;40:631-635.
52. Güler M. Irrigation effects on quality characteristics of durum wheat. Can J Plant Sci. 2002;83:327-331.
53. Gonzalez JA, Bruno M, Valoy M, Prado FE. Genotypic variation of gas exchange parameters and leaf stable carbon and nitrogen isotopes in ten quinoa cultivars

grown under drought. J. Agron. Crop. Sci. 2010;0931-2250.

54. Dhanpal, AH. Simulation of soil moisture regime: Application of the SWATRE model to a maize crop on the reddish brown earths in the dry zone of Sri Lanka. Agric. Sys. 1992;38:61-73.
55. Jefferies EC, Hook JE, Blair SL, Using crop models to plan water withdrawl for irrigation in drought years. Agric. Sys. 1992;45(3):271-289.
56. Lilley JM, Kirkegaard JA. Benefits of increased soil exploration by wheat roots. Field Crops Res. 2011;122:118-130.
57. Dardanelli JL, Ritchie JT, Calmon M, Andriani JM, Collino DJ. An empirical model for root water uptake. Field Crops Res. 2004;87:59-71.
58. Black CA. Methods of soil Analysis (Part 2nd) Agronomy Monograph of American Society of Agronomy. Madison Wisconsin, 1965;149-159.
59. Busse, MD,Ratcliff AW, Shestak, CJ, Powers RF. Glyphosate toxicity and the effects of long-term vegetation control on soil microbial communities. Soil Bio. and Biochem. 2001;33:1777–1789.
60. Gomez KA. Techniques for field experiments with rice. International Rice Research Institute Manila, Philippines; 1972.
61. Gregory PJ, Bengough AG, Grinev D, Schmidt S, Thomas WTB, Wojciechowski T, Young IM. Root phenomics of crops: opportunities and challenges. Funct Plant Biol. 2009;36:922-929.
62. Zarkti H, Ouabbou H, Udupa SM, Hilali A. Agro-morphological variability in durum wheat landraces of Morocco Australian Journal of Crop Science. 2012;6(7):1172-1178.
63. Passioura JB. The role of root system characteristics in the drought resistance of crop plants. In: Drought Resistance in Crops with Emphasis on Rice. International Rice Research Institute, Los Banos, Philippines. 1982;71-82.
64. Asseng S, Ritchie JT, Smucker AJM, Robertson MJ. Root growth and water uptake during water deficit and recovering in wheat. Plant Soil. 1998;201:265-273.
65. Palta JA, Chen X, Milroy SP, Rebetzke GJ, Dreccer MF, Watt M. Large root systems: Are they useful in adapting wheat to dry environments? Funct. Plant Biol. 2011; 38:347-354.
66. Reynolds M, Dreccer F, Trethowan R. Drought-adaptive traits derived from wheat wild relatives and landraces. J. Exp. Bot. 2007; 58:177-186.
67. Hamid DJ, Karim NN, Mohsen A. Effect of reduced irrigation regimes on yield, yield components and some quality traits of three bread wheat cultivars (*Triticum aestivum* L.) International J. Agric. Crop Scie. 2012;4(5):234-237.
68. Reynolds MP, Kazi MA, Sawkins M. Prospects for utilizing plant-adaptive mechanisms to improve wheat and other crops in drought- and salinity-prone environments. Ann Appl Biol. 2005; 146:239-259.
69. Manschadi AM, Hammer GL, Christopher JT, Voil Pd. Genotypic variation in seedling root architectural traits and implications for drought adaptation in wheat (*Triticum aestivum* L.). Plant Soil. 2008;303:115-129.
70. Trethowan RM, Mahmood T. Genetics options for improving the productivity of wheat in water-limited and temperature-stressed environments. In: Yadav SS, Redden RJ, Hatfield JL, Lotze-Campen H, Hall AE (eds) Crop Adaptation to Climate Change, John Wiley & Sons Ltd, Ames, Iowa. 2011; 218-237.
71. Sinclair TR. Limits to crop yield? In: Boote KJ, Bennett JM, Sinclair TR, Paulsen GM (eds.) Physiology and determination of crop yield, American Society of Agronomy, Madison USA. 1994;509-532.
72. Fukai S, Cooper M. Development of drought-resistant cultivars using physio-morphological traits in rice. Field Crops Res. 1995;40:67-86.
73. Kamoshita A, Wade LJ, Ali ML, Pathan MS, Zhang J, Sarkarung S, Nguyen HT. Mapping QTLs for root morphology of a rice population adapted to rainfed lowland conditions. Theor. Appl. Genet. 2002; 104:880-893.
74. Manschadi AM, Christopher J, Voil Pd, Hammer GL The role of root architectural traits in adaptation of wheat to water-limited environments. Funct. Plant Biol. 2006;33:823-837.
75. O'Toole JC. Adaptation of rice to drought-prone environments. In: Drought resistance in crops with the emphasis on rice, Manila, IRRI. 1982;195-213.
76. Turner NC, Wright GC, Siddique KHH. Adaptation of grain legumes (pulses) to water-limited environments. In: Sparks DL

(ed) Advances in Agronomy. 2001;71:193-231.

77. Ober ES. Breeding for improved drought tolerance and water use efficiency. In: Arable cropping in a changing climate. Proceedings of the HGCA Conference, UK. 2008;28-37.
78. Sanguineti MC, Li S, Maccaferri M, Corneti S, Rotondo F, Chiari T, Tuberosa R. Genetic dissection of seminal root architecture in elite durum wheat germplasm. Ann. Appl. Biol. 2007; 151:291-305.
79. Schweiger P, Petrasek R, Hartl W. Root distribution of winter wheat cultivars as affected by drought. In: International Symposium Root Research and Applications, Root RAP, Boku, Vienna, Austria. 2009;345.
80. Yamauchi A. Significance of root system structure in relation to stress tolerance in cereal crop. In: Low-input sustainable crop production system in Asia. Korean Soc. Crop Sci., Korea. 1993;347-360.
81. Iijima M, Kono Y. Interspecific differences of the root system structures of four cereal species as affected by soil compaction. Jpn. J. Crop Sci. 1991;60:130-138.
82. Masle J. High soil strength: Mechanical forces at play on root morphogenesis and in root: shoot signaling. In: Waisel Y, Eshel A, Kafkafi U, Eds., Plant Roots the Hidden Half. Marcel Dekker Inc, New York, Basel. 2002;807-819.
83. Agnew ML, Carrow RN. Soil compaction and moisture stress preconditioning in Kentucky bluegrass. II. Stomatal resistance, leaf water potential and canopy temperature. Agron. J. 1985;77:878-884.
84. Lipiec J, Ishioka T, Szustak A, Pietrusiewicz J, Stępniewski W. Effects of soil compaction and transient oxygen deficiency on growth, water use and stomatal resistance in maize. Acta Agric. Scand. Sect. B: Soil Plant Sci. 1996;46: 186-191.
85. Ginkel Mv, Ogbonnaya F. Novel genetic diversity from synthetic wheats in breeding cultivars for changing production conditions. Field Crops Res. 2007;104:86-94.
86. Munoz-Romero V, Benitez-Vega J, Lopez-Bellido L, Lopez-Bellido RJ. Monitoring wheat root development in a rainfed vertisol: Tillage effect. Eur. J. Agron. 2010; 33:182-187.
87. Spielmeyer W, Dolferus R. Breeding for improved water productivity in temperate cereals: Phenotyping, quantitative trait loci, markers and the selection environment. Functional Plant Biology. 2010;37:85–97.
88. F. A. O. Deficit irrigation Practices. FAO Water Report No. 22. Food and Agricultural Organization of the United Nations. Rome; 2002.

14

Effects of Planting Dates on Leaf and Grain Yield of Black-Eyed Bean Cowpea Type in Mashonaland East Province in Zimbabwe

A. Matikiti[1*]

[1]*Crop Sciences Department, University of Zimbabwe, Zimbabwe.*

Author's contribution

The sole author designed, analyzed and interpreted and prepared the manuscript.

Editor(s):
(1) Masayuki Fujita, Dept. of Plant Sciences, Faculty of Agriculture, Kagawa University, Japan.
(2) Marco Aurelio Cristancho, National Center for Coffee Research, CENICAFÉ, Colombia.
Reviewers:
(1) Hendrie Nunes, University of São Paulo, Brazil.
(2) Anonymous, University of Arizona, USA.
(3) Anonymous, University of Toronto, Canada.

ABSTRACT

Aims: To determine the effect of planting date on leaf and grain yield of Black-eyed bean (BEB) of cowpea type,
Study Design: A field experiment was carried out in a factorial arrangement in randomised complete block design with three replications.
Place and Duration of Study: The study was conducted at both on-station (University of Zimbabwe, Crop Sciences Department experimental blocks and on-farm in Mashonaland East Province, Mutoko District, Katsukunya village during the 2005 – 6 and 2006 - 7 cropping seasons.
Methodology: Two sowing dates (14 December and 16 January) were used. The leaf harvesting treatments were started four WACE for the cowpea to set a sufficient framework on both stations. Leaf harvesting was done every week on Fridays for on-station and Wednesdays for on-farms sites and terminated at the 50% flowering stage for BEB after harvesting for four weeks (7 WACE) at the UZ site and for three weeks (6 WACE) in Mutoko a total of twenty-four plants were harvested per each treatment.
Results: The results showed that planting date had significant effect on leaf, grain and biomass yield. Highest (1492 kg/ha) grain yield was obtained with late planting in January while highest

*Corresponding author: E-mail: amatikiti@agric.uz.ac.zw;

(1225.5 kg/ha) leaf yield was obtained with early planting in December
Conclusion: Planting date is an important factor influencing both leaf and grain yield of cowpea.

Keywords: Planting date; black-eyed bean; yield; harvesting intensity and interval.

1. INTRODUCTION

Cowpea (*Vignau niguiculata* L. Walp) (Fabaceae) is a very important food legume in the world. It contains about 22% protein and constitutes a major source of protein for resource-poor rural and urban people [1]. Worldwide the crop is highly valued for its leaf and grain for human consumption and for forage and therefore often has a dual utility [2]. Black-eyed bean (BEB) was introduced into the country by non-governmental organisation to offer farmers an early maturing, highly nutritious cowpea type and relatively low input crop which has also had a market appeal. Planting and production practices are similar to the usually grown local landraces; however farmers had little knowledge on the effects of some of these practices on BEB.

The great importance of BEB crop is not only from its ability to grow in the new reclaimed areas as an economic crop, but also for producing high leaf and grain yield under the stressful conditions as compared to the local landraces. BEB is a as short season crop, and its growth period is almost half of the local landraces [3]. It is drought tolerant, making it suitable for the marginalized areas where most rural areas are situated.

It is known that high productivity of any crop is as a result of many factors and operations. Moreover, the quantity and quality of the crop is affected by the cultivation management practices prevailing in the agro ecological zone where the crop is grown, hence, the pronounced role of the agronomical processes such as planting dates, sowing density, soil fertility and rainfall requirements play very important role on the total performance of the crop [4].

Planting date of a crop is considered important among other factors that influence growth and productivity. It plays a vital role in germination, growth, yield and quality of the crop [5]. Therefore, edaphic factors vary under the Zimbabwean Natural Regions conditions. Also planting date is a great factor in organising and securing work schedule for farmers. Planting BEB on a suitable date according to environmental conditions of the region is therefore the best method to maximise yield and quality. There is therefore a great need to determine the responses of BEB that has been released to the smallholder farmers in Zimbabwe to different planting dates, hence the objectives of this study were:

- To determine the direct influence of planting dates on the performance of BEB cowpea type.
- To develop recommendations for BEB production practises under semi-arid conditions in Zimbabwe.

2. MATERIALS AND METHODS

2.1 Experimental Sites

The experiment was conducted across four sites, three sites in Mutoko, Nyadire area and the other site at the University of Zimbabwe, Department of Crop Science experimental plots.

2.1.1 University of Zimbabwe, department of crop science

The experiment was carried out at the University of Zimbabwe, Department of Crop Science in the 2005/6 and 2006/7 seasons. The Department of Crop Science is located in the suburb of Mount Pleasant, 10 km north of the city centre of Harare. University of Zimbabwe is in Natural Farming Region IIa and receives about 750 – 1000 mm rainfall per annum. The area lies at 1500 metres above sea level, latitude 17°31^1 east and longitude 30°50^1 south. The arable experimental plots are flat with slopes of 2% or less. The soils are red fersiallitic clay with more than 30% clay soil using the Zimbabwean classification system [6]. Fertility of these soils is maintained through regular application of inorganic fertilizers, crop rotations and good management.

2.1.2 Mutoko Nyadire area

This area is in Mashonaland East Province, the experiment was carried out in Mutoko District, Katsukunya village under Chief Kanyongo 15 km from Mutoko centre in 2005/6 and 2006/7 seasons. The area is situated at latitude 18°

10[1]east and longitude 032°17[1] south and its 812 meters above sea level. It is in Natural Region III which is characterized by miombo woodland with three marked seasons: a warm wet summer (November-April), a cool dry winter (May-August) and a short hot dry spring (September-October). Soils are mainly coarse-grained sands and sandy loams over sandy clay loams [7]. Rainfall is confined to summer, and is in the range 450 – 650 mm per annum. There are periodic seasonal droughts and severe dry spells during the rainy season hence drought tolerant crops are recommended. The area was chosen because it simulates conditions under which most small-holder farmers are located and also cowpea varieties do very well under the conditions in the area. The soils are derived from granite rock and are sandy, light textured and of fair agricultural potential. However, due to their structural and textural characteristics, the soils are subject to high levels of erosion and leaching under high rainfall conditions. Fertility of these soils is rapidly lost through poor management practices. The main climatic constraint to dryland crop production in Mutoko is rainfall. Temperatures also play an important role in the area high temperatures results in high potential evaporation rates leaving the soils dry [7].

2.2 Description of Varieties Used in the Experiment

2.2.1 Black-eyed bean (BEB)

BEB is a cowpea type with large white grain with a black patch around the hilum, hence the name. It has been recently introduced into the country by non-governmental organizations mainly due to its attractive colour that raises the market potential, as well as its high nutritional value that has potential to raise health levels in the smallholder farming sector. BEB is determinate and is well adapted and combines excellent stable yields with very good tolerance to drought. The variety is short seasoned, requiring less than 100 days to reach physiological maturity, needing about 70 – 75 days of warm weather to mature, hence allowing its production twice in any one rainy season or can be planted late in the season in higher rainfall areas [3].

2.2.2 Local landrace *(Chigwa)*

It is an indeterminate variety which has been grown by smallholder farmers for a long time which can be trailing, semi-trailing or twining. The tolerance of the crop to dry conditions and its ability to fix nitrogen makes it an ideal crop for smallholder farmers who farm in marginal areas, but its main drawback has been low grain yields. This landrace takes approximately five months to mature. Local landraces provide tender leaves for an extended period [8].

2.3 Soil Analysis

At each experimental site soil samples were taken on both on-farm and on-station sites for physical and chemical analysis at the Chemistry and Soils Research Institute of Zimbabwe (Table 1).

2.4 Planting Dates

The first season experiments were planted on the 14th of December 2005 for on-station and the on-farm was planted on the 15th of December 2005 and the second experiments were planted on the 16th of January 2006on the on-station and the 17th of January 2006for the on-farm. The second season experiments were planted on the 14th of December 2006 for on station and on the second experiments were planted on the 16th of January 2007 on the on-station and the 17th of January 2007 for the on-farm.

2.5 Experimental Design and Treatments

2.5.1 On-Station

The experiment was set up as 2 x 2 x 6 x 2 factorial, with planting dates as first planting (14 December and second planting (16 January), cowpea variety (BEB and *Chigwa*), leaf harvesting intensity (one leaf, two leaves and three leaves), and leaf harvesting frequency (weekly and bi-weekly) as factors. A control in which no leaf was harvested was included for each variety. Hence the experiment was made up of 12 treatment combinations.

Treatments were arranged as a randomized complete block design with three replications. The gross plot size was 2.7 m x 2.5 m consisting of 6 rows, 2.5 m long, spaced 0.45 m apart. The net plot size was 0.9 m x 1.90 m, consisting of two middle rows, 1.90 m long and a total of twenty-four (24) plants were sampled for both leaf and grain yield. A basal dressing of Single Super Phosphate (20.5%, P2O5, S 12.9%) was dribbled into open planting furrows at 40 kg ha^{-1}. Cowpea seeds were sown by hand. Two weeks after crop emergence (WACE) seedlings were

Table 1. Summary of soil analysis results for all the experimental sites in the 2005 – 6 and 2006 – 7 seasons respectively

	UZ sites		Mutoko sites		
	Block 9	Block 13	Chinyanga	Chirimanzu	Munetsi
Soil texture	MG/SC	MG/SCL	MG/S	MG/S	MG/S
Soilcolour	Dark Brown	Dark Brown	Pale Brown	Pale Brown	Light Brown
pH ($CaCl_2$)	5.9	5.6	4.9	4.6	4.7
Mineral Nitrogen (ppm)	37	35	3	9	11
Initial Nitrogen (ppm)	17	13	14	6	5
Phosphorus (ppm)	36	37	10	13	11
Potassium (ppm)	0.28	0.19	0.09	0.11	0.14
Calcium (ppm)	7.51	6.15	0.76	0.89	0.90
Magnesium (ppm)	2.69	2.87	0.38	0.35	0.34
Total Chromium (ppm)	9.8	8.3	1.4	1.5	1.2

MG- Medium-grained SCL- Sand clay loam SC - Sand Clay S – Sand

thinned to an in-row spacing of 15 cm to achieve the recommended plants/ha population.

2.5.2 On-farm

VECO organization provided the established links to the community and the intrinsic knowledge of farmers' concerns. This was done to improve the timeliness of sowing, trial supervision and contact with farmers. The selected farmers received free of charge cowpea seed; agro-chemicals for recommended practices and all the land preparations expenses. Agronomic details were the same as on-station. The treatment combinations were as follows:

1. no leaf harvested
2. 1 leaf harvested weekly
3. 1 leaf harvested 2 weekly interval
4. 2 leaves harvested weekly
5. 2 leaves harvested 2 weekly interval
6. 3 leaves harvested weekly (this was left out because there were no leaves,)

2.6 Establishment and Maintenance of Experiments

2.6.1 On-station experiment

On-station expermients was planted on land from which a crop of maize had been grown as green mealies previously. After removal of stover, the existing weed growth was killed by application of Gramoxone (paraquat 25%, 75%. inert ingredients) applied at the rate of 189.39 litres/ha. The herbicide was applied using a knapsack sprayer calibrated to apply 200 litres of spray liquid ha^{-1}. After the scorching and collapse of weeds, three days after herbicide application, remaining green weeds were cleared by hand hoeing.

The second experiment was planted after a crop of green maize. A basal dressing of Single Super Phosphate (20.5% P_2 O_5 S 12.9%) was dribbled into open planting furrows at 40 kg ha^{-1}. Two weeks after crop emergence (WACE) seedlings were thinned to an in-row spacing of 15 cm to achieve the recommended 150000 plants/ha population. To reduce snail damage in the first two WACE, broadcasting of snail and slug killer (metaldehyde 2.0%, carbaryl 2.0%, captan 0.5%, and inert ingredients) was done at two, three and four WACE at the rate of 8 kg ha^{-1}. Dimethoate (40% emulsifiable concentrate, 60% inert ingredients) was applied at 0.760 litres ha^{-1} on three occasions at four, six and seven WACE to control aphids. Plots were kept weed-free throughout the experiment by hand hoeing. On-station, BEB was attacked by *Ascocyta phaseolorum* during pod formation in the first growing season, and was controlled with alternating sprays of Benomyl (50% butylcarbomyol, 50% inert ingredients), copper oxychloride (85% Cu, 15% inert ingredients) and Dithane M45 (80% mancozeb, 20% inert ingredients).

2.6.2 On-farm experiments

Experiments were established by the researcher with participation of farmers who provided land, labour and made observations. Land preparation was done by farmers using ox-drawn mould board ploughs and harrowed to a fine tilth before planting furrows were opened using hoes. Planting and general management of the fields were as on-station.

2.6.3 Harvesting and measurements

The leaf harvesting treatments were started four WACE for the cowpea to set a sufficient framework on both stations. Leaf harvesting was done every week on Fridays for on-station and Wednesdays for on-farms sites and terminated at the 50% flowering stage for BEB after harvesting for four weeks (7 WACE) at the UZ site and for three weeks (6 WACE) in Mutoko a total of twenty-four plants were harvested per each treatment. Harvesting of the *Chigwa* continued for ten weeks (up to 14 WACE) at the UZ site and eight weeks (up to 12 WACE) at Mutoko sites; because of its indeterminate characteristics, *Chigwa* grows vegetatively for a longer duration than the BEB. Fresh weight of leaves was determined using a balance (Sartorius 1507 model) soon after harvesting. Leaf dry weight was determined by oven-drying leaves at 70% for 48 hours until there was no further weight loss, and then weighing the leaves.

Grain for the BEB was harvested at harvest maturity as indicated by the turning of pod colour from green to brown and rattling of grain in pods, at 14 WACE for on-station experiments and 12 WACE for on-farm experiments. Harvesting of the local landrace was done 17 WACE at both on-station and on-farm sites. Moisture content was determined using a digital moisture meter. Above-ground biomass was harvested on the same day as the pods.

2.7 Statistical Analysis

Collected data was subjected to analyses of variance (ANOVA) using GENSTAT (Version 6.0, 2005). Assumptions of ANOVA for normal data were carried out using GENSTAT (Version 6.0, 2005). Leaf and grain yield data across on-farm sites were subjected to homogeneity of variance using Bartlett's test using the formula (used when degrees of freedom are equal) outlined in Gomez and Gomez, [9]:

$$\frac{(2.3026)\ (f)\ (k \log s^2p - \sum_{i=1}^{k} \log si^2}{1 + [(k + 1)/3kf]}$$

Where s^2 = error mean square
f = degree of freedom of each s^2
k = number of variances
log = logarithm base 10

3. RESULTS

3.1 Leaf Yield

3.1.1 Effects of planting dates, leaf harvesting intensity, leaf harvesting interval and variety on dry leaf yield

Planting dates had highly significant ($p<0.001$) effect on cowpea variety performance on leaf yield. Local landrace had the highest (1225.5kg/ha) leaf yield than BEB in all the seasons, on all the planting dates (Fig. 1). Leaf harvesting intensity was significantly ($p<0.001$) affected by planting dates at both on-station and on-farm sites for the 2005 – 6 and 2006 – 7 seasons (Fig. 1). Leaf harvesting interval had a significant ($p<0.001$) effect on the leaf yield on planting dates.

There was a significant interaction ($p<0.001$) in leaf harvesting intensity, leaf harvesting interval, cowpea variety and planting dates for all the seasons. The interaction is shown by the response of the cowpea varieties to the different planting dates, intensities and interval. For the planting dates, leaf yield decreased with late planting. However, for leaf harvesting intensity, leaf yield increased with increase in intensities. For leaf harvesting interval leaf yield increased with weekly harvests and was decreased for every fortnightly harvest, except for three leaves where it was done fortnightly only as there were no enough leaves for weekly harvests. Early planting dates produced the highest leaf yield for both varieties on all the sites for all the seasons (Fig. 1).

3.2 Grain Yield

3.2.1 Effects of planting dates, leaf harvesting intensity, leaf harvest interval and variety on grain yield

Grain yield responded significantly to planting dates, hence there was significant effect ($p<0.001$) of planting dates on grain yield on all the sites (Fig. 2). Cowpea variety had significant ($p<0.001$) difference grain yield on all the sites.

There was a significant ($p<0.001$) interaction in planting dates, leaf harvesting intensity, leaf harvesting interval and variety on grain yield on both sites as shown in the Fig. 2. The interaction is shown by the differential response of the two cowpea varieties on grain yield to planting dates,

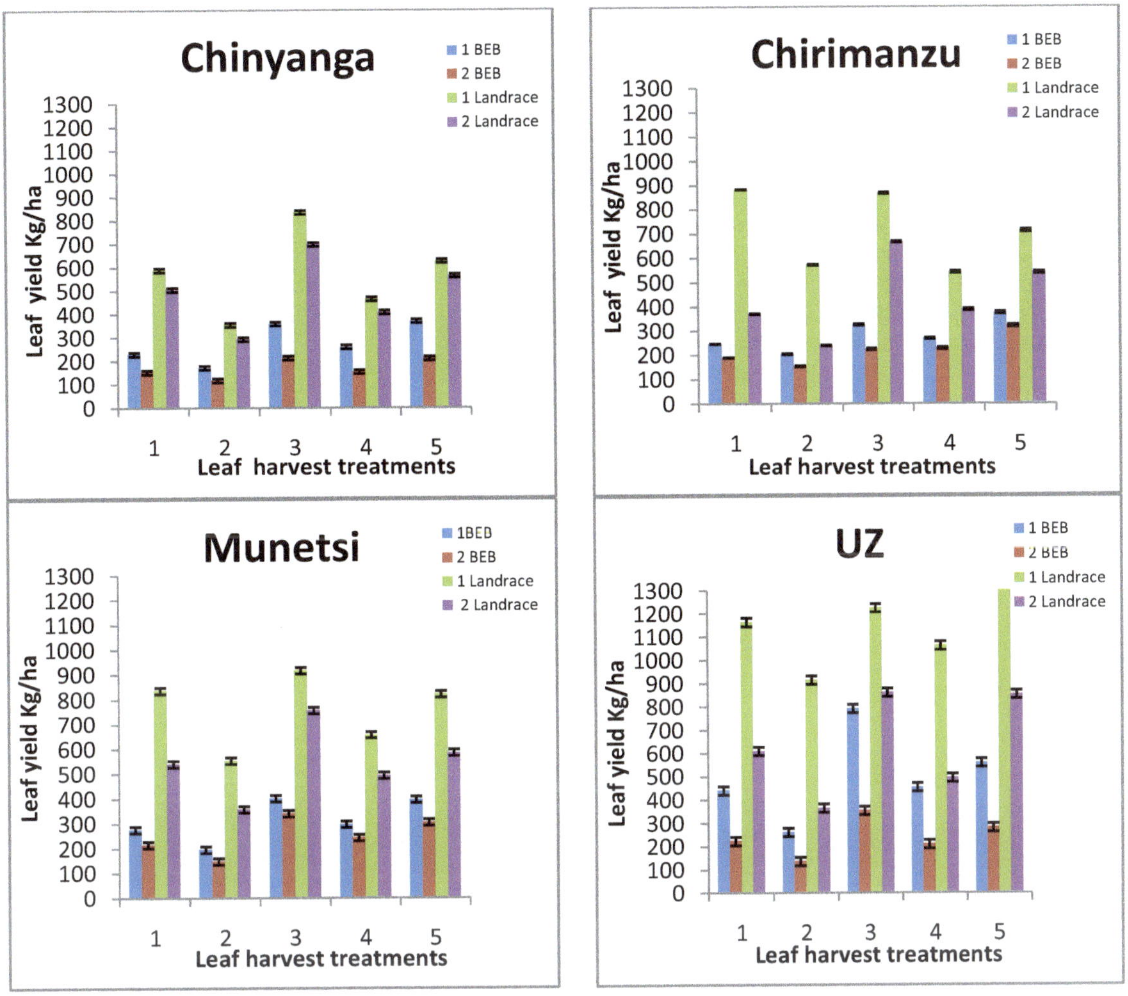

Fig. 1. Combined effects of planting dates on leaf yield of BEB and local landrace for both on-farm and on-station sites. Planting dates were 1BEB – 14 December, 2BEB – 16 January, 1Landrace – 15 December, 2Landrace – 17 January. Leaf harvesting treatments 1, 2, 3, 4 and 5 represents one leaf harvested weekly, one leaf harvested fortnightly, two leaves harvested weekly, two leaves harvested fortnightly, three leaves harvested fortnightly respectively.

leaf harvesting intensity and interval. Late planting dates in January produced the highest grain (1492kg/ha) yield than early planting dates in December (964kg/ha). BEB produced the highest grain yield in all the planting dates in all the leaf harvesting intensities and interval on all the sites in all the seasons. Harvesting two leaves produced the lowest grain yield in all the sites whereas the control and harvesting one leaf fortnightly produced the highest yield as there were significant differences on the two treatments. Leaf harvesting interval had highly significant ($p<0.001$) effect on grain yield. Harvesting weekly produced less grain than harvesting fortnightly (Fig. 3).

3.3 Aboveground Biomass

3.3.1 Effect of planting dates, leaf harvesting intensity, leaf intensity interval and variety on aboveground biomass

Similarly like leaf yield, planting dates had significant effect ($p<0.001$) effect on aboveground biomass yield hence, yield was significantly influenced by planting dates. Early planting in December resulted in more aboveground biomass than planting late in January. Cowpea variety responded significantly to aboveground biomass yield with the local landrace (17219 kg/ha) producing the highest biomass yield than BEB (9111 kg/ha) in both sites. Generally, the UZ site produced the highest aboveground biomass (Fig. 3).

4. DISCUSSION

The study revealed that planting dates influences cowpea grain yield. Hence, there is need to grow the crops on the best utilization of moisture, nutrients and solar radiation on an optimum planting date in order to obtain the recommended yields. Therefore, by choosing the best planting date determines the different stages of plant growth at which the plants can come in accordance with preferred environmental conditions that lead to increased photosynthesis output resulting in increased yield. In this study, leaf and aboveground biomass high yields obtained from the early or first planting date suggests that the growth conditions were optimum for the vegetative growth and conversely the low grain yields in this phase indicate that the conditions in the late or second planting dates could be sub-optimal. The vegetative growth tendency observed in this experiment implies that they are enabling growth conditions available, therefore the advantage of early planting lead to high biomass yield. In [10], authors Faukner and Mackie, showed that cowpeas planted early in the season in Southern

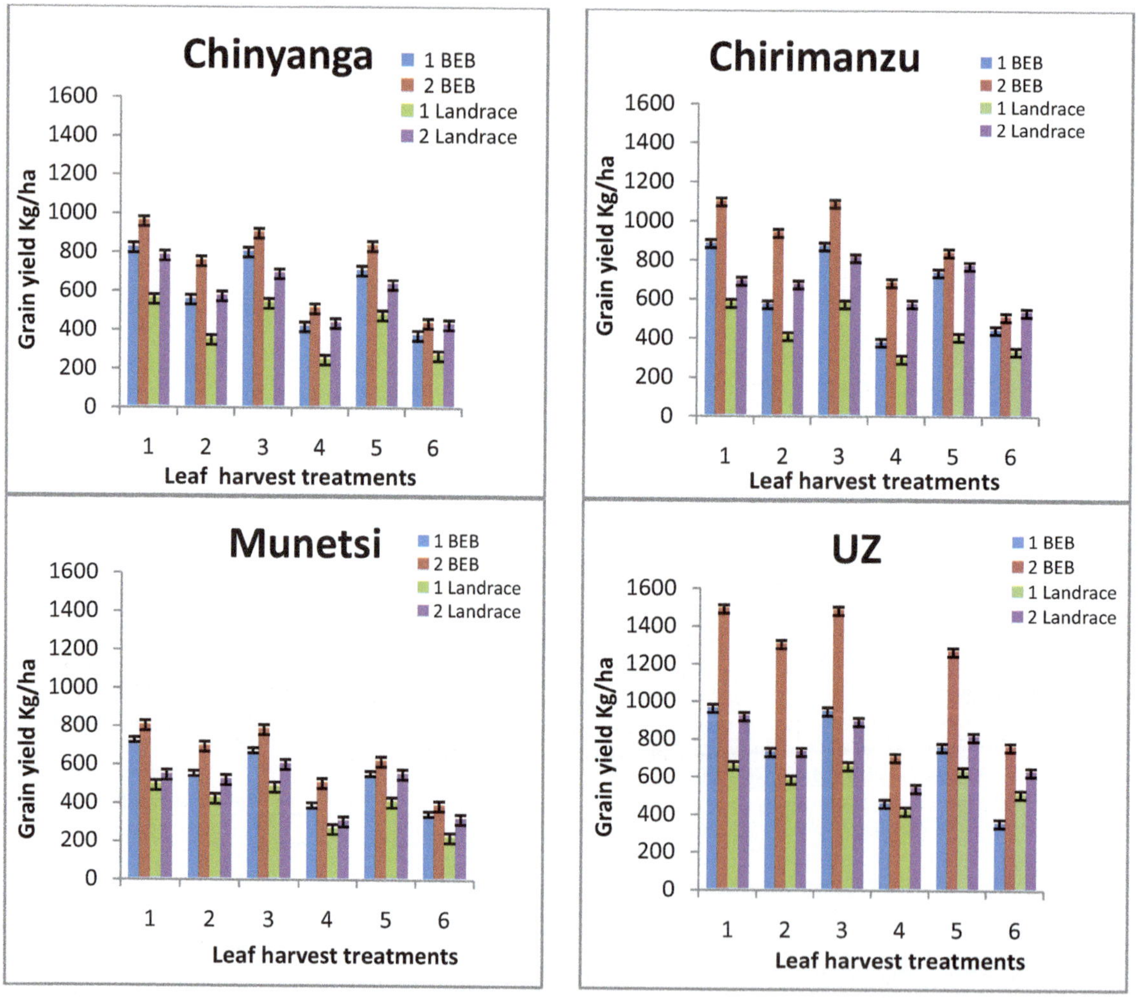

Fig. 2. Combined effects of planting dates on grain yield of BEB and local landrace for both on-farm and on-station sites. Planting dates were 1BEB – 14 December, 2BEB – 16 January, 1Landrace – 15 December, 2Landrace – 17 January. Leaf harvesting treatments 1, 2,3, 4, 5and 6 represents no leaf harvesting, one leaf harvested weekly, one leaf harvested fortnightly, two leaves harvested weekly, two leaves harvested fortnightly, three leaves harvested fortnightly respectively.

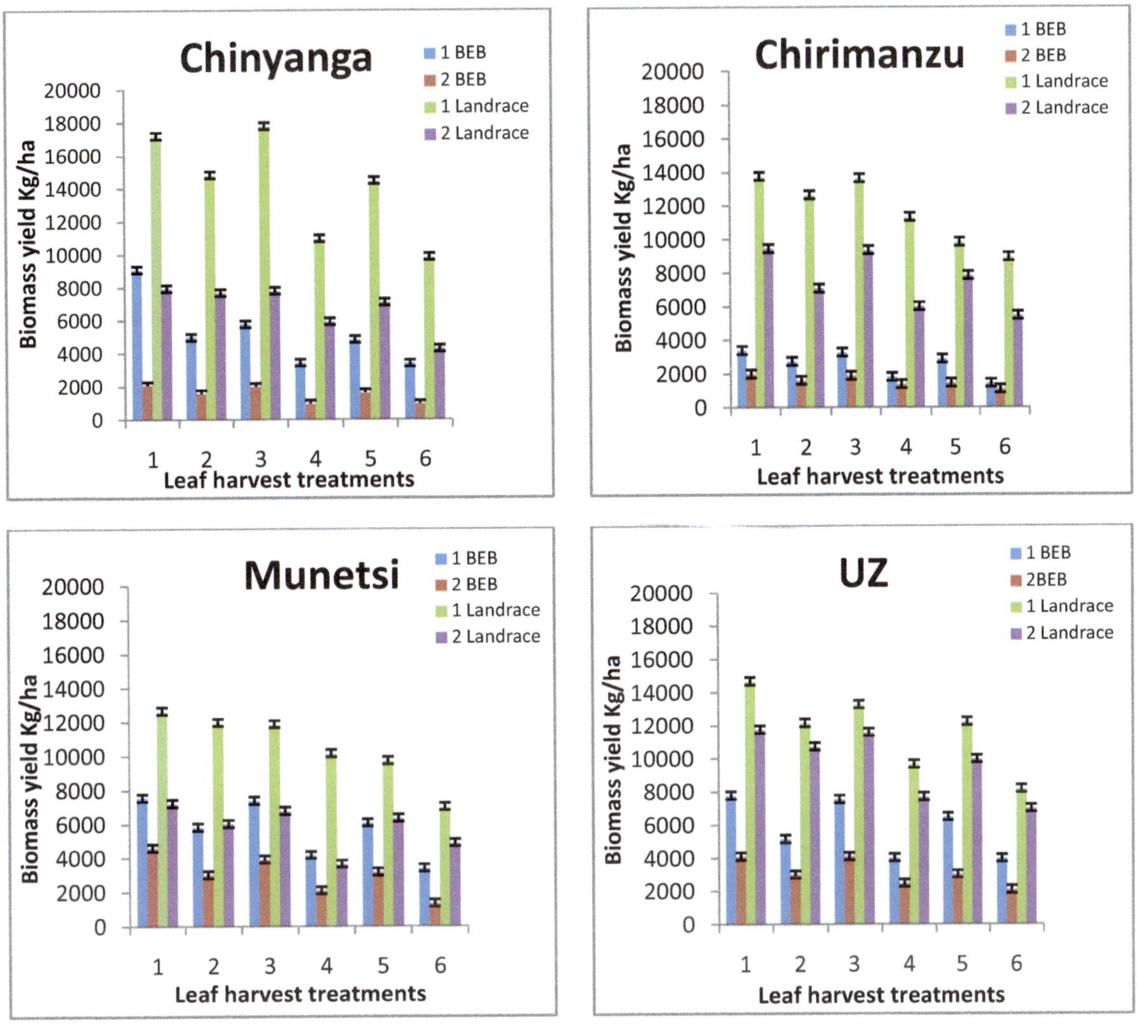

Fig. 3. Combined effects of planting dates on biomass yield of BEB and local landrace for both on-farm and on-station sites. Planting dates were 1BEB – 14 December, 2BEB – 16 January, 1Landrace – 15 December, 2Landrace – 17 January. Leaf harvesting treatments 1, 2,3, 4, 5and 6 represents no leaf harvesting, one leaf harvested weekly, one leaf harvested fortnightly, two leaves harvested weekly, two leaves harvested fortnightly, three leaves harvested fortnightly respectively.

Nigeria produced little or no seed due to the production of dense foliage at the expense of pod formation, then later on another researcher attributed the production of dense foliage to good soil and too much rainfall. It was also observed that weather conditions prevailing at any given time can be more conducive to the vegetative growth than pod formation [11]. The results are in agreement with [5] who also observed that there are numerous factors that contribute to the realization of a successful crop yield and one of them was planting date management which has a profound effect on the development and final outcome of the crop. However, in this study it was found that selection of a specific variety will have a large impact on the way in which planting date should be managed. As was observed in the experiment the selection of the local landrace was ideal as the early planting produced more leaf yield than late planting and this also occurred in the BEB crop. Hence, the time frame in which the crop can be planted due to weather and or other circumstances should have a large impact on the selection of a suitable variety [12]. Therefore, early plantings result in higher vegetative growth tendencies at the expense of the grain yield in both the local landrace and BEB. Previous reports also showed that in late planting situation, higher yields potentials are often realised from a more determinate, shorter season variety as it was observed from this study on BEB, whereas the indeterminate local landrace produced low grain yields with late planting as they were not able to mature the pods formed late in the season. For grain yield, the results of the study revealed that the late planting date offered better growing conditions due to temperatures coincidences of flowering and insemination flowers with more activity of insects and appropriate weather, more nutrients and photosynthetic materials are transmitted to the plants [13].

However, on leaf harvesting, the study revealed that leaf harvesting reduces cowpea grain yield. Whilst increasing leaf harvest intensity increased leaf yield significantly, this occurred at the expense of grain yield as the reduction of grain yield was presumably related to leaf harvest intensity. In this study, the more drastic reduction in grain yield could be explained by the fact that leaves are the major determinants of source strength, their harvesting will result in reduced source strength and subsequently lower yields by reducing the amount of photo-assimilates partitioned on the grain during grain fill and new vegetative growth. The effect of leaf harvesting on grain yield observed in this study also confirms the report by [14], which produced similar results where limited harvest of leaves had a detrimental effect on yield of cowpea harvested at maturity because of reduction in photo-assimilates.

In addition, leaf harvesting interval had also an effect on the yield performance of the cowpea where harvesting weekly resulted in more leaf yield but less grain yield. With longer harvest intervals specifically for one leaf harvested at fortnightly intervals for both the determinante (BEB) and the determinate (local landrace) the resultant leaf yield was less and did not affect the grain yield because of compensatory effect in terms of cowpea biomass. This also confirms the findings by [15], who revealed that removal of few leaves in cowpea resulted in proportionally little leaf area removal, hence the negligible negative effect in yield at moderate harvest intensities and extended intervals.

Regardless of planting dates, the results from this study also indicated that indeterminate types due to their vegetative characteristics tend to produce more leaves than determinant types. This was in agreement with earlier reports [16]. Determinant types have no time to recover from defoliation because they have a short flowering period whereas indeterminate types have a long flowering period, hence have enough time to recover from defoliation and channel photo-assimilates produced to pods resulting in higher grain yields.

5. CONCLUSION

In this study, it can be concluded that both BEB and the local landrace, although very well adapted to the region may not well be used for early planting dates as they faced a decline in grain yield. Hence, they can only be planted in January for quantity and quality enhancement. The results of this study also suggest that some varieties could be grown for grain harvest, while others could be grown for vegetable harvesting as the yield efficiency was suppressed by the combination of leaf and grain harvest.

ACKNOWLEDGEMENTS

The special thanks to RUFORUM for sponsoring this study and the farmers in Mutoko namely: Munetsi, Chirimanzu and Chinyanga.

COMPETING INTERESTS

Author has declared that no competing interests exist.

REFERENCES

1. Muleba N, Dabire C, Suh JB, Drabo I, Ouedraogo JT. Technologies for cowpea production based on genetic and environmental manipulations in the semi-arid tropics. 1997;23(2):195-206.
2. Akundabweni LS, Peter-Paul C, Singh BB. Evaluation of elite lines of cowpea (*vigna unguiculata (L)* Walp.) for leaf /fodder plus grains (i.e. dual purpose). Tropical Agriculture (Trinidad). 1990;15(2)133-136.
3. Madamba R. Evaluating leaf, grain yield and nutritive value of leaves of different cowpea varieties. Department of Research and Specialist Services. Crop Breeding Institute. Box CY550. Causeway, Harare, Zimbabwe; 2000.
4. Akande SR, Olakojo SA, Ajayi SA, Owolade OF, Adetumbi JA, Adeniyan ON, Ogunbodede BA. Planting date affects cowpea seed yield and quality at Southern

Guinea, Nigeria. Seed Technology. 2012;34(1):51–60.

5. Shiringani RP, Shimelis HA. Yield response and stability among cowpea genotypes at three planting dates and tests environments. African Journal of Agricultural Research. 2011;6(14):3259-3263.
6. Nyamapfene K. The Soils of Zimbabwe. Nehanda Publishers. Harare. Zimbabwe; 1991.
7. Mariatou D, Kwaramba RJ. Review of vegetable IPM potential in Zimbabwe: A case study of Mashonaland East Province. Ministry of Agriculture Harare; 1999.
8. Johnson DT. The cowpea in the African areas of Rhodesia. Rhodesia Agricultural Journal. 1970;6(7):61-64.
9. Gomez KA, Gomez, AA. Statistical Procedures for Agricultural Research. 2nd ed. IRRI. Wiley – Interscience Publication; 1984.
10. Enyi BAC. A Spacing/Time of Planting Trial with Cowpea *(Vigna unguiculata (L) Walp).* Ghana Journal of Science. 1972;13:1.
11. Ezendima FOC. Seasonal variations in the growth of cowpea seedlings in the humid tropical environment of Southern Nigeria. Journal of West African Science Association. 1967;(12):45-49.
12. Dhlamini MS, Edje OT. Effects of Defoliation on cowpea yield responses of cowpea (Vigna uniguiculata (L) Walp to nitrogen fertilizers and leaf plucking frequencies. University of Swaziland; 2003.
13. Sobhani MA, Rahmikhdoev G, Mazaheri D, Majidian M. Effects of sowing date, cropping pattern and nitrogen on crop growth rate, yield and yield component summer sowing buckwheat *(Fagopyrum esculentum Moench).* Journal of Applied Environmental and Biological Sciences. 2012;2(1):35–46.
14. Bubenheim DL, Mitchell CA. Evaluation of new candidate crop species For CELSS, in proceedings, Space Life Sciences Symposium. Three decades of life sciences research in space, Washington, DC; 1987.
15. Nielsen SS, Ohler TA, Mitchell CA. Cowpea leaves for human consumption; Production, Utilization and Nutrient Composition in Advances of Cowpea Research. 1st ed. Devon. United Kingdom; 1997.
16. Thalji T, Shalaldeh G. Effect of Planting Date on Faba Bean *(Vicia faba L.)* Nodulations and Performance under Semiarid Conditions. World Journal Agricultural Sciences. 2006;2(4):477-482. ISSN 1817-3047. IDOSI Publications.

15

Dry Bean Response to Fertilization Using Minjingu Phosphate Rock and Composted Tughutu (*Vernonia subligera* O. Hoffn)

Patrick A. Ndakidemi[1*]

[1]*School of Life Sciences, Nelson Mandela African Institution of Science and Technology, P.O. Box 447, Arusha Tanzania.*

Author's contribution

This whole work was carried out by author PAN.

Editor(s):
(1) Masayuki Fujita, Dept. of Plant Sciences, Faculty of Agriculture, Kagawa University, Japan.
Reviewers:
(1) Anonymous, Animal Science Research Institute, Karaj, Iran.
(2) Tunira Bhadauria, Department of Zoology, Feroze gandh Post Graduate College, Kanpur University, Uttar Pradesh, India.
(3) Anonymous, IAPTA - Pólo Regional Centro Sul, Brazil.

ABSTRACT

Aims: The objectives of the study were toevaluate the effects of organic materials from Tughutu (*Vernonia subligera* O. Hoffn) and locally available Minjingu phosphate rock or commercial Triple super phosphate fertilizer on the plant growth and amount of phosphorus (P) and calcium (Ca) content in bean shoots (*Phaseolus vulgaris* L.) and their residual effects in the soil after harvest.
Study Design: The experiments were set up as a randomized complete block design
Place and Duration of Study: Field experiments were conducted in famers' fields in the Western Usambara Mountains in Tanzania.
Methodology: Each farmer's field was assigned as a replicate. The experimental treatments consisted of six plots as follows: control, Minjingu Phosphate Rock (MPR) or Triple Super Phosphate (TSP) each at 26 kg P ha^{-1}, Tughutu at 2.5 t dry matter ha^{-1} and Tughutu at 2.5 t dry matter ha^{-1} combined with Minjingu Phosphate Rock or Triple Super Phosphate at 26 kg P ha^{-1}.
Results: The dry matter yield and amount of P and Ca in bean shoots significantly increased by supplying the organic and inorganic fertilizers above the control treatment. Addition of P fertilizers alone or in combination with Tughutu also modified the soil pH and significantly increased the concentration of P and Ca in the soil at harvest.
Conclusion: Overall, combination of Minjingu phosphate rock or Triple super phosphate with

**Corresponding author: Email: ndakidemipa@gmail.com;*

Tughutu enhanced the effectiveness of these fertilizers and resultedin better growth and greater P and Ca amount in the shoots as well as their concentration in the soil at harvest.

Keywords: Dry matter; pH; triple super phosphate; minjingu phosphate rock.

1. INTRODUCTION

The densely populated areas of the Western Usambara Mountains (WUM) in northern Tanzania suffer from low soil nutrient levels [1,2] owing to intensive cultivation of crops without the application of external inputs. This practice has of late resulted in the depletion of essential plant nutrients, including P and Ca that are important for plant growth [2]. These constraints, amongst others, have been reported to be the cause of poor growth of bean and a major threat to the sustainability of the small-scale production systems in Western Usambara Mountains [2,3] Phosphorus deficiency is known to impair early nodule functioning [4], retard plant growth and reduce the ability of the plant roots to release protons that may be useful in the solubilization of organic and inorganic P sources in the soil [5,6]. Calcium deficiency in legumes is also known to impair N_2 fixation and retard plant growth [7-12].

The majorities of small-scale farmers living in Western Usambara Mountains are poor and cannot afford to purchase expensive agricultural inputs such as Triple super phosphate or lime that could supply phosphorus and calcium to the commonly grown crops in the area, including beans [13]. Therefore, there is a great need to explore the available indigenous sources of nutrients that will ultimately be affordable to the poorly resourced farmers.

Incorporating some organic materials in the soil may contribute significantly in providing mineral nutrients needed by the plants [14,15] and hence increase the crop yields. The organic matter may consist of a wide range of mineral nutrients, upon the process of decomposition and mineralization, they may be made readily available for plant uptake and growth [16,17]. Additionally, the use organic plant materials may improve other important soil characteristics such as their water retention and the microbiological activities in the soil [18].

The exploratory work by [17] revealed that traditional shrub Tughutu contained more P and Ca than *Tithonia diversifolia*, the most researched shrub species in East Africa [18-21]. A number of studies also reported that Minjingu phosphate rock fertilizer has higher contents of water soluble P and Ca [22-24]. From this background, it is evident that these two indigenous resources could be used to boost agricultural production and improve people's livelihood in poor soils of the Western Usambara Mountains.

A few studies conducted in East Africa have shown improved growth and increased concentration of P and Ca nutrition in plants following the application of organic materials and phosphate rocks as fertilizers [1,15,16,24-33]. Significant synergistic effects were reported in growth or yield by combining the organic and inorganic P fertilizers such as rock phosphates or Triple Super Phosphate [1,14-16,29-31,34].

This study was conducted with the objective of evaluating the effects of supplying Tughutu alone, or in combination with locally available Minjingu phosphate rock or commercial Triple Super Phosphate fertilizer on plant growth, and amounts of P and Ca in bean shoots. Furthermore, soil chemical properties such as pH, P and Ca availability were assessed at harvesting.

2. MATERIALS AND METHODS

2.1 SiteLocation and Description

Six sites in farmers' fields were planted to beans in Western Usambara Mountains (04°47.976 S, 38°17.623' E) during the 2000 - 01 cropping season. The Western Usambara Mountains are located in an altitude ranging from 900 to 2300 m. a.s.l. and the major soils are classified as Humic and Chromic Acrisols, Luvisols and Lixisols [28]. Before planting, soil samples were collected from each site, bulked and analysed separately for pH, P, and Ca content for each site. Details of soilcollection and analysis are described in [2].

2.2 Field Trials

The fields were ploughed by a hand hoe. The experiments were set up as a randomized complete block design. Each farmer's field was assigned as a block. The experimental treatments consisted of six plots assigned to each block as follows: control, Minjingu phosphate rockor Triple Super Phosphate each at 26 kg P ha^{-1}, Tughutu at 2.5 t dry matter ha^{-1} and Tughutu at 2.5 t dry matter ha^{-1} combined with Minjingu phosphate rock or Triple Super Phosphate at 26 kg P ha^{-1}. P levels from Minjingu phosphate rock and Triple Super Phosphate were calculated considering the water soluble P in each fertilizer. Each treatment was planted in 25 m^2 plot. Inter-row and intra-row spacing of 50 cm and 20 cm respectively were used. Tughutu used in this study was collected from the Magamba forest reserve in Western Usambara Mountains. Fresh Tughutu leaves were surface applied and incorporated into the soil to 15 cm depth by hand hoe one week before sowing and allowed to decompose [17]. Recommended that Minjingu phosphate rock be broadcasted and incorporated into the soil by a hand hoe at sowing. Triple Super Phosphate was applied into the rows at planting. Two bean seeds coated with *Rhizobium tropici* strain CIAT 899 (10^9 g cellsseed^{-1}) purchased from the Nairobi MICERN, University of Nairobi, Kenya, were planted in each hole and allowed to grow to maturity. The inoculants packets were supplied with gum arabic for sticking as many cells as possible onto the seeds. Bean variety Lyamungu 90 used in this study was supplied by the breeder of the Selian Agricultural Research Institute, Arusha, Tanzania.

2.3 Plant Harvest and Analysis

Twenty plants were sampled 45 days after planting Di Ammonium Phosphate (DAP) from the two inner rows (1 m^2) in each plot for the above-ground biomass assessment. Bean yields from this study are reported elsewhere by [1]. The above ground shoots were cut with a knife at about 1 cm above the soil surface. Plant samples were dried to constant weight at 70°C in an oven for 48 h and ground into fine powder for the analysis of P by dry ashing method [35]. Phosphorus in the digest was measured by the molybdenum blue method as described by [36]. Calcium concentrations were determined by atomic absorption spectrophotometer as described by [37]. The measured concentrations of nutrients were multiplied by plant dry matter to obtain the above ground biomass amounts of the mineral. At harvest, soil adhering to the roots was collected from around 10 bean plants from each plot and analysed for pH, P and calcium as described above.

2.4 Statistical Analysis

Dry matter yield, P and Ca concentration and amounts in the tissues were analyzed using one-way analysis of variance (ANOVA). These computations were done with the software program STATISTICA [38]. The Fisher least significance difference (LSD) was used to compare treatment means at $P \leq 0.05$.

3. RESULTS AND DISCUSSION

3.1 Soil Analysis

Resultsof soil analysis are shown in Table 1 and reported elsewhere by [2]. Generally, P was low at all the sites with reference to optimum level of 11 mg P kg^{-1} soil as proposed by [39] for bean production. The exchangeable Ca ranged from 3.39 - 5.59 cmol kg^{-1} (Table 1). The proposed optimum level of Ca for the majority of crops is 5.0 cmol kg^{-1} [40] Four out of the six sites contained Ca levels lower than the proposed optimum levels and hence should receive some form of amelioration.

Table 1. Selected chemical properties of soils collected from the Western Usambara mountains in the Lushoto District of the Tanga Region, Tanzania

Site	Chemical properties of soils*		
	pH H_2O	Bray 1 P	Ca
		mg kg^{-1}	cmol kg^{-1}
Magamba	4.9	4.20	3.39
Ubiri	5.8	3.16	3.99
Ngulwii	5.5	2.80	5.59
Soni	5.9	1.68	5.59
Mbuzii	5.9	2.80	4.79
Mkuzi	5.2	4.20	4.79
Average	5.5	3.14	4.59

**Source: Ndakidemi and Semoka (2006)*

3.2 Dry Matter Yield

There was a small variation in dry matter yield across sites (Table 2) suggesting that to a greater extent, the six farmers' fields were fairly uniform. The dry matter (DM) yield of bean responded strongly ($P \le 0.001$) to all amendments used in this study (Table 3) indicating that P was among the limiting nutrients in the study area. Dry matter yields also correlated significantly with their shoot amounts P (Table 3; $r = 0.97$, $P = .05$) and Ca ($r = 0.86$, $P = .05$) suggesting that the dry matter was related to their amounts. Moreover, there was significant correlation between the amendment treatments and dry matter ($r = 0.83$, $P = .05$) indicating insufficient P and probably Ca in soils used in the study (Table 1). The dry matter increased by more than twofold by adding Triple Super Phosphate alone or Tughutu combined with Minjingu phosphate rock as compared with the control treatment. Addition of Tughutu or Minjingu phosphate rock alone increased the dry matter by more than 50% as compared to the control treatment. However, combining Tughutu and Triple super phosphate gave the best results by increasing the dry matter more than threefold as compared with the control. In general, Minjingu phosphate rock or Tughutu supplied alone produced comparable dry matter yield, whereas there was no significant dry matter yield difference between Triple Super Phosphate or Tughutu combined with Minjingu phosphate rock (Table 3). The observed variations in dry matter are partly explained by the chemical composition of the materials used in this study [28,23]. Yield from Minjingu phosphate rock or Tughutu treatment were generally lower than those of Triple Super Phosphate owing to low P contents in these materials [28,23]. Generally there was a positive effect on dry matter yield by combining Minjingu phosphate rock or Triple Super Phosphate with Tughutu. This work is consistent with the work done by [34,15,29] that reported yield increases in maize by combining *Tithonia* or *Tephrosia vogelii* with Triple Super Phosphate or Minjingu phosphate rock.

3.3 Phosphorus and Calcium Amount in Shoots

Significant variation on P and Ca amounts in shoots were observed ($P \le 0.001$) in all plots that were supplied with the amendments (Table 3). The amount of P and Ca in this study almost paralleled the dry matter yield (Table 3). This suggests that P and Ca were among the major limiting elements for bean growth in the Western Usambara Mountains (Table 3; [2,3]). Combining Tughutu and Triple Super Phosphate resulted in significantly higher P amount compared with all other treatments tested and was followed by Tughutu combined with Minjingu phosphate rock (Table 3). The good response of Triple Super Phosphate as compared with Minjingu phosphate rock could be due to the higher and faster dissolution rate of Triple Super Phosphate than Minjingu phosphate rock that improved their concentration in the soil (Table 4), in shoots (data not shown) and made them available for plant uptake (Table 3). Similar results have also been reported by [24,33] in maize.

Table 2. Estimates of micro-site variations on dry matter yield of common bean in WUM, northern Tanzania

Site	Dry matter yield** g $plant^{-1}$
Magamba	17.95
Ubiri	18.24
Ngulwii	17.55
Soni	19.17
Mbuzii	16.95
Mkuzi	20.94
Average	**18.47**

***Dry matter was determined by taking the average yield across all treatments from each site.*

Calcium amount in shoots almost followed the trend of P, but combining Tughutu with Minjingu phosphate rock or Triple Super Phosphate produced the best results that were not different (Table 2). A comparative study on the properties of Tughutu and other organic sources from Western Usambara Mountains revealed that Tughutu was superior to most of them in N, P and K content [17] Furthermore, mineralization patterns of Tughutu were faster and higher for N, P and K as compared with those of Tithonia and other useful shrubs. Therefore, high qualities of this plant render it a good source of nutrient for bean growers in Western Usambara Mountains.

In this study, combining organic material with Triple Super Phosphate or Minjingu phosphate rock promoted the uptake of P and Ca in bean. This was also reflected in the dry matter yield (Table 2). Research evidence suggests that Ca plays a crucial role in plant growth and development. This is responsible for structural rigidity of the cell wall, controlling the cell membrane structures and functions, and in many different signal transduction processes [41], which ultimately may contribute significantly in regulating plant growth and development as observed in this study (Table 2). Results from this study showed that the amount of P and dry matter yield in beans was significantly enhanced with the addition of Tughutu, Minjingu phosphate rock and the combination of Tuguthu with Minjingu phosphate rock or Triple Super Phosphate (Table 2 and 3). Phosphorus plays a key role in a plant's energy transfer mechanism (high energy phosphate), an essential intermediary in the metabolism of carbohydrate, protein and phospholipid transformations in plants and final plant growth [6]. Taken together, these results suggest that small-scale farmers in Western Usambara Mountains have the potential to benefit from the biomass transfer technology combined with a cheap inorganic resource such as Minjingu phosphate rock in improving their crop production.

3.4 Soil pH, Residual P and Ca at Harvest

The treatments supplied significantly (*P*<.05) changed the soil pH, P and Ca concentration in the soil solution at harvest (Table 4). Compared with control, Triple Super Phosphate supplied alone significantly decreased the soil pH more than with all other treatments. [42,43] also reported significant pH decreases with Triple Super Phosphate supply. However, the addition of Tughutu, Minjingu phosphate rock and the combination of Tuguthu with Minjingu phosphate rock or Triple Super Phosphate increased the soil pH as compared with Triple Super Phosphate and the control treatment (Table 4). The increase in soil pH owing to Minjingu phosphate rock can be owing to reasonably higher Ca content in this material [23]. Minjingu phosphate rock in this context provided a liming effect to the soil and hence raised its pH. Similar increases in soil pH with Minjingu phosphate rock applications and use of organic amendments have been reported by [16,15,30,31,34].

The effect of Tughutu, Minjingu phosphate rock, Triple Super Phosphate and/or their combination on residual P in the soil at harvest is shown in Table 4. Their application in the soil significantly (*P*=.05) increased the residual P in the soil. In this study, the use of Tuguthu combined with Triple Super Phosphate or Minjingu phosphate rock or Triple Super Phosphate applied alone elevated the residual P above the optimum level of 11 mg P kg^{-1} soil for bean production [39]. Addition of Tuguthu alone doubled the residual P in the soil as compared with control, but the level was still low. The observed small increase in soil P with Tughutu application was probably from the decay and the release into the soil of the organically bound P held in this material. These results suggest that Tughutu supplied alone has lower P input in the soil. Addition of Minjingu phosphate rock alone was not sufficient to raise the residual P value to the proposed optimum level (Table 4). It is possible that soil conditions were not favourable for Minjingu phosphate rock solubility and release of enough P into the soil in the first year. Amorphous rock phosphate could be weathered and released for plant growth over time [44,45]. Combining Minjingu phosphate rock with Tughutu released P above the proposed optimum level of 11 mg P kg^{-1} soil for bean production (Table 4). Studies have shown that organic acids released during the decomposition of organic materials improved the availability of soluble P from rock phosphates [6,14-16,30,31]. These findings suggest that repeated addition of inorganic phosphorus applied alone or in combination with Tughutu could build soil P reserves to reasonable levels and sustain crop productivity in Western Usambara Mountains.

The addition of Tughutu, Minjingu phosphate rock, Triple Super Phosphate and/or their combination influenced the residual exchangeable Ca in the soil (Table 4). The proposed optimum level of exchangeable Ca in the soil is 5.0 cmol kg^{-1} [40]. Except Triple Super Phosphate and control, other treatments increased the exchangeable Ca above the proposed optimum levels in the following order: Tuguthu + Minjingu phosphate rock>Minjingu phosphate rock>Tuguthu + Triple Super Phosphate>Tughutu>Triple Super Phosphate> control (Table 4). All treatments involving Minjingu phosphate rock were superior in increasing residual soil Ca as compared with others. Similar observations in which the addition of Minjingu phosphate rock in soils increased the residual Ca as compared with Triple Super Phosphate have been reported by [32]. Minjingu phosphate rock has a reasonably higher Ca content than Triple Super Phosphate [23] and Tughutu [17].

Table 3. Effects of Tughutu application, Minjingu phosphate rock or triple super phosphate supply on dry matter yield (DMY), P and Ca amount in shoots of *Phaseolus vulgaris* L. measured at 45 days after planting

Treatments	Dry matter yield	P	Ca	P	Ca
	$g.plant^{-1}$	$mg.plant^{-1}$	$mg.plant^{-1}$	$kg.ha^{-1}$	$kg.ha^{-1}$
$P_0 + T_0$ (Control)	9.68±0.26d	12.31±0.80d	116.81±4.02d	2.46±0.16d	23.36±0.80d
T_0 + MPR	15.58±0.66c	26.66±2.12c	302.38±23.34bc	5.33±0.42c	60.48±4.67bc
T_0 + TSP	19.88±0.69b	38.86±2.46b	326.34±21.21b	7.77±0.49b	65.27±4.24b
T	15.27±0.81c	24.06±1.88c	216.34±18.97c	4.81±0.38c	43.27±3.79c
T+ MPR	22.03±1.13b	45.07±3.75b	557.51±61.05a	9.01±0.75b	111.50±12.21a
T+ TSP	28.35±1.92a	66.80±6.86a	529.80±41.23a	13.36±1.37a	105.92±8.25a
F - Statistics	37.84***	28.97***	26.48***	28.97***	26.48***

***** *Significant at $P \leq 0.001$. a) P_0 = zero level of phosphorous; T_0 = zero level of Tughutu; Minjingu phosphate rock 26 kg P ha-1 = 26 kg P ha^{-1} of Minjingu phosphate rock; Triple Super Phosphate 26 kg P ha^{-1} = 26 kg P ha^{-1} of Triple Super Phosphate; T2.5 t ha^{-1} = 2.5 t of Tughutu ha^{-1} b) Mean values within a column followed by different letter(s) differ significantly at P=.05 according to Fishers least significance difference*

Table 4. Effects of Tughutu application, Minjingu phosphate rock or Triple Super Phosphate supply on pH, residual P and Ca in the soil collected from rhizosphere of *Phaseolus vulgaris* L. plants at harvest

Treatments	pH H_2O	P	Ca
		$mg.kg^{-1}$	$cmol.kg^{-1}$
$P_0 + T_0$ (Control)	5.32±0.16bc	3.28±0.41d	4.52±0.40b
T_0 + MPR	5.86±0.18a	10.59±0.47b	5.99±0.45a
T_0 +TSP	5.23±0.16c	14.76±0.80a	4.72±0.41b
T	5.79±0.20ab	6.93±0.51c	5.01±0.41ab
T+ MPR	5.98±0.20a	14.61±1.23a	6.14±0.41a
T+ TSP	5.71±0.17abc	16.79±1.06a	5.05±0.37ab
F - Statistics	2.92*	41.77*	2.68*

** Significant at P=.05. a) P_0 = zero level of phosphorous; T_0 = zero level of Tughutu; Minjingu phosphate rock 26 kg P ha^{-1} = 26 kg P ha^{-1} of Minjingu phosphate rock; Triple Super Phosphate 26 kg P ha^{-1} = 26 kg P ha^{-1} of Triple Super Phosphate; T2.5 $t.ha^{-1}$ = 2.5 t of Tughutu ha^{-1} b) Mean values within a column followed by different letter(s) differ significantly at P□.05 according to Fishers least significance difference*

4. CONCLUSION

The inherently low soil fertility status in most parts of the Western Usambara Mountains makes sustaining agricultural production difficult. This is the first work to evaluate Tughutualone and/or in combination with phosphate rock or Triple super phosphate on their influence on growth, amount of phosphorous and calcium content in bean shoots. The application of Tughutu, Minjingu Phosphate Rock or triple super phosphate and their combination showed synergistic effects in plant growth, amount of phosphorous and calcium and improvement in selected chemical soil properties measured at the end of the crop season. In conclusion, the locally available shrub (Tughutu) and Minjingu Phosphate Rock can constitute a very valuable and viable supplement to the industrial fertilizer sources in the Western Usambara Mountains. Future research could also focus on the relative cost (collecting, purchasing, supply of the products, transport and farm labour) of applying Tughutu and Minjingu Phosphate Rock in relation to that of triple super phosphate. There would appear to be no negative effects of Tughutu and Minjingu Phosphate Rock application on these soils or the bean crop.

ACKNOWLEDGEMENTS

This work was supported by the Directorate of Research and Development in the Ministry of

Agriculture, Food Security and Cooperatives of Tanzania, through TARP II World Bank project.

COMPETING INTERESTS

Author has declared that no competing interests exist.

REFERENCES

1. Ndakidemi PA. Agronomic and economic potential of tughutu (*Vernonia subligera* O.Hoffn) and minjingu phosphate rock as alternative P sources for bean growers. Pedosphere. 2007;17(6);732-738.
2. Ndakidemi PA, Semoka JMR. Soil fertility survey in western Usambara Mountains, northern Tanzania. Pedosphere. 2006;16; 237-244.
3. Ndakidemi PA, Semoka JMR. Soil fertility constraints to bean production in Lushoto District and their amelioration by application of lime, phosphorous and potassium. In Msumali GP, Mabagala RB (eds.). Bean Research. Proceedings of 12th Bean Research Workshop, Sokoine University of Agriculture, Morogoro, Tanzania. 1993;8:32-50.
4. Vance CP. Symbiotic nitrogen fixation and phosphorus acquisition. Plant nutrition in a World of Declining Renewable Resources. Plant Physiol. 2001;127:390-377.
5. Tang C, Hinsinger P, Drevon JJ, Jaillard B. Phosphorus deficiency impairs early nodule functioning and enhances proton release in roots of *Medicago truncatula* L. Ann. Bot. 2001;88:131-138.
6. Vance CP, Uhde-Stone C, Allan DL. Phosphorus acquisition and use: critical adaptations by plants for securing a nonrenewable resource. New Phytol. 2003;157:423-447.
7. Greenwood EAN, Hallsworth EG. Studies on the nutrition of forage legumes. II. Some interactions of Ca, P, Cu and Mo on the growth and chemical composition of *Trifolium subterraneum* L. Plant Soil. 1960;12:97-127.
8. Banath CL, Greenwood EAN, Loneragan JF. Effects of calcium deficiency on symbiotic nitrogen fixation. Plant Physiol. 1966;41;760-763.
9. Lowter WL, Loneragan JF. Calcium and nodulation in subterranean clover (*Trifolium subterraneum* L.). Plant Physiol. 1968;43:1362-1366. on P availability. Soil Biol. Biochem. 2003;35:207-216.
10. Muns DN. Nodulation of *Medicago sativa* in solution culture. V. Calcium and pH requirements during infection. Plant Soil. 1970;32:90-102.
11. Blevins DG, Barnett NM, Bottino PJ. The effects of calcium and the ionophore A23187 on nodulation, nitrogen fixation and growth of soybeans. Physiol. Plant. 1977;41:235-238.
12. Streeter JG. Effect of elevated calcium concentration in infected cells of soybean (*Glycine max* L. Merr.) nodules on nitrogenase activity and N input to the plant. J. Exp. Bot. 1998;49:997-1003.
13. Lyamchai CJ, Lyiom SD, Ndondi RV, Owenya MZ, Ndakidemi PA. Massawe NF. Participatory rural appraisal for Kwalei Catchment, Lushoto District. Selian Agricultural Research Institute, Arusha, Tanzania. 1998;71.
14. Ikerra TWD, Mnkeni PNS, Singh BR. Effects of added compost and farmyard manure on P release from minjingu phosphate rock and its uptake by maize. Norw. J. Agr. Sci. 1994;8:13-23.
15. Ikerra ST. Use of minjingu phosphate rock combined with different organic inputs in improving phosphorus availability and maize yields on a Chromic Acrisol in Morogoro, Tanzania, Ph.D Thesis. Sokoine University of Agriculture, Morogoro, Tanzania. 2004;262.
16. Savini I. The effect of organic and inorganic amendments on phosphorous release and availability from two phosphate rocks and triple super phosphate in phosphorous fixing soils, MSc Thesis. University of Nairobi, Kenya. 2000;104.
17. Wickama JM, Mowo JG. Using local resources to improve soil fertility in Tanzania. Managing Africa's Soils Working Paper International Institute for Environment and Development (IIED). London. 2001;21:24.
18. Kwabiah AB, Palm CA, Stoskopf NC, Voroney RP. Response of soil microbial biomass dynamics to quality of plant materials with emphasis. 2003;35(2):207–216.
19. George TS, Gregory PJ, Wood M, Read D. Buresh RJ. Phosphatase activity and

organic acids in the rhizosphere of potential agroforestry. 2002;34(10):1487–1494.
20. Kwabiah AB., Stoskopf NC, Palm CA, VOoroney RP. Soil availability as affected by the chemical composition of plant materials: Implications for P-limiting agriculture in tropical Africa. Agr. Ecosyst. Environ. 2003;100:53-61.
21. Kwabiah AB, Stoskopf NC, Palm CA, Voroney RP, Rao MR, Gacheru E. Phosphorus availability and maize response to organic and inorganic fertilizer inputs in a short-term study in western Kenya. Agr. Ecosyst. Environ. 2003;95:49-59.
22. Weil RR. Soil and Plant Influences on Crop Response to Two African Phosphate Rocks. Agron. J. 2000;92:1167-1175.
23. Szilas C. The Tanzanian Minjingu phosphate rock – possibilities and limitations for direct application. PhD Thesis, Royal Veterinary and Agricultural University, Copenhagen; 2002.
Available:http://www.igv.kvl.dk/-casz
24. Msolla MM. Agronomic evaluation of maize performance using hard and soft Minjingu phosphate rocks and in mixtures with triple phosphate on selected Tanzanian soils.
25. Mnkeni PNS, Semoka JMR, Buganga BBS. Agronomic effectiveness of Minjingu phosphate rock as a source of phosphorus for maize in some soils of Morogoro District, Tanzania. Zimbabwe J. Agric. Res. 1991;30:27-37.
26. Semoka JMR, Mnkeni PNS, Ringo HD. Effectiveness of Tanzanian phosphate rocks of igneous and sedimentary origin as sources of phosphorus for maize. Zimbabwe J. Agric. Res. 1992;30:127-136.
27. Haque I, Lupwayi NZ, Ssali H. Agronomic effectiveness of unacidulated and partially acidulated Minjingu rock phosphate on *Stylosanthes guianensis*. Trop. Grasslands. 1999;33:159-164.
28. Mowo JG. Effectiveness of phosphate rock on Ferralsols in Tanzania and the influence of within field variability, Ph.D Thesis. Wageningen Agricultural University, The Netherlands. 2000;64.
29. Mkangwa CZ. Influence of Minjingu Phosphte rock application on *Tephrosia vogelii* fallow productivity and subsequent maize response on Ferralsol in Morogoro, Tanzania. PhD Thesis, Sokoine University of Agriculture, Morogoro, Tanzania. 2004;227.
30. Savini I, Smithson P, Karanja N. Effects of added biomass, soil pH and calcium on the solubility of Minjingu phosphate rock in a Kenyan Oxisol. Arch. Agron. Soil Sci. 2006;52:19-36.
31. Savini I, Smithson PC, Karanja NK, Yamasaki H. Influence of *Tithonia diversifolia* and triple superphosphate on dissolution and effectiveness of phosphate rock in acidic soil. J. Plant Nutr. Soil Sci. 2006b;169;593-604.
32. Szilas C, Semoka JMR, Borggaard OK. Can local Minjingu phosphate rock replace superphosphate on acid soils in Tanzania? Nutr. Cycl. Agroecosys. 2006;77:257-268.
33. Msolla MM, Semoka JMR, Szilas C, Borggaard OK. Crop (Malze) response to direct application of local phosphate rock on selected acid soils of Tanzania. Comm. Soil Sci. Plant Anal. 2007;38:93-106.
34. Osunde AO, Zapata F, Sangiga N. Agronomic evaluation of rock Phosphate as a phosphorus sources for *Leucaena leucocephala* grown on an altisol. In Proceedings of the African Association of Biological Nitrogen Fixation (AABNF) held at the International Institute of Tropical Agriculture (IITA), Ibadan Nigeria, 24-28 September, 1990: 1992.
35. Chapman HD, Pratt PF. Methods of Analysis for Soils, Plants and
36. Murphy J, Riley JP. A modified single solution method for the determination of phosphate in natural waters. Anal. Chim. Acta. 1962;27:31-36.
37. Hesse PR. A text book of soil chemistry analysis. John Murray Ltd. London. 1971;120-121and 309.
38. STATISTICA. StatSoft, Inc. 2300 East 14th Street, Tulsa, OK 74104, USA; 2008.
39. Schwatz HF, Corralles MAP. Nutritional disorders in beans. In Schwatz HF, Corrales MAP (eds.). Bean Production Problems in the Tropics. CIAT, Cali, Columbia. 1989;57-604.
40. Marx ES, Hart J, Steven RG. Soil test interpretation guide. Oregon State University extension services. Oregon State University, Corvallis. 1996;7.
41. Helper PK, Calcium: A central regulator of plant growth and development. Plant Cell. 2005;17:2142-2155.
42. Robinson JBD. The influence of fertilizer and manure on the pH reaction of a coffee

soil. E. Afr. Agric. and For. J. 1956;22:76-80.

43. Huffman EO. Reaction of phosphate in soils: Recent research by TVA. Proceeding 71. The Fertilizer Society, London; 1962.

44. Chieh SH, Menon RG. Factors affecting the agronomic effectiveness of phosphate rock for direct application. Fert. Res. 1995;41:227-234.

45. Smalbarger SA. Singh U, Chien SH, Henao J, Wilkens PW. Development and validation of a phosphate rock decision support system. Agron. J. 2006;98:471-483.

16

Dietary Inclusion of Ethanolic Extracts of *Jatropha curcas* on the Performance and Carcass Characteristics of Broiler Chickens

G. O. Adeyemo[1*] and F. A. Oluyede[1]

[1]Department of Animal Science, Faculty of Agriculture, University of Ibadan, Nigeria.

Authors' contributions

This work was carried out in collaboration between both authors. Author GOA designed the study, wrote the protocol and all drafts of the manuscript. Author FAO reviewed the experimental design, managed the analyses of the study, performed the statistical analysis and wrote the first draft of the manuscript. Both authors read and approved the final manuscript.

Editor(s):
(1) Javier Alarcon Lopez, School of Engineering, University of Almería La Cañada de San Urbano, Spain.
(2) Anonymous.
Reviewers:
(1) Aliyu Abdullahi Mohammed, Department of Animal Science, Usmanu Danfodiyo University, Sokoto, Nigeria.
(2) Anonymous, Nigeria.

ABSTRACT

A study was conducted to evaluate the effect of *Jatropha curcas* leaf extract on the performance and carcass characteristics of broiler chickens. One hundred and fifty day old Arbor Acre broiler chicks were used. They were divided into five treatments and each treatment was replicated five times. Treatment one, (Negative control, no antibiotics), Treatment two (positive control, with antibiotics) while treatments three, four and five had varying inclusion levels of the *Jatropha curcas* leaf extract (0.25 g/100 kg, 0.50 g/100 kg and 0.75 g/100 kg of feed respectively). The birds were raised for a total number forty two days during which, they were fed *ad libitum* on a deep litter system of management. Weekly weight gained and feed consumed were recorded. Evisceration of carcass was done after forty two days; each part was weighed and recorded. Performance characteristics indicated no significant ($P>0.05$) difference in feed consumed and feed conversion ratio, but treatment 4 with 0.50 g/100 kg of *Jatropha curcas* extract has the best feed conversion ratio, but there were significant ($P<0.05$) differences in total weight gain.

**Corresponding author: E-mail: gbemiadeyemo@yahoo.com, gbemiadeyemo7@gmail.com;*

Keywords: Broilers; broiler carcass; Jatropha curcas; performance.

1. INTRODUCTION

Antibiotics are used in poultry production as non-nutritive feed additives to increase overall performance of poultry and improve their immune status. Such antibiotics are used as growth promoting agents, prophylactic, chemotherapeutic and other medicinal purposes. Non-stop use at sub therapeutic levels as done in animal feeding may result in drug resistant microorganisms in humans because of drug residues in animal products finding their way into man through consumption.

Antibiotic growth promoters have been banned in first world countries. This ban on the use of synthetic feed additives has triggered intensive research to develop alternative strategies to maintain health and performance in poultry production system. Researchers have identified several metabolites in medicinal plants, which play beneficial roles in the production and immune system of birds against many diseases, have excellent medicinal worth and could be use as natural growth promoters to in place of antibiotics and other non natural feed additives.

Bacterial diseases are tackled world-wide with anti-microbial agents that kill microorganisms or inhibit their growth. Antimicrobial agents that reversibly inhibit growth of bacteria are called bacteriostatic whereas those with irreversible lethal action on bacteria are known as bactericidal [1]. Microbial processes or structures are disrupted by antimicrobial agents making them different from those of the host. They may damage pathogens by hampering cell wall synthesis, inhibiting microbial protein and nucleic acid synthesis, disrupting microbial membrane structure and function, or blocking metabolic pathways through inhibition of key enzymes [2]. It has been reported that plant extracts and their products provide active curative ingredients in herb remedies which are used in many parts of the world [3].

Herbs were used and are still been used locally in the treatment of infections, many centuries before scientific studies were discovered. Before an antimicrobial agent is accepted for use in human beings it passes through several stages among which are the following (a) has selective toxicity (it should act on bacteria without damaging the host tissues); (b) be bactericidal rather than bacteriostatic; it should be effective against a broad range of bacteria; [3] should not be allergic; (d) should remain active in plasma, body fluids etc.; (e) should be stable and preferably water soluble; (f) desired levels should be reached rapidly and maintained for adequate period of time; (g) it should not give rise to resistance in bacteria; (h) it should have long shelf life; (i) it should not be expensive [1].

The chemical treatment of diseases which has gain acceptance world over also have conditions for its successive use, some of which include; the route of administration and location of the infection, the presence of interfering substances, the concentration of the drug in the body, the nature of the pathogen, the presence of drug allergies and drug resistance by microorganisms. Increasing number of drug resistance pathogens is of serious public health concern. Bacteria often become resistant in several different ways. Unfortunately, a particular type of resistance mechanism is not confined to a single class of drugs. Two bacteria may use different resistance mechanisms to withstand the same chemotherapeutic agent. Furthermore, resistant mutants arise spontaneously and are then selected for in the presence of the drug. Bacteria can become resistant to a drug by excluding it from cell, pumping the drug out of the cell, enzymatically altering it, modifying the target enzymes or organelles to make it less drug sensitive and effective [3]. Thousands of species of different medicinal plants exist and are used globally for the cure of different diseases. These plants are used as antimicrobial agents and several studies have been done by researchers to find out their scientific bases [4]. Some of these plants include; *Anacardium occidentale* (cashew), *Pilostigma recticulatum*, *Anogeissus leiocarpa*, *Enantia chlorantha*, *Senna occidentalis*, *Jatropha curcas* ("Lapalapa"), *Azadirachta indica* ("Dogonyaro, Nigeria local name") *etc.* The presence of some substances in plants such as alkaloids, glycosides, volatile oils, gums, tannins, steroids, saponins, phlobatannins, flavonoids and a host of other chemical compounds referred to as secondary metabolites. The actions of these plants on microorganisms have been found to be due to these secondary metabolites that are present in them [5-7]. Medicinal plants like *J. curcas* have played a major role in the treatment of various diseases including bacterial and fungal infections. The extracts of *Jatropha* species including *J. curcas* displayed potent cyto-toxic, anti-tumor and anti-microbial activities in different assays. The latex of *J. curcas* also showed antibacterial activity against *Staphylococcus aureus* [8], however the antimicrobial activity of the other parts have not been fully investigated.

Jatropha curcas Linn. (Family Euphorbiaceae) Physic nut, is *J. curcas* plant has been initially considered a traditional herb in many parts of the world [9]. This is partially true of this drought resistant shrub because it is widely grown in Africa, India Central and South America and South-East Asia,. It has gained importance in Malaysia as a source of biofuel production. *J. curcas* have been used in different forms for the treatment of different types of infection. The leaves part has been used as an antiseptic substance during birth, the root part is also being used to treat sexually transmitted diseases and the seed is used to treat skin diseases [9] However, its application as a remedy for many of these ailments has not been fully established by identifying the phyto-chemical compounds responsible for the various treatments they are used for. Hence, this study was conducted to evaluate the performance and carcass characteristics of broiler chickens fed *Jatropha curcas* and substantiate the effect of the phyto-chemicals present in it on broiler production.

2. MATERIALS AND METHODS

The study was carried out at the poultry unit of the Department of Animal Science Teaching and Research farm, University of Ibadan, Nigeria (latitude 7 73" N and longitude 3 5' to 3 36 E). Ibadan is located 228 m above sea level and has a mean annual rainfall of 1289.2 mm based on 27 year of records [10]. The experimental pens were thoroughly cleaned, washed and disinfected. The condition of housing and management of birds were the same in all groups.

2.1 Processing of Leaf Extract

Fresh leaves of *Jatropha curcas* were collected within Ibadan, South-Western Nigeria. The leaves collected per time was rinsed with clean water to remove any foreign matter, chopped and air-dried. The dried leaves were milled using a roller mill. One litre of an 80% ethanol extraction fluid was mixed with 200 g of powdered plant material. The mixtures were kept for 2-5 days in tightly sealed vessels at room temperature protected from sunlight, and mixed several times daily with a sterile glass rod. This mixture was filtered through muslin cloth and the residue, adjusted to the required concentration (500 ml of 80% ethanol for the residue of 200 g of powdered plant material) with the extraction fluid for further extraction. Further extraction of the residue was repeated 3-5 times until a clear colorless supernatant extraction liquid was obtained indicating that no more extraction from the plant material was possible [9].

The extracted liquid was subjected to rota-evaporator (Model # R) to remove the ethanol. Rota evaporation was used to concentrate the smaller quantity of extract. A 250 ml aliquot of extracted liquid was subjected to rota evaporatoration for 3-4 h. Extract from this method was then weighed and stored in desiccators until further use.

2.2 Experimental Birds and Management

A total number of 150 day old arbor acre broilers were purchased and randomly allocated to 5 treatments and 6 replicates per treatment containing 6 birds in a completely randomized design. Water and feed were supplied *ad libitum* while a standard medication and vaccination programs was adopted according to the recommendation of [11]. Weekly feed intake and individual bird weights were recorded for six weeks. Weekly weight gains and feed conversion ratio were calculated from the data obtained.

2.3 Experimental Diets

Diet 1- basal diet without Antibiotics
Diet 2– basal diet with Antibiotics
Diet 3– basal diet with 0.25 g *Jatropha curcas* leaf extract
Diet 4– basal diet with 0.50 g *Jatropha curcas* leaf extract
Diet 5– basal diet with 0.75 g *Jatropha curcas* leaf extract

The composition of experimental diet is shown in Table 1

2.4 Performance Characteristics

Feed and water were given *ad libtum*. Feed consumed (g/bird) was recorded daily, leftover was weighed and discarded. Individually, body weight (BW, g)/bird was taken on pen basis before offering feed the initial day and then at weekly intervals up to 7 weeks. Feed conversion ratio was calculated as feed intake per unit gain (g/intake/g BW gain).

2.5 Carcass Characteristics

At day 42, two birds were randomly selected from each of the replicate for carcass analysis. The selected birds were starved over night and their live weight recorded. The birds were slaughtered by severing the jugular vein and fully bled before scalding in hot water. The birds were de-feathered after scalding and their plucked weight was taken. The eviscerated weights were recorded. The head, neck, shanks, back, breast, gizzard, and organs such as liver, kidney, heart, spleen, lungs, and pancreas were removed, weighed and recorded. The different cut parts were expressed as percentage of dressed weight.

Table 1. Gross composition of experimental diets

Ingredients	Starter	Finisher
Maize	56.00	59.00
Soyabean meal	36.50	28.00
Fish meal	2.20	1.30
Wheat offal	1.60	7.50
Dicalcium phosphate	1.50	1.50
Methionine	0.10	0.10
Salt	0.25	0.25
Premix	0.25	0.25
Total	100	100
Calculated values		
Crude Protein (%)	23.02	20.00
Metabolizable energy (kcal/kg)	2995.10	2950.05
Crude Fibre (%)	3.51	3.79
Calcium (%)	1.03	1.00
Available phosphorus(%)	0.63	0.61
Lysine (%)	1.14	1.09
Methionine (%)	0.57	0.55

2.6 Chemical Analysis

Feed were subjected to proximate analysis according to [12] method. Crude Protein, Ether Extract, Ash, Dry Matter and Nitrogen Free Extracts were the nutrients analyze for.

2.7 Statistical Analysis

Data generated were subjected to one-way analysis of variance (ANOVA) using the [13] package and means were separated using Duncan multiple range test of the same software.

3. RESULTS AND DISCUSSION

The chemical composition of the experimental diets is shown in Table 2. Performance indices of the experimental birds fed *Jatropha curcas* extract supplemented diets are shown in Table 3 The final body weight(kg) obtained (1.73, 1.84, 1.61, 1.75 and 1.67) for birds on diets 1,2,3,4, and 5 respectively varied significantly ($p<0.05$). Values obtained for weight gain (kg) (1.72, 1.87, 1.56, 1.72 and 1.65 for birds on diet 1,2,3,4 and 5 respectively) varied significantly ($p<0.05$). Values obtained for feed intake (kg/bird) 4.11, 4.39, 3.85, 4.02 and 3.94 respectively) for treatment 1,2,3,4 and 5 were not statistically different ($p>0.05$). Also the values obtained for Feed Conversion Ratio (FCR) were 2.37, 2.43, 2.52, 2.35 and 2.41 respectively were not statistically different ($p>0.05$) but birds on diet 4 had the best FCR.

Table 2. Chemical composition of experimental diets

Parameter	Starter	Finisher
Dry matter	90.20	90.50
Crude protein	23.05	20.90
Crude fibre	3.70	5.10
Ether extract	2.70	2.50
Ash	6.50	6.00

Various reports [14-16], indicated that plant extracts has the ability to improve FCR and increase BWG (Body Weight Gain). However, plant extracts are also known to exert anti-microbial actions in vitro against important pathogens including fungi [17]. This is due to the abundance of very complex mixture of bioactive components [17].

Results for weight gain showed that T2 had the highest value, followed by T4. This high value may be due to the ability of antibiotics to suppress sensitive populations of bacteria in the intestine which was in line with the findings of [18]. Also, diversion of nutrients from being used by the gut microorganism for the use of birds may be responsible for values observed. Treatment 4 values can be said to be due to growth promoting effect of plant extract as explained by [14,15].

The higher feed intake observed in treatment 4 among the treatments with *Jatropha curcas* plant extract may be due to the effect of plant bioactive on digestion as this was in agreement with the work of [19] who believes that some plant extract may stimulate the production of saliva and of gastric and pancreatic juice favoring the secretion of enzymes and therefore improving nutrient digestibility. Once there is increase in pancreatic and gastric juices the animal tends to consume more because feed is quickly degraded. Also it can be said that the synergistic effect of all primary and secondary component may be responsible for the higher feed intake.

There is improvement in feed conversion ratio of T4 when compared with other treatments with different level of *Jatropha curcas* leaf extract which further confirms the antibacterial and antifungal properties of *Jatropha curcas*. It can be said that the low body weight gain (BWG) recorded in treatment 3 may be due to low inclusion rate of the extract in the feed of the birds on that treatment. This is in support with [20] who stated that discrepancy may be attributed to differences in the inclusion level of products containing plant extracts and essential oils or in the main components of the essential oil. Also the level of active ingredient in supplements, the environmental conditions and the composition of the basal diet [20,21] has effect on the performance of the birds.

Table 4 shows the summary of different parts of broiler and organ weight as expressed in percentage of dressed weight. The average live weight of the birds at the end of the experiment ranges from 1710 g to1880 g in T3 and T4 respectively. Value obtained showed, there were no significant ($p>0.05$) differences.

The average dressed weight of the birds at the end of the experiment ranges from 1390 g to 1540 g in T2 and T5. Values obtained for T1 and T4 are 1500 g and 1510 g respectively, no significant ($p\geq0.05$) difference across the treatment.

The carcass and organ parts were expressed as the percentage of the dressed weight. The values obtained for breast, thigh, drumstick, back, neck, liver, gizzard, heart, feet, spleen, lungs kidney and head were not statistically different across the treatment. Values for breast obtained for T2, T3 and T4 are 26.34, 26.78 and 27.62 respectively. Also values obtained for Thigh ranges from 14.04 to 12.88 in both T3 and T4 respectively. Values obtained for T2 and T5 are 13.20 and 13.00.

Table 3. Performance characteristics of broilers fed *Jatropha curcas* leaf extract diets

Parameters	1	2	3	4	5	SEM
Initial body weight (g)	0.40	0.40	0.41	0.39	0.40	0.11
Final body weight (kg)	1.73^{a}	1.84^{a}	1.61^{b}	1.75^{a}	1.67^{b}	0.06
Total weight gained (kg)	1.33^{ab}	1.44^{a}	1.20^{b}	1.34^{ab}	1.27^{b}	0.14
Total feed consumed (kg)	4.11	4.39	3.85	4.02	3.94	0.25
Feed conversion ratio	3.09	3.05	3.21	3.00	3.10	0.21

ab: Means in the same row with different superscripts are significantly(P<0.05) different

Table 4. Carcass characteristics and organ weight of broilers fed *Jatropha curcas* leaf extract diets

Parameters	1	2	3	4	5	SEM
Live weight(kg)	1.84	1.80	1.71	1.88	1.84	0.097
Dresses weight (kg)	1.50	1.54	1.39	1.51	1.54	0.08
Thighs(% of dressed wt)	14.04	26.34	26.78	27.62	25.46	0.37
Drumsticks (√√)	11.64	12.32	12.68	12.07	12.25	0.36
Wings (√√)	9.69^{ab}	10.08^{ab}	10.4^{a}	9.26^{b}	9.88^{ab}	0.67
Backs (√√)	16.59	17.23	16.60	17.91	18.51	0.67
Breasts (√√)	27.62	26.34	26.78	27.85	25.46	0.85
Neck (√√)	5.31	5.80	5.38	6.04	5.53	0.29
Livers (√√)	3.50	3.24	3.32	2.92	3.16	2.28
Gizzards (√√)	4.46	4.25	3.99	4.19	4.37	0.34
Hearts (√√)	0.58	0.61	0.57	0.53	0.61	0.06
Feet (√√)	3.77	4.97	5.23	4.39	4.85	0.49
Pancreas (√√)	0.43^{a}	0.36^{ab}	0.23^{b}	0.25^{b}	0.42^{a}	0.05
Spleens (√√)	0.21	0.16	0.20	0.12	0.11	0.04
Lungs (√√)	0.78	0.72	0.65	0.59	0.78	0.42
Kidneys (√√)	0.81^{a}	0.89^{ab}	0.71^{a}	0.44^{b}	0.85^{ab}	0.10
Heads (√√)	3.04^{ab}	3.19^{ab}	3.66^{a}	2.91^{a}	3.11^{a}	0.22

ab:Mean in the same row with different superscripts are significantly(P<0.05) different; Key: % of dressed weight = √√

Values obtained for wing in T3 and T4 shows significant differences with the value ranges from 10.4 to 9.26. The pancreas also showed significant differences which ranges from 0.43 to 0.23 respectively in T1 and T3. Results of the carcass analysis and organ weights are shown in Table 4. Best performance in terms of live weight and dressed weight were recorded for birds on treatment 1, 2, 4 and 5. Treatment 4 had the highest among the treatment especially in terms of live weight, thigh, and breast .Treatment 2 and 5 had the highest value for dressed weights while the highest values for drumstick and wing were observed in treatment 3. The highest value for breast were also seen in T4. This means the treatments with extract had the best total edible meat yield compared to birds from the control.

Although, the result showed that treatment 2 had the highest body weight values, treatments with leaf extract competes favourably with birds on control treatments (T1 and T2) in terms of dressed weight which means there were better distribution and location of meat on several parts especially the thigh, drumstick and breast.

Wing values differ significantly between treatments while values for liver, gizzard, heart, spleen and lungs did not differ significantly. This development showed there were reduced concentration of anti-nutritional factors and leaf extract in diets fed to the birds. This is because higher physiological values in organs and offals are stimulated by the presence of anti-nutritional factors and other harmful secondary metabolites. The lower values observed in treatments 3 and 4 for pancreas and treatment 4 for kidney showed that plant bio-actives present in *Jatropha curcas* leaf extract enhance the functioning of pancreas and kidney as digestive and excretory organs.

4. CONCLUSION

Most performance characteristics measured indicated no differences between the treatments. But broilers fed diet with 0.50 g/100 kg of *Jatropha curcas* extract gave the best feed conversion ratio and total weight gain.

COMPETING INTERESTS

Authors have declared that no competing interests exist.

REFERENCES

1. Rajesh B, Rattan LI. Essentials of Medical Microbiology. 4th Edition. Jaypee Brothers Medical Publishers (P) Ltd. New Delhi 110002, India. 2008;500.
2. Willey JM, Sherwood LM, Woolverton CJ. Prescott, Harley and Klein's Microbiology. 7th Edition; 2008.
3. Haslam E, Lilley TH, Martin Y, Magnolotto R. Traditional Herbal Medicine; the Roles of Polyphenols. Plants Medica. 1989; 4(2):1-8.
4. Omotayo AE. Antibacterial activity of some antimalarial plants. Proceeding of Nigerian Society for Microbiology. 39:69-72c Graw Hill, New York, NY 10020. 1998;1088.
5. Kochlar SL. Tropical crops. In: A Textbook of Economic Botany. Macmillan Publishers Ltd, London and Basingstoke. 1986; 21(25):33-34
6. Sofowora EA. Medicinal plants and traditional medicine in Africa. Spectrum Books Ltd, Ibadan, Nigeria. 1993;289.
7. Oyagade JO, Awotoye OO, Adewumi JT, Thorpe HT. Antimicrobial activities of some Nigerian medicinal plants. Screening for Antimicrobial Activity. Bioscience Research Communication. 1999;11(3):193-197
8. Thomas OO. Re-examination of the antimicrobial activities of *Xylopia aethiopica, Carica papaya, Ocimium gratissimum* and *Jatropha curcas.* Fitoterapia. 1989;60:147-161.
9. Gubitz GM, Mittelbach M, Trabi M. Exploitation of the tropical oil seed plant *Jatropha curcas* Linn. Bioresour. Technol. 1999;67:73-82.
10. Alabi RT, AG Ibiyemi. Rainfall in Nigeria and food crop production. In Agronomy in Nigeria. University of in Ibadan, Nigeria; 2000.
11. Oluyemi, Robert. Poultry Production in warm wet climate. Macmillian Press Ltd. 2000;197.
12. AOAC. Assosiation of Official Analytical Chemists. Official Method Analysis, 15th Edn. Washington, D. C; 1990.
13. SAS. Statistical Analysis System. Version. 8. Cary, North Carolina, SAS Institute. 1999;2.
14. Esteve-Garcia E, Brufau J, Perez-Vendrel A, Miquel T, Duven K. Bioefficacy of European Commission. Regulation (EC) No. 1831/2003 of the European Parliament and of the council of 22 September 2003 on additives for use in animal nutrition. Off. J. Eur. Union L. 2003;268:29–43.
15. Bafundo KW, Cox LA, Bywater R. Review lends perspective to recent sciectific findings on virginiamycin, antibiotic

resistance debate. Feed Stuffs. 2002;75:26 -7.

16. Van Campenhout L, Van hemel J, Vandenkerckhove J, Mollen K, Sas B. Performance of an alternative to antibiotics in broilers with high intestinal counts of *Clostridium* perfringens. Proc. 13th Eur. Symp. on Poult. Nutr. Oct. Blankenberge, Belgium. 2001;127 - 8.
17. Windisch W, Kroismayr A. The effects of phytobiotics on performance and gut function in monogastrics. Accessed in 2006. Available:www.feedinfo.com. www.milliyet.com.tr/extra/venus/vitamin/vit001/axvitamin02.html
18. Janssen AM, Scheffer JJC, Svendsen AB. Antimicrobial activity of essential oils: A 1976-1986 literature review. Aspects of the test methods. Planta Medica. 1987;53:395-398.
19. Mellor S. Alternatives to antibiotic. Pig Progress. 2000;16:18-21
20. Giannenas I, Florou Paneri P, Papazahariadou M, Christaki E, Botsoglou NA, Spais AB. Effect of dietary supplementation with oregano essential oil on performance and the intestinal mucosa, and their consequences on digestive health in young non-ruminant animals. Anim. Feed Sci. Technol. 2003;108:95–117.
21. Lee KW, Everts HJ, Kappert H, Frehner R, Losa R, Beynen AC. Effects of dietary essential oil components on growth peformance, digestive enzymes and lipid metabolism in female broiler chickens. Brit. Poultry Sci. 2003;44:450-457.

17

Growth Response of Guinea Grass (*Panicum maximum*) to Cutting Height and Poultry Manure

C. C. Nnadi[1*], C. C. Onyeonagu[1] and S. C. Eze[1]

[1]Department of Crop Science, University of Nigeria Nsukka, Enugu State, Nigeria.

Authors' contributions

This work was carried out in collaboration between all authors. Authors CCO and CCN designed the study, wrote the protocol and wrote the first draft of the manuscript. Authors CCO and CCN reviewed the experimental design and all drafts of the manuscript. Authors CCO and CCN managed the analyses of the study. Author CCN performed the statistical analysis. All authors read and approved the final manuscript.

Editor(s):
(1) Anonymous.
Reviewers:
(1) Valentin Ivanov Kosev, Institute of Forage Crops, Pleven, Bulgaria.
(2) Lucas R. Petigrosso, National University of Mar del Plata, Argentina.
(3) Anonymous, Brazil.

ABSTRACT

A field experiment conducted in 2014 at Nsukka, Nigeria was aimed to study the effect of cutting height and poultry manure on the growth of *Panicum maximum.* Nine treatments were evaluated as factorial combinations of poultry manure applied at 0, 5 and 10 ton ha^{-1} and cutting heights of 5, 10 and 15 cm. The treatments were fitted into in a randomized complete block design (RCBD), replicated three times. The research was conducted during the period of March to September, 2014 at the Teaching and Research Farm of the Department of Crop Science, Faculty of Agriculture, University of Nigeria, Nsukka. Grass cover and tiller number were significantly increased with 10 ton ha^{-1} poultry manure compared with the control. Dry matter content and bare ground area reduced with increased application of poultry manure. Cutting at 15 cm significantly produced taller plants than 5 cm but statistically the same with 10 cm height of cut. A combination of 10 cm cutting height and 10 ton ha^{-1} of poultry manure gave greater tiller population per m^2 than 5 cm with zero poultry manure. However application of 10 ton ha^{-1} of poultry manure with harvest at either 10 or 15 cm produced higher growth establishment of the grass.

**Corresponding author: E-mail: nnadichineducollins@gmail.com;*

Keywords: *Panicum maximum; poultry manure; cutting height; growth; Nigeria.*

1. INTRODUCTION

Panicum maximum is an extremely variable species; loosely to densely tufted, shortly rhizomatous, erect or geniculately ascending, rooting at the lower nodes [1]. Leaf blades linear to narrowly lanceolate; panicle open, oblong or pyramidal, with secondary branches well developed and flexuous; spikelets oblong, blunt or acute, rounded on the back [1]. It remains one of the best cultivated grasses for the tropics and is used extensively for forage purposes [2]. In the last 10 years, clashes have occurred between the herdsman and farmers in many parts of the country Nigeria. Incessant clashes between Fulani herdsman and local farmers from North to south of the country has claimed many lives [3]. Climate change has partly been blamed. About 35 percent of land that was cultivable 50 years ago is now desert in 11 of Nigeria's northern states [4]. The root of the conflict has been path of grazing / encroachment of grazing land. Shortage of pastures has compelled these herdsmen to allow their cattle, graze in cultivated farmlands. Most pastures in Nigeria lack proper management in terms of optimum defoliation and fertilization. They are often subjected to burning and extreme grazing, which have been shown to result in serious deterioration of pastures [5]. The ultimate panacea to this problem is the establishment of sown and maintained pastures. There is also the need to improve pastures through good management. Such management practices include the use of fertilizers and cutting management [6]. Improved management practices require many inputs which are not commonly used in natural grassland livestock production system. Therefore to be more readily acceptable, the system must not be capital intensive and must require little new technology [7]. Under such circumstances, improvement of pastures through application of poultry manure and cutting management could be useful alternatives.

The regrowth after defoliation is one of the most important physiological processes and it determines the sward structure [8]. Grasses form the foundation of forage–livestock systems around the world because they can be consumed and converted by animals into useful products. Consumption assumes the grass is harvested directly through grazing, or by machine for green chop, silage, or hay. Knowing when and how to harvest for optimal forage quantity and quality while safeguarding the persistence of stands require an understanding of grass growth and regrowth mechanism [9]. Grasses can be most productive when clipped with mowers or bitten by animals. Defoliation can be productive or destructive. The effect on grass productivity of differential cutting heights is conflicting. For optimum growth following defoliation there must be cell division and expansion in certain meristem systems. Knowledge about the location and specific function of these meristems is critical for successful forage management [10].

Poultry manure is a valuable source of plant nutrients and organic matter and when used as a fertilizer, will improve crop and forage production and soil quality [11]. The use of inorganic fertilizer has not been helpful under intensive agriculture because it is often associated with reduced crop yield, soil acidity and nutrient imbalance [12-14]. Soil degredation which is brought about by loss of organic matter accompanying continuous cropping becomes aggravated when inorganic fertilizer are applied repeatedly. This is because crop response to applied fertilizer depends on soil organic matter [15]. The rising cost of inorganic fertilizers coupled with their inability to recondition the soil has directed attention to organic manures in recent times [16]. Poultry manure is very cheap and effective as a good source of N for fertilization, but its availability remain an important issue due to its bulky nature, while inorganic fertilizer is no longer within the reach of poor resource farmers due to its high cost [17]. Moreover, the need and utilization of poultry manure has overtaken the use of other animal wastes, because of its high content of nitrogen, phosphorus and potassium [18]. Furthermore, poultry manure is preferred amongst other animal wastes because of its high concentration of macro nutrients [19-21]. The quantity of organic material which can be introduced into the soil either by natural returns through roots, stubbles, slough off root nodules and root exudates or by artificial application in the form of organic fertilizer, determines crop growth response [22].

Nevertheless poultry manure differs in quality and pasture response to poultry manure fertilization also differs to soils and among location. Consequently, recommendation for organic fertilizer based on result from a particular experiment may not always be suitable for all growing conditions in every location [23]. Therefore, the objective of this study is to determine the effect of cutting height, poultry manure and the best combination of the two on the growth of *Panicum maximum* grown in Nsukka, Nigeria.

2. METHODOLOGY

2.1 Study Area and Sampling

The experiment was conducted at the Teaching and Research Farm of the Department of Crop science, faculty of Agriculture, University of Nigeria, Nsukka. Nsukka is located at latitude 06°52'N and longitude 07°24'E, and altitude of 447.2m above sea level. It is characterized by the low land humid tropical conditions. The soil is classified as an ultisol [24]. The guinea grass (*Panicum maximum*) experiment was established in 2014 (March-April). The experiment was a 3x3 factorial laid out in a Randomized Complete Block Design and was replicated three times. Treatments comprised three cutting heights of 5, 10, and 15 cm and three levels of poultry manure (PM) at 0, 5, and 10 ton ha-^{1}. The land measuring 22.5 meters long by 10 meters wide with an area of 225 m2 was marked out into three blocks of 20.5 m x 2 m each. Each block was further divided into 9 plots of 2 m x 1.8 m. Each block was separated by 1 m pathway, while the plots were demarcated by 0.5 m pathway within the blocks. The three levels of poultry manure were applied to their respective plots and rooted cuttings of *Panicum maximum* with height of 15 cm planted with the spacing of 20 cm x 30 cm. Manure treatment effect alone was considered during the establishment period (first general harvest). Plant scoring of the plots was done before any harvest to determine the extent of cover by the grass species, the weed species and bare ground cover. Scoring was done using five grading score for subjective evaluation [25]. Cutting was carried out every 5 weeks with shears, aided by a wooden block at varying heights. A quadrant of 1 m x1 m square was used to determine the sampling area.

2.2 Data Collection Technique

Plant height was taken using the mean of three readings taken at random from the sample area in each plot. Tiller counts were made in each plot using a 25 cm square quadrant. The mean of three throws per plot was used to calculate tiller population per m^2. Sub samples of the grass species weighing 100 g were put in a paper envelop and dried in a force air oven set at 80°C and weighed after attaining constant dry weight. These were used to calculate the dry matter content of grass species as follow:

$$\text{Dry matter content \%} = \frac{\text{Dry weight of sample (g)}}{\text{Sample fresh weight (g)}} \times \frac{100}{1}$$

2.3 Statistical Analysis

All data collected were statistically analyzed using GENSTAT (2008) statistical package [26]. Separation of treatment means for statistical significance was done using the least significant difference (LSD) as outlined by Obi [27].

3. RESULTS

Increased application rate of poultry manure gave higher grass height and tiller number of *Panicum maximum* during the first general harvest (Table 1). The extent of weed cover was significantly ($P < .05$) high with no application of poultry manure. Bare ground cover was highest at 0 ton ha-^{1} compared to 5 or 10 ton ha-^{1} of poultry manure. Grass cover increased with higher rates of poultry manure (Table 1).

Grass dry matter content was not significantly influenced by either cutting height or poultry manure application in the second and third harvest periods (Table 2). However, interaction of cutting height at 10 cm with 5 ton ha-^{1} of poultry manure, significantly gave highest dry matter content compared to harvest at 5 cm with application of poultry manure at 10 ton ha-^{1}.

Table 1. Treatments effect on the growth parameters measured at the first general harvest

PM (tonha^{-1})	PH	TN	GDMC	GC	WC	BG
0	72.67	130.2	20.6	2.4	1.67	1.4
5	91.9	180.6	23.7	3.2	1.44	1.0
10	89.9	198.0	30.3	4.0	1.56	1.0
$LSD_{0.05}$	11.87	35.54	2.88	0.54	NS	0.33

PM = poultry manure; PH = plant height; GDMC = grass dry matter content; TN = tiller number; GC = grass cover; WC = weed cover; BG = bare ground cover; NS = Non- Significant F- test at 5% probability level

Increase in height of cut above ground significantly increased grass height in the second and third harvest periods (Table 3). Cutting height of 15 cm gave higher grass height over 5 cm. However harvest at 15 and 10 cm were statistically the same. Grass height increased with higher application of poultry manure. Furthermore, Poultry manure and cutting height interaction significantly affects the grass height. Harvesting at height of either 10 or 15 cm with application of poultry manure at 10 ton ha^{-1} significantly increased height of grass compared to harvest at 5 cm above ground with no poultry manure application. Moreover, during the second and third harvest period, the grass heights were highest when harvested at 15 cm above ground and poultry manure applied at rate of 10 ton ha^{-1}.

Grass tiller number per square meter was reduced (P > .05) with lower cutting height (Table 4). There was a significant increase in the number of tillers per square meter with application of poultry manure compared to control. Meanwhile, the highest tiller number was obtained in both periods when harvest was done at 10 cm above ground and poultry manure applied at 10 ton ha^{-1} and was significant compared to cutting height of 5 cm with no application of poultry manure.

Bare ground and weed cover were not significantly affected by either cutting height, poultry manure or their interaction (Tables 5 and 6).

Table 2. Treatments effect on grass dry matter content

	Poultry manure (ton ha^{-1})			
Cutting height (cm)	**0**	**5**	**10**	**Mean**
	2nd harvest period (4 – 7 – 2014)			
5	23.2	18.8	20.3	20.7
10	21.8	20.5	22.1	21.5
15	21.0	22.9	21.7	21.9
Mean	22.0	20.7	21.3	21.4
	3rd harvest period (8 – 8 – 2014)			
5	26.2	25.3	21.1	24.2
10	26.4	27.8	25.1	26.4
15	27.3	24.8	24.7	25.6
Mean	26.7	26.0	23.6	21.4
			2nd harvest period	**3rd harvest period**
$LSD_{0.05}$ between two cutting height mean (C):			NS	NS
$LSD_{0.05}$ between two poultry manure mean (P):			NS	NS
$LSD_{0.05}$ between two (C) x (P):			NS	5.69

NS, Non- Significant F- test at 5% probability level

Table 3. Height response of guinea grass to the treatments

	Poultry manure (ton ha^{-1})			
Cutting height (cm)	**0**	**5**	**10**	**Mean**
	2nd harvest period (4 – 7 – 2014)			
5	103.6	117.7	127.3	116.2
10	128.0	111.9	139.4	126.4
15	117.0	142.9	149.8	136.6
Mean	116.2	124.2	138.8	126.4
	3rd harvest period (8– 8 – 2014)			
5	69.3	82.3	77.9	76.5
10	80.4	77.9	87.6	82.0
15	84.9	88.1	89.5	87.5
Mean	78.2	82.8	87.5	82.0
			2nd harvest period	**3rd harvest period**
$LSD_{0.05}$ between two cutting height mean (C):			17.23	9.07
$LSD_{0.05}$ between two poultry manure mean (P):			17.23	9.07
$LSD_{0.05}$ between two (C) x (P):			29.84	15.71

NS, Non- Significant F- test at 5% probability level

Harvesting at 15 cm above ground other than at 5 cm gave the highest grass cover (Table 7). Meanwhile, higher poultry manure rate significantly increased the grass cover. However, interaction of either 10 or 15 cm height of cut with application of poultry manure at 10 ton ha^{-1} gave more grass cover during the second harvest period than where the 5 cm cutting was combined with zero poultry manure.

4. DISCUSSION

Plant height decreased (P > .05) with lower cutting height but increased (P > .05) progressively with higher cutting heights. Adams et al. [28] reported that close cutting of grassland above ground level reduced the ability of the grass to replenish leaf area, set seeds and store food reserves in their root, thereby reducing grass growth. Increase in application of poultry manure increased the grass height. This could be attributed to the increase in the nutrient content of the soil with essential nutrient elements from poultry manure such as nitrogen, phosphorus and potassium in plants, leading to formation of plant metabolites that helped to build the plant tissue. This is in accordance with Feisal et al. [22] with Sorghum bicolor, and in line with the findings of Farhed et al. [29] who attributed the increase of plant height in maize to the increase in the rate and availability of poultry manure throughout the growing season.

Table 4. Plant tiller number (per m2) calculated for each treatment during the harvest periods

	Poultry manure (ton ha^{-1})			
Cutting height (cm)	**0**	**5**	**10**	**Mean**
	2nd harvest period (4 – 7 – 2014)			
5	241.6	341.3	387.7	323.6
10	323.7	263.5	468.8	352.0
15	318.4	448.0	457.1	407.8
Mean	294.6	350.9	437.9	361.1
	3rd harvest period (8 – 8 – 2014)			
5	206.4	293.3	301.9	267.2
10	252.8	216.0	323.7	264.4
15	272.5	304.0	298.7	291.7
Mean	243.9	271.3	308.1	274.4
			2nd harvest period	**3rd harvest period**
$LSD_{0.05}$ between two cutting height mean (C):			NS	NS
$LSD_{0.05}$ between two poultry manure mean (P):			97.90	60.89
$LSD_{0.05}$ between two (C) x (P):			169.57	105.46

NS, Non- Significant F- test at 5% probability level

Table 5. Bare ground cover estimated for the treatments at the harvest periods

	Poultry manure (ton ha^{-1})			
Cutting height (cm)	**0**	**5**	**10**	**Mean**
	2nd harvest period (4 – 7 – 2014)			
5	1.0	1.0	1.0	1.0
10	1.0	1.0	1.0	1.0
15	1.3	1.0	1.0	1.1
Mean	1.1	1.0	1.0	1.0
	3rd harvest period (8 – 8 – 2014)			
5	1.0	1.0	1.0	1.0
10	1.0	1.0	1.0	1.0
15	1.0	1.0	1.0	1.0
Mean	1.0	1.0	1.0	1.0
			2nd harvest period	**3rd harvest period**
$LSD_{0.05}$ between two cutting height mean (C):			NS	NS
$LSD_{0.05}$ between two poultry manure mean (P):			NS	NS
$LSD_{0.05}$ between two (C) x (P):			NS	NS

NS, Non- Significant F- test at 5% probability level

Table 6. Estimated weed cover in response to the treatments

	Poultry manure (ton ha^{-1})			
Cutting height (cm)	**0**	**5**	**10**	**Mean**
	2nd harvest period (4 – 7 – 2014)			
5	2.3	2.0	2.0	2.1
10	2.0	2.3	2.0	2.1
15	2.0	2.0	2.0	2.0
Mean	2.1	2.1	2.1	2.1
	3rd harvest period (8 – 8 – 2014)			
5	2.0	1.7	2.0	1.9
10	1.7	2.0	2.0	1.9
15	1.7	1.7	1.7	1.7
Mean	1.8	1.8	1.9	1.8
			2nd harvest period	**3rd harvest period**
$LSD_{0.05}$ between two cutting height mean (C):			NS	NS
$LSD_{0.05}$ between two poultry manure mean (P):			NS	NS
$LSD_{0.05}$ between two (C) x (P):			NS	NS

NS, Non- Significant F- test at 5% probability level

Table 7. *Panicum maximum* response to the treatments on grass cover

	Poultry manure (ton ha^{-1})			
Cutting height (cm)	**0**	**5**	**10**	**Mean**
	2nd harvest period (4 – 7 – 2014)			
5	3.0	4.0	4.0	3.7
10	4.0	3.0	5.0	4.0
15	3.3	5.0	5.0	4.4
Mean	3.4	4.0	4.4	4.0
	3rd harvest period (8 – 8 – 2014)			
5	2.0	3.0	2.3	2.4
10	2.3	2.0	2.7	2.3
15	2.7	3.0	3.0	2.9
Mean	1.3	2.7	2.7	2.6
			2nd harvest period	**3rd harvest period**
$LSD_{0.05}$ between two cutting height mean (C):			NS	0.60
$LSD_{0.05}$ between two poultry manure mean (P):			0.76	0.60
$LSD_{0.05}$ between two (C) x (P):			1.32	NS

NS, Non- Significant F- test at 5% probability level

Grass tiller number per square meter was not significantly influenced by height of cutting as was reported by Wijitphan et al. [30]. This also concurred with Onyeonagu and Ugwuanyi on *Panicum maximum* [31]. Tillering in grasses is a very dynamic process, and time of sampling can strongly influence the situation encountered [32]. However, Carlassare and Karsten reported increase in grass tiller population with reduction in height of cut above ground in orchad grass (*Dactylis glomerata*) [33]. This contradicts with result observed with *Panicum maximum* in the study, though a number of studies have found that when orchard grass, a tall growing bunch – grass was grazed or cut frequently close to the ground level, its tiller density, stand persistence and productivity were limited by short grazing height regimes [34]. The present study revealed increase ($P<.05$) in grass tiller number with poultry manure application. Barbosa et al. [35] obtained a similar result with *Panicum maximum.* They reported that application of nitrogen a major component of poultry manure significantly increased grass tiller numbers. The same result was obtained in perennial ryegrass as reported by Wiliam and Asiegbu [36].

Dry matter content generally decreased with higher poultry manure application. Onyeonagu and Asiegbu revealed a decrease in dry matter content of herbage with nitrogen application compared with where no nitrogen (a major nutrient element of poultry manure) was applied in *Panicum maximum* pasture [37]. Dry matter content of forage is the fraction or remains after all moisture content has been removed. This contains the nutritional components of energy, protein, fibre, minerals and vitamins. Fresh pasture has high water content and will have a lower percentage of dry matter than an equivalent weight of dryer feed. However, cutting height had no effect on dry matter content.

Weed cover were found to decrease with higher application of poultry manure and vice versa for grass cover. This could be attributed to nitrogen fertilizer application effect in stimulating greater growth of desired species and denser vegetative cover than broad–leaved weeds thus limiting weed invasion [37].

5. CONCLUSION

The higher plant height observed with the 15 cm cutting height compared with 5 or 10 cm height indicates an advantage in a 15 cm cutting height. Grass cover, tiller number, as well as plant height were generally increased with application of poultry manure. Increase in poultry manure application resulted to decrease in the dry matter content of grass and bare ground area. A combination of 10 cm cutting height with 10 ton ha^{-1} poultry manure significantly produced greater tiller population per m^2. The 10 ton ha^{-1} manure combined with either the 10 cm or 15 cm cutting significantly increased the extent of grass cover compared with where the 0 ton ha^{-1} was combined with the 5 cm cutting.

ACKNOWLEDGEMENTS

We humbly appreciate the breathtaking effort of all that contributed immensely to the successful completion of this work.

COMPETING INTERESTS

Authors have declared that no competing interests exist.

REFERENCES

1. Cook BG, Pengelly BC, Brown SD, Donnelly JL, Eagles DA, Franco MA, Hanson J, Mullen BF, Partridge IJ, Peters M, Schultze – Kraft R. Tropical Forages: An interactive selection Tool; {CD-ROM}, CSIRO, DPI&F (Qld), CIAT and ILRI, Brisbane, Australia; 2005. Accessed 16 June 2014. Available:http://www.tropicalforages.info/../Html/
2. Motta MS. *Panicum maximum*. Empire Journal of Experimental Agriculture. 1953; 21:33–41.
3. Abbah T. Fulani / farmers clashes claim 300 lives in 5 months; Sunday trust newspaper, Category: News Published on Sunday; 2013.
4. Yammama K. Normad – farmer clashes increases as pasture shrinks. IRIN Humanitarian News and Analysis, Nigeria; 2009. Accessed on 14 February 2014. Available:http://www.irinnews.org/report/84761/nigeria-normad-farmer-clashes-increase-as-pasture-shrinks
5. Dev I. Problems and Prospects of Forage production and utilization of Indian Himalaya. ENVIS Bulletin: Himalayan Ecology and Development. 2001;9(2):1–13.
6. Bamikole MA, Akinsoyinu AO, Ezenwa I, Babayemi OJ, Akinlade J, Adewumi K. Effect of six-weekly harvests on the yield, Chemical Composition and Dry Matter Degradability of *Panicum Maximum* and *Stylosanthes hamata* in Nigeria. Grass and Forage Science. 2004;59(4):357–363.
7. Omaliko CPE. Stock piling of three tropical forage grass species. Agronomy Journal. 1982;75(2):677–979.
8. Mathew CG, Lemaire NR, Sackville AH, Hernandez G. A modified self thining equation to describe size / density relationships for defoliated swards. Annals of Botany. 1995;76;579–587.
9. Waller S, Moser L, Reece. Understanding Grass Growth: The key to profitable livestock production. Trabon printing Co., Inc. Kansas city, Missouri; 1985.
10. David BH, Kimberly JH, Patti S, Sara G, Craig G, Merle T. Grass growth and regrowth for Improved management. Oregon state university; 2000. Accessed on 14 February 2014. Available:http://www.fsl.orst.edu/../regrowth/
11. Neil G. Best practice guideline for using poultry litter on pasture. New South Wales, Department of Primary Industries; 2007. Accessed on 09 March 2014. Available:http://www.dpi.nsw.gov.au
12. Obi ME, Ebo PO. The effect of different management practices on the soil physical properties and maize production in

severely degraded soil in southern Nigeria. Biological resource technology. 1995;5: 117–123.
13. Ojeniyi SO. Effect of goat manure on soil nutrients and okra yield in a rain forest area of Nigeria. Applied Tropical Agriculture. 2010;5:20–30.
14. Ayoola OT, Adeniyan ON. Influence of poultry manure and NPK fertilizer on yield and yield compoment of crops under different cropping systems in south west Nigeria. African Journal of Biotechnology. 2008;5(15):1386-1392.
15. Agboola AA, Omueti JA. Soil fertility problem and its management in tropical Africa. Paper Presented at the International Institute of Tropical Agriculture Ibadan, Nigeria.1982;25.
16. Olatunji OO, Aderinola OA, Babajide PA, Ewetola EA, Lawal BA, Owoade FM, Oyeyiola YB, Olayiwola AO. Effect of poultry manure on soil physio chemical properties, aggregate stability and biomass yield of *Panicum maximum.* International Journal of Current Reseaech. 2012;5(4): 13–16.
17. Rahman SA. The place of organic manure in sustaining agricultural development in Nigeria. Paper presented at science technology and society national workshop in Lafia, Nasarawa State; 2004.
18. Schjegel AJ. Effect of composted manure on soil chemical properties and nitrogen use by grain sorghum. Journal of Production Agriculture. 1992;5:153–157.
19. Warman PR. The effect of fertilizer, chicken manure and dairy manure on Timothy yield, tissue composition and soil fertility. Agricultural Wastes. 1986;18:289–298.
20. Duncan J. Composting chicken manure. WSU cooperative extension, King county master gardener and cooperative extension livestock advisor, Washington State, Pullman; 2005.
21. Oagile D, Namasiku M. Chicken manure enhanced soil fertility and productivity: Effects of application rates. Journal of Soil Science and Environmental Management. 2010;1(3)46–54.
22. Feisal MI, Awad OA, Ahmed ME. Influence of chicken manure on growth and yield of forage sorghum (*Sorghum bicolor* L. Moench). International Journal of Agriculture and Forestry. 2012;2(2):56–60.
23. Ahmed M, Tswanya MN, Nda SL, Yisa PZ. Effect of varieties and organic manure on the vegetative growth and yield of tomato (*Lycopersicon lycopersicum* MILL.). First National Conference of the Crop Science Society of Nigeria (CSSN) – Nsukka. 2013; 9-12.
24. Asiegbu JE. Response of onion to lime and fertilizer N in a tropical ultisol. Tropical agriculture (Trinidad). 1989;(2)66:161-166.
25. Snedecor GW, Cochram WG. Statistical method 6th edition, Iowa State University Press America. 1967;246.
26. Genestat Release 10.3 DE, Discovery Third Edition, Lawes Agricultural Trust Rothamsted Experimental Station; 2008.
27. Obi IU. Statistical method for detecting differences between treatment means. SNAAP press limited, Enugu, Nigeria. 1986;45.
28. Adams BW, Ehlert G, Robertson A. Grazing System For Public Grazing Lands. Range notes no 10 alberta forestry lands division. Leth-Brdge, Alberta. 1991;1-8.
29. Farhad W, Saleem MF, Cheema, Hammad HM. Effect of poultry manure levels on the productivity of spring maize (*Zea mays* L). The Journal of Animal and Plant Sciences. 2009;19(3):122–125.
30. Wijitphan S, Lorwilai P, Arkaseang C. Effect of cutting heights on productivity and quality of King Napier grass (*Pennisetum purpureum* cv. King Grass) under irrigation. Pakistan Journal of Nutrition. 2009;8:1244–1250.
31. Onyeonagu CC, Ugwuanyi BN. Influence of cutting height and nitrogen fertilizer on plant height and tiller production of guinea grass (*Panicum maximum* Jacq) pasture. African Journal of Agricultural Research. 2012;7(48):6401–6409.
32. Gutman M, Noy–Meir I, Pluda D, Seligman NA, Rothman S, Sternberg M. Biomass Partitioning following defoliation of annual and perennial Mediterranean grasses. 2001; In: Onyeonagu CC. and Ugwuanyi BN. Influence of cutting height and nitrogen fertilizer on plant height and tiller production of guinea grass (*Panicum maximum* Jacq) pasture. African Journal of Agricultural Research. 2012;7(48):6401–6409.
33. Carlassare M, Karsten HD. Species population dynamics in a mixed pasture under two rotational sward grazing height

regimes. Agronomy Journal. 2013;95:844–854.
34. Murphy WM, Silman JP, Mccroy LE, Flack SE, Schmitt AL, Mzamane NM. Management of neutral kenttuchy bluegrass-white clover pasture. American Journal of Alternative Agriculture. 1997; 12:140-142.
35. Barbosa MA, Cecato U, Onorato VM, Coellio EM. Estudo de perfithamento de capim mombaca (*Panicum maximum*). In: reuniao annual da sociedade brasileira de zootecnia 34. Juiz de for a, Anias ..... Juiz De for A: SBZ. 1997;114.
36. Wilman D. Asiegbu JE. The effect of clover variety, cutting interval and nitrogen application on herbage yields, proportions and heights in perennial ryegrass – white clover swards. Grass and Forage Science. 1982;37:1-13.
37. Onyeonagu CC. Asiegbu JE. Effects of cutting management and N fertilizer application on plant height, tiller production and percentage dry matter in a run-down *Panicum maximum* pasture. Agro-Science. 2005;4(1):74–77.

18

Field Evaluation of Levo Botanical Insecticide for the Management of Insect Pests of Eggplant (*Solanum melongena* L.)

John Peter Nyaaba Aetiba[1] and Enoch Adjei Osekre[1*]

[1]Department of Crop and Soil Sciences, Kwame Nkrumah University of Science and Technology, Kumasi, Ghana.

Authors' contributions

This work was carried out in collaboration between both authors. Author JPNA performed the experiment, collected data, performed the analyses and did the write-up. Author EAO designed the experiment, guided data collection, analyses and write up. Both authors read and approved the final manuscript.

Editor(s):
(1) Marcelo Daniel Preite, Department of Organic Chemistry, Pontifical Catholic University of Chile, Chile.
Reviewers:
(1) Mohammed Suleiman, Department of Biology, Umaru Musa Yar'adua University, Katsina, Nigeria.
(2) Anonymous, South Africa.

ABSTRACT

Renewed interest in the use of botanical insecticides for the management of insect pests necessitated field experiments to be carried out during the major and minor cropping seasons in 2013 at the plantation crops section of the Department of Crop and Soil Sciences of the Kwame Nkrumah University of Science and Technology (KNUST), Ghana to evaluate the insecticidal potency of Levo botanical insecticide for the management of insect pests of eggplant (*Solanum melongena* L.). The study comprised the following treatments: (i) Levo (a.i. Oxymatrine) at 1.68 ml /0.5 litre water and (ii) Lambda-super (a.i. Lambda-cyhalothrin) (a check) at 1.5 ml /0.5 litre water; an untreated control (water only) was also maintained. *Leuninodes orbonalis* (Guen), *Bemisia tabaci* (Gennadius), *Aphis gossypii* (Glover), and *Eublemma olivacea* (Walker) were collected on eggplant in the study area. Significant differences (P<0.5) were observed among the treatments with respect to the abundance of *A. gosspyii*, L. *orbonalis, B. tabaci* and *E. olivacea* on eggplant during the major season. Similar results were obtained in the minor season. Significantly higher yields were obtained from the insecticide-treated eggplant plots. The study showed that Levo was

**Corresponding author: E-mail: osek652001@yahoo.co.uk;*

as effective as Lambda super and can be a substitute in the management of insect pests of eggplant.

Keywords: Botanical; insecticide; eggplant; insect pests.

1. INTRODUCTION

Vegetables are important sources of vitamin, minerals and plant proteins in human diets throughout the world [1] and are rapidly becoming an important source of income for rural population [2]. Eggplant (*Solanum melongena* L.) is a popular vegetable crop grown in the subtropics and tropics [3]. According to [4], more than two million hectares are devoted to the cultivation of eggplant in the world. In Ghana it is one of the most important vegetable crops extensively cultivated especially in the forest zone, during both the major and minor seasons [5]. Eggplant is a potential export crop for Ghana, but only a very small share of the total production in Ghana is exported. However, its exports are on the ascendency. Exports of vegetables, including eggplant, have increased from 2,033 Mt in 2010 to 3,695 Mt by 2010 [6].

Despite its importance in terms of nutrition and export potential for foreign exchange, increasing damage by arthropod pest is affecting eggplant cultivation [2]. Eggplant is infested by a plethora of insect pests throughout the world. A survey of vegetable pests conducted by the world vegetable center (AVRDC) indicated *Leuninodes orbonalis* (Guen) is the most destructive pest in most major eggplant producing countries [2]. Other insect pests that severely constrain eggplant production in the tropics are; leaf lopper, whiteflies, thrips, aphid, spotted beetles, leaf rollers, stem borers and blister beetles [1].

The application of synthetic insecticides is the primary control strategy against insect pests of the crop. For instance, farmers in certain areas of Philippines spray chemical insecticides up to 56 times during a cropping season; the total quantity of pesticide used per ha of eggplant was about 41 litres of different brands belonging to the four major pesticide groups [7,8]. In Bangladesh, some farmers spray about 180 times during a cropping season [9].

Although synthetic pesticides application remains the primary agricultural pest control strategy, it is evident that society cannot continue to tolerate their harmful effects on the environment and non-target organisms. One way to manage this menace is to develop pest management systems that are based on judicious application of synthetic insecticides. Thus there is the urgent need for the development of alternative control strategies [10].

Recently, studies have been intensified on the use of naturally occurring pesticides for pest control. Many investigators isolated, identified and screened chemical compounds from leaves and seeds of many botanical families for insect deterrence and growth inhibition. Some of the anti-pest plants documented included Neem, Chrysanthemum, Annona, Mahogoni, Albizziaetc [11].

According to [12], botanicals are one of the groups of safe insecticides which have a broad spectrum of anti-pest activity, relatively to specific mode of action, low mammalian toxicity and more tendency to disintegrate, in nature or metabolic in a biological system. Moreover, their preparation and application at farm level are more convenient for the farmers and are quite incorporable into integrated pest management programs. It is based on this that Levo (a. i. Oxymatrine) botanical was evaluated for its insecticidal activity against insect pests of eggplant.The specific objectives were to(i) identify the major insect pest species affecting eggplant in the study area, (ii) determine the effect of Levo botanical insecticide on the incidence of insect pests of eggplant, (iii) determine the effect of the insecticide on damage caused by the insects to eggplant, and (iv) determine the effect of the insecticide on fruit yield of eggplant.

2. MATERIALS AND METHODS

2.1 Experimental Site

The study was conducted in both major and minor crop growing seasons at the Faculty of Agriculture plantation site of the Kwame Nkrumah University of Science and Technology (KNUST), Kumasi, Ghana. The major growing season lasted June to September, 2013 while the minor season spanned late September to November 2013.

2.2 Source of Seeds

The variety of eggplant used was the local "*Aworoworo*" which was obtained from the Crops Research Institute of the Council for Scientific and Industrial Research (CSIR-CRI), Kwadaso, Kumasi, Ghana. This variety is improved but susceptible to insect pests and diseases.

2.3 Field Layout and Experimental Design

The field was laid out in a randomized complete block design (RCBD) with three treatments. Each plot measured 5 m x 5 m with inter and intra-row spacing of 1 m x 0.5 m. The total plot size was 468.75 m^2.A 1.5 m alley was allowed between the plots so as to minimize spray drifts to adjacent plots.

2.3.1 Treatments

Three treatments, each with four replications were used.

(i) Levo (a.i. Oxymatrine) at 1.68 ml / 0.5 litre water (recommended dose)
(ii) Lambda super (a.i. Lambda-cyhalothrin) at 1.5 ml / 0.5 litre water (recommended dose)
(iii) Control (water only)

2.4 Transplanting of Seedlings

Eggplant seedlings were transplanted on 11th July, 2013 and 9th October, 2013 during the major and minor seasons, respectively at one seedling per stand. Replacement of dead seedlings was carried out a week after transplanting. Each plot contained five rows with 10 plants per row, giving a total of 50 plants per plot for both crops.

2.5 Sampling for Insect Pest Infestation on Eggplant Fields

Sampling for insect pests was done weekly between 0800 – 1100 h when the insects' build-up was high. Sampling was done on five plants randomly selected from the three middle rows per plot. For *B. tabaci*, sampling involved visual examination of each plant and the number of insects on two leaves were recorded. This was done for the first two weeks of sampling but thereafter three leaves were examined with the aid of a magnifying lens. For the other insects, namely *A. gossypii*, *E. olivacea*, and *L. orbonalis*, three leaves from both the upper and lower canopies were collected into high density polyethylene bottles containing 70% ethanol. These were transported to the insectary for processing and identification using a stereo microscope. Insecticide application was done after sampling.

2.6 Cultural Practices on Eggplant and Okra Fields

The normal agronomic practices (e. g. weeding, irrigation, fertilization) were carried out.Manual weeding was done at three weeks interval. The field was irrigated as and when necessary since rainfall was erratic. Fertlllzer was applied in two splits. The first dose of N: P: K 15: 15: 15 was applied three weeks after transplanting of eggplant at a rate of 10/g per plant while Urea (46% N) at 2.2 g per plant was applied three weeks later.

2.7 Estimation of Fruit Yield

Eggplant fruits were harvested every three or four days when they reached maturity and then weighed using a Switzerland-made Metler Toledo PB302 electronic weighing scale in the laboratory. The results obtained for each treatment were then extrapolated to kilograms per hectare (kg/ha) using the formula:

$$\text{Fruit yield/ha} = \frac{1000}{\text{area harvested}} \times \text{fruit yield/plot [13].}$$

2.8 Yield and Yield Components Assessments

The following data were also recorded; number of fruits per plant and weight of fruits per plant and yield. Damaged fruits included eggplant fruits that had injuries or blemishes apparently caused by insects.

2.9 Data Analysis

Insect and other count data were transformed using square root transformation and percentages by arcsine transformation and then subjected to Analysis of Variance (ANOVA) using statistix software, version (9.0). The treatment means were separated using Tukey at 5% probability.

3. RESULTS

3.1 Insect Pests Collected on Eggplant in the Major Season in 2013

There were significant differences among treatments with respect to abundance of *B. tabaci* and *A. gossypii* (Table 1). There was also significant difference between insecticide-treated plots and the control in the densities of *L. orbonalis* and *E. olivacea*.

3.2 Insect Pests Collected in the Minor Season on Eggplant in 2013

The effect of the different treatments on the abundance of insect pests on eggplant in the minor season is presented in (Table 2). There were significant differences among treatments with respect to densities of *A. gosspyii*, *B. tabaci*, *L. orbonalis*, and *E. olivacea*.

3.3 Yield of Eggplant as Affected by Various Insecticide Treatments in the Major and Minor Seasons in 2013

Significantly more fruits per plant and fruit yield were obtained in the Levo-treated plots than the Lambda super and control plots in the major season (Table 3). The mean fruit weight of the fruits from the Levo-treated plots was significantly more than that from the Lambda Super and control plots. Significantly more damaged fruits were recorded from the Lambda Super and control plots than the Levo-treated plots. In the minor season, there were no significant differences among the treatments with respect to the mean number of fruits per plant and the fruit weight. Significantly more fruit yield was obtained from the insecticide-treated plots than the control plots. However, significantly more fruits from the control plots were damaged than from the insecticide-treated plots (Table 4).

Table 1. Mean number of insect pests collected on eggplant (*Solanum melongena* L) as affected by insecticides treatment in the major season in Kumasi, Ghana in 2013

	Mean number (± SEM) of insect per leaf			
Treatment	***A. gossypii***	***B. tabaci***	***L. orbonalis***	***E. olivacea***
Levo	1.17±0.07 c	1.67±0.05 c	0.88±0.28 b	0.80±0.02 b
Lambda Super	2.75±0.21 b	2.29±0.09 b	0.80±0.21b	0.85±0.25 b
Control	5.32±0.32 a	2.97±0.12 a	1.39±0.45 a	1.39±0.04 a

Means with the same letter in a column are not significantly different from each other (P< .05, Tukey test)

Table 2. Mean number of insect pests collected on eggplant (*Solanum melongena* L.) as affected by insecticides treatments in the minor season in Kumasi, Ghana in 2013

	Mean number (± SEM) of insect per leaf			
Treatment	***A. gossypii***	***B. tabaci***	***L. orbonalis***	***E. olivacea***
Levo	1.14±0.61 c	1.61±0.04 c	0.05±0.03 c	0.89±0.03 c
Lambda Super	1.88±0.11 b	2.22±0.04 b	1.32±0.05 b	1.06±0.04 b
Control	3.45±0.16 a	2.67±0.05 a	1.50±0.05 a	1.50±0.05 a

Means with the same letter in a column are not significantly different from each other (P<.05, Tukey test)

Table 3. Yield, yield components and mean damaged fruit of eggplant (*Solanum melongena* L.) as affected by various treatments in the major season in Kumasi, Ghana in 2013

Treatment	Mean no. of fruits plant $^{-1}$	Mean % damaged fruits	Mean fruit weight (g)	Mean yield (kg ha $^{-1}$)
Levo	56.50±5.04 a	2.55±0.95 b	2.44±0.27 a	997.38±109.06 a
Lambda super	42.50±1.32 b	7.75±0.63 a	1.61±0.05 b	643.73±21.55 b
Control	40.00±2.31 b	9.50± 0.50 a	1.41±0.06 b	562.60±25.35 b

Means with the same letter in a column are not significantly different from each other (P<0.05, Tukey test)

Table 4. Yield, yield components and mean damaged fruits of eggplant (*Solanum melongena* L.) as affected by various treatments in the minor season in Kumasi, Ghana in 2013

Treatment	Mean no. of fruits plant $^{-1}$	Mean % damaged fruits	Mean fruit weight (g)	Mean yield (kg ha $^{-1}$)
Levo	71.75±7.93 [a]	7.00±0.82 [b]	2.77±0.50 [a]	1107.30±198.79 [a]
Lambda Super	71.75±9.84 [a]	9.25±1.25 [b]	2.44±0.35 [a]	974.00±139.17 [a]
Control	80.00±6.87 [a]	19.00±2.55 [a]	1.60±0.17 [a]	638.25±67.60 [b]

Means with the same letter in a column are not significantly different from each other (P<0.05, tukey test)

4. DISCUSSION

4.1 Effect of Insecticide Treatments on Insect Pests of Eggplant

Although synthetic pesticides are target specific and effective, their effects on the environment are mostly deleterious. Plant-based pesticides contain active ingredients with low half-life period and their effects on the environment are not too detrimental making them more acceptable for pest management [14].

On eggplant, the results from the studies in both the major and minor seasons showed that Levo was effective against *B. tabaci* by keeping the mean number to about 1.6 per leaf, comparable to that of Lambda super (Tables 1 and 2). The effectiveness of Levo against whiteflies may be attributable to its possession of anti-feeding and repelling properties. [15] tested 2 g and 5 g leaf extracts of custard apple (*Annona squamosa*) on *Tribolium castaneum* (Herbst) and reported that they were successful in controlling the infestation of *T. castaneum*, which, they explained, could be attributed to the repellent properties of the acetogenins in the leaf extracts. [16] also reported that the seed extracts of *A. squamosal* had repellent and anti-oviposition properties against *Ceratitis capitata* (Wiedemann).

The abundance of *A. gossypii* on eggplant was significantly higher in the control plots than insecticide-treated plots in both the major and minor seasons. Comparing the insecticides, lower numbers of *A. gossypii* were recorded on Levo-treated plots as compared to Lambda super (Tables 1 and 2). This could be due to the over-dependence on Lambda super by farmers within the experimental area for the control of pests which might have contributed to the insects developing resistance to the insecticide. [17] observed that *A. gossypii* populations on eggplant are becoming resistant to commonly used insecticides in Ghana.

Levo against *A. gossypii* was very effective (Tables 1 and 2). The effectiveness of botanical insecticides in controlling aphids was reported by [18], who detected that aqueous neem seed extract was very effective in reducing the number of cotton aphids infesting okra, particularly when applied as a foliar spray. According to [19], who also used neem extract to control aphids, the control of aphids resulted from the failure of nymphs to moult and from inhibition of adults to reproduce.

The results from the experiment for both crops in the major and minor seasons showed that *L. orbonalis* and *E. olivacea* densities throughout the experiment were lower compared with that recorded on the control plots. Thus Levo was as effective as lambda super in reducing *L. orbonalis* and *E. olivacea* densities on eggplant. [18] reported that lepidopterous insects are highly sensitive to neem compounds and that the order Lepidoptera, to which these insects belong, was classified as one group of the insects well controlled with many botanical insecticides. The report concluded that, the results held a promise for the management of the other members of this order, which attack okra and cotton.

All the insecticide-treated plots had increased yield of eggplant compared to the control. Levo-treated plots had yield of 997.38 kgha $^{-1}$ followed by Lambda super, 643.73 kgha $^{-1}$ and the control, 562.60 kgha $^{-1}$. The relatively higher incidence of insects might have accounted for the low yield on the untreated, control plots. Thus the significantly higher yields obtained from the insecticide-treated plots could be attributed to the marked decrease in the densities of insects collected from those plots in addition to reduced defoliation.

It is evident from the study that Levo application contributed to a yield similar to Lambda super and can therefore be a good substitute for the management of insect pests on the crop.

5. CONCLUSION

The study has shown that *A. gosspyii, B. tabaci, L. orbonalis,* and *E. olivacea,* were the most important insect pests of eggplant in the study area. The incidence and abundance of these pests were found to have been reduced in all insecticides-treated plots while yield increases were also recorded in all treated plots of eggplant. The study also showed that Levo was as effective as the Lambda Super and could be a substitute for the latter insecticide in the management of insect pests of eggplant.

COMPETING INTERESTS

Authors have declared that no competing interests exist.

REFERENCES

1. Srinivasan R. Insect and mite pests on eggplant: A field guide for indentification and management. AVRDC–The World Vegetable Center, Shanhua, Taiwan. AVRDC Publication. 2009;09:64.
2. Alam SN, Rashid MA, Rouf FMA, Jhala RC, Patel JR, Satpathy S, Shivalinga swamy TM, Rai S, Wahundeniya I, Cork A, Ammaranan C, Talekar NS. Development of an integrated pest management strategy for Eggplant fruit and shoot borer in South Asia. Technical Bulletin 28. AVRDC – The World Vegetable Center, Shanhua, Taiwan. 2003;66.
3. Owusu J. The effect of Gardona (Tetrachlovinphos) on the insect fauna of eggplant, *Solanum melongena* L. A B.Sc. (Agric.) Dissertation. Faculty of Agriculture, University of Ghana, Legon; 1980.
4. FAO. Food and Agriculture Organization of the United Nations, Rome; 2007. Last Accessed Nov. 2013.
   Available:http://faostat.fao.org/faostat/
5. Akpabi IGK. A guide to garden egg production in Ghana, In: Horticultural Crops. A report by the commodity committee of NARP, CSIR, Accra, Ghana. 1989;3:20-40.
6. Ministry of Food and Agriculture. Agriculture in Ghana: Facts and Figures. MoFA annual Report. 2010; 2011:53.
7. Gapud VP, Canapi BL. Preliminary survey of insects of onions, eggplant and string beans in San Jose, Nueva Ecija. Philippines Country Report, IPM CRSP – First Annual Report; 1994.
8. Orden MEM, Patricio MG, Canoy VV. Extent of pesticide use in vegetable production in Nueva Ecija: Empirical evidence and policy implications. Research and Development Highlights 1994, Central Luzon State University, Republic of the Philippines. 1994;196-213.
9. SUSVEG-Asia. Brinjal integrated pest management (IPM); 2007. Accessed June, 2013.
   Available:http://susvegasia.nri.org/susvegasiabrinjalipm4.html
10. Osei Owusu D. Effect of crude ethanolic leaf extract of sours op (*Annona muricata* L.) on eggplant shoot and fruit borer. MSc Thesis. Kwame Nkrumah University of Science and Technology Kumasi, Ghana; 2010.
11. Reed DK, Freedman B, Ladd TL. Insecticidal and antifeedant activity of neriifolin against codling moth, striped cuncumber beetle and Japanese beetle. Journal of Economic Entomolgy. 1982;75: 1093-1097.
12. Muhammed A. Antixenotic and antibiotic impact of botanicals for organic stored wheat insect pests, University of Agriculture Faisalabad Pakistan; 2009.
13. Asante SK, Tamo M, Jackai LEN. Integrated Management of cowpea insect pests using elite cultivars, date of planting and minimum insecticide application. Journal of Africa Crop Science. 2001; 9(4):655-665.
14. Sharma SK, Dua VK, Sharma VP. Field studies on the repellent action of neem oil. Southeast Asian. Journal of Tropical Medicine of Public Health. 1995;26:180-182.
15. Hussain MM, Ali SH, Rahim A, Mondal KAMSH. Studies on the repellent effect of two indigenous plants, biskaan tali (*Polygonum hydropiper*) and ata (*Annona squamosa*) leaf on *Tribolium castaneum* Herbst. Bangladesh Journal of Science and Industrial Research. 1995;30(1):81-85.
16. Epino PB, Chang F. Insecticidal activity of *Annona squamosal* L. seed extracts against the Mediterranean fruit fly, *Ceratitis capitata* (Diptera: Tephritidae). Philippine Entomologist. 1993;9:737-747.

17. Owusu EO. Preliminary biochemical evidence of insecticide resistance development in Ghanaian populations of cotton aphid, *Aphis gossypii* Glover (Homoptera: Aphididae).Ghana Journal of Science. 1997;37:67-70.
18. Ahmed SMS. Laboratory evaluation of some biological activity of *Ricinus communis* seed extracts against the rice weevil, *Sitophilus oryzae* (L.). Arab University Journal of Agriculture Science. 2000;8(3):853-861.
19. Lowery DT, Isman MB, Bhard NL. Laboratory and field evaluation of neem for the control of Aphids (Homoptera: Aphididae). Journal of Economic Entomology. 1993;6(3):864-870.

19

Mechanism of Nanosilicon-Mediated Alleviation of Salinity Stress in Faba Bean (*Vicia faba* L.) Plants

Amira M. S. Abdul Qados[1*]

[1]*Department of Biology, Faculty of Science, Princess Nora bint Abdul Rahman University, Saudi Arabia.*

Author's contribution

The sole author designed, analyzed and interpreted and prepared the manuscript.

Editor(s):
(1) Vladimir K. Mukhomorov, Agrophysical Institute, 14 Grazhdanskiy Ave., Russia.
(2) Mintesinot Jiru, Department of Natural Sciences, Coppin State University, USA.
Reviewers:
(1) Myrene Roselyn D'souza, Department of Chemistry (P.G. Biochemistry), Mount Carmel College, India.
(2) Anonymous, Vietnam.
(3) Anonymous, Malaysia.
(4) Anonymous, Saudi Arabia.

ABSTRACT

Aims: In the current study we try to clarify the mechanism that might be involved in the ameliorating effects of Nano-silicon (NSi) and Silicon (Si) on faba bean (*Vicia faba*) plants grown under different levels of salinity stress that we found in a previous study.
Study Design: Factorial completely randomized design Pot experiments were used with NSi and Si applied at 4 concentrations each (0, 1, 2 and 3 mM) and NaCl (0, 50, 100 and 200 mM) were studied.
Place and Duration of Study: Experiments were carried out in the greenhouse of the Experimental Station, Faculty of Science, Princess Nora Bint Abdulrahman University, Kingdom of Saudi Arabia during winter season of 2012/2013.
Methodology: Effects of NSi and Si on membrane characteristics, photosynthetic pigments, sugar content, free proline, antioxidant enzymes and mineral elements were investigated in NaCl stressed and non-stressed faba bean plants. The experiment was arranged in a factorial design with 4 replications at NaCl levels of 0, 50, 100 and 200 mM. The tested NSi and Si concentrations were 0, 1, 2 and 3 mM for each.
Results: NaCl treatments caused an increase in proline content and in some enzyme activities, Chl a, b and carotenoids were decreased. Application of NSi caused a significantly increase in the activity of ascorbate peroxidase (APX), catalase (CAT) and peroxidase (POD) in plant leaves, but

**E-mail: Amira_m.s@hotmail.com;*

caused a decrease in the activity of superoxide dismutase (SOD) as compared to unstressed plants. Oxidative damage, produced by salinity stress, seemed to decrease in accordance with the increase in antioxidant enzymes activity under NSi and Si treatments, thus tolerance against salt stress was observed. The improvement of salt tolerance resulted from NSi and Si treatments was accompanied with improved membrane stability, chloroplast formation and sugar accumulation.

Conclusion: Nano-silicon treatments can reduce the adverse effects of salinity on *V. faba* plants by enhancing the activity of antioxidant enzymes.

Keywords: Nanosilicon; Vicia faba; antioxidant enzymes; salt tolerance.

1. INTRODUCTION

Faba bean (*Vicia faba* L.) is cultivated in the many regions of the world as a source of food for humans and animals. It is also traditionally used as a cover crop to recover nitrogen content and prevent erosion of the soil, and is appreciated for its good agronomic characteristics [1]. The high percentages of protein, carbohydrate, minerals and vitamins in faba bean seeds make it an important source of nutrition in many countries. Being rich in calcium, iron, thiamin, niacin tocopherols, and folic acid faba bean is recommended to strengthen the immune system [2]. In addition, faba bean plants increase the concentration of nitrogenous compounds in the soil and make it more fertile [3]. However, growth and metabolism of faba bean plants are affected by many abiotic stresses such as salinity and water deficit.

Salinity is considered one of the main abiotic stresses which reduce growth and production of many crops [4]. There is a large area of the world adversely affected by high salt concentrations that have a negative impact on the yield of cultivated crops [5]. Therefore, most plants growing in saline conditions have various strategies to allow them to overcome salinity stress. The deleterious effect of salinity on plant growth is attributed to osmotic effect that makes it difficult for a plant to absorb water from saline soils, and/or to ionic effect resulting from accumulation of toxic salt ions to a level at which plants cannot grow well [6]. Both of these components, osmotic and ionic effects, are harmful to growing plants because they can cause changes in cell membrane characteristics, water status, enzyme activities, protein synthesis and gene expression [7]. In addition, reactive oxygen species (ROS) such as superoxide, hydrogen peroxide and hydroxyl radicals, can be produced in saline conditions to a high levels and accumulate in the chloroplasts and other organelles, leading to a disruption in the metabolism within plant cells through lipid peroxidation, protein oxidation and enzyme inhibition [8].

In order to overcome the harmful effect of ROS, produced at salt stress conditions, an efficient antioxidant system, including antioxidant enzymes, should be developed in plant cells. Among these antioxidants are the following enzymes: catalase, peroxidase superoxide dismutase and glutathione reductase [9]. In a previous study by Hernandez et al. [10] they found that salt stress conditions stimulated the stressed plants to activate their antioxidant enzymes in order to endure salt effect. It was found that ROS could harm the photosynthetic apparatus through the effect on chlorophyll structure, plasma membranes and protein formation [11,12]. The accumulation of compatible solutes such as sugars, amino acids, proline and other soluble compounds permits plants to maintain cell turgor pressure necessary for cell expansion during stress conditions; such compounds act usually as osmo-protectants [13]. Proline, in particular, is considered on of the protectants because it has the ablility to scavenge ROS and other free radical compounds, thereby ensuring membrane stabilization and preventing protein denaturation during severe osmotic stress [14]. The protective role of proline in plants under abiotic stress such as salinity or water deficit has been reported in several plant species [15].

The properties of nano-materials and their potential applications are increasing in various fields, especially agricultural biotechnology [16]. However, the investigations on behavior of nanometals such as nanosilica in plants and the mechanism of interaction, its influence, and agricultural application are still under rudimentary stage [17]. Based on current knowledge, Liang et al. [6] concluded that Si is not inert, but acts as a physical or mechanical barrier in plants. It is not only deposited in the cell wall, but is also actively involved in the metabolic and/or physiological activities, especially in plants subject to environmental stresses [6]. Generally, the ameliorative effect of bulk silicon sources varies genotypically among plant species and particle size [18]. Si (Atom No =14, Atom mass = 28.09, MP 1,410°C, density 2.33 g cm^{-3}) is a metalloid, exists in two forms either crystal plate or non crystal brown powder. Si is a relatively inactive element at room temperature. At higher temperatures, however, silicon becomes much more reactive and it combines with oxygen, nitrogen, sulfur, phosphorus, and other elements.

Nanomaterials, because of their tiny size, may show unique characteristics. For example, they can change physico-chemical properties compared to bulk materials. They have greater surface area than bulk materials, and due to this larger surface area, their solubility and surface reactivity tend to be higher. Si nano-particles were found to increase the growth of different species i.e., maize (16), rice (17) and *V. faba* (19) plants. Therefore, we focus on the effect of nanosilica and its influence on faba bean physiology and metabolic defense compounds. Our previous study [19] revealed improved growth parameters and germination attributes in faba bean plants treated with Si and NSi, but the physiological and biochemical responses to Si and NSi were not well studied. Therefore, the current study was established to examine the effect of nanosilicon (NSi) and silicon particles (Si), as a promising plant development regulatory substance to increase the salt tolerance of faba bean plants grown under salinity stress, and to clarify the mechanism that might be involved in the ameliorating effects of NSi and Si on faba bean plants grown under salt stress conditions.

2. MATERIALS AND METHODS

Pot experiments were carried out in the greenhouse during winter season to investigate the mechanisms that might be involved in the ameliorating effects of NSi and Si on faba bean plants grown under salinity stress conditions. The effect of NSi and Si on on characteristics of cell membrane, chlorophyll a and b, carotenoids, carbohydrates, antioxidant enzymes, free proline and mineral elements were investigated in NaCl stressed and non-stressed faba bean plants. The experiments were arranged in a factorial design with 4 replications at NaCl levels of 0, 50, 100 and 200 mM. The tested Si and NSi concentrations were 0, 1, 2 and 3 mM.

2.1 Preparing and Sowing of Seeds

Seeds of faba bean (*Vicia faba* L.) were obtained from authorized agriculture company and were sterilized with 10% sodium hypochlorite solution for 10 minutes, washed three times with distilled water, and coated with N-fixing bacteria (*Rhizobium leguminosarum*) using Tween 20 agent as an adhesive and scattering material. Identical seeds were then sown in plastic pots (30 cm inner diameter) filled with 10 kg sandy soil. Physical and chemical properties of the soil used in the study were recorded in Table 1. After sowing, irrigation was applied to supply seedlings with 100% available water, at two-day intervals until the seedlings reached the third leaf stage. Seedlings were then thinned to 3 plants/pot and pots were divided into four main groups for salt-stress treatments, with each group divided into seven subgroups for Si and NSi foliar application. Plants were fertilized with Sangeral complete fertilizer (Sinclair Horticulture LTD, England) in two equal portions; the first during the seedling stage and the second at the start of flowering stage.

At the fourth leaf stage, irrigation solutions containing one of four levels of NaCl (00, 50, 100 or 200 mM) were used and were repeated after two weeks. To avoid osmotic shock due to high concentrations, plants were started with lower salt concentrations, then, salt concentration was increased on a daily basis until each group reached the concentration determined for it. Plants were then irrigated with tap water every three days with the addition of sodium chloride to the irrigation water every two weeks (each pot received 300 ml). Each pot was washed with 500 ml distilled water 2 days before the irrigation with saline solutions to prevent the increase in osmotic potential resulting from the accumulation of salts by the succession of irrigation procedures. Sodium silicate (Na_2SiO_3), as source of Si, and silicon nano-particles (NSi) were purchased from Sigma-Aldrich Company, St. Louis, MO, USA with a purity of 99.5%, through SOMATCO chemical company in Riyadh, Saudi Arabia. Just at the end of each salt treatments, foliar application of Si (35 µm), or NSi (40 nm) in concentrations of 0, 1, 2, 3 mM was performed using a small pressure pump (60 psi) after adding Tween 20 (0.5%) as a wetting agent. The nano particles or Si were suspended directly in distilled water and dispersed by ultrasonic vibration (100W, 40 kHz) for 30 min. Small magnetic bars were placed in the suspension for stirring before use to avoid aggregation of the particles. Different doses of particles suspensions were prepared as (mM). Each plant received 100 mL of deionized water as well as the solutions of Si and NSi. The experiment consisted of 28 treatments (4 salt treatments with 3 Si treatments, 3 NSi treatments and water as control) and arranged in a factorial completely randomized design with 4 replicates for each treatment to make a sum of 112 pots. Si treatments were represented by Si1, Si2 and Si3; while NSi treatments were represented by NSi1, NSi2 and NSi3.

2.2 Measurements

Three weeks after the second application of Si or NSi (about 100 days after germination), three replicates were taken from each treatment, and the following parameters were measured:

2.2.1 Determination of membrane characteristics

Lipid peroxidation (LP), electrolyte leakage (EL), and membrane stability index (MSI) were determined as followes:

2.2.1.1 Lipid peroxidation

Lipid peroxidation was determined by measuring the amount of MDA according to Unyayar et al. [20]. About 0.5 g of leaf tissues from control and treated groups were cut into small pieces and homogenized by the addition of 5 ml of 5% trichloroacetic acid (TCA) solution (using UP200 Ultra homogenizer; Hielscher Ult. Tech., Germany). The homogenates were then transferred into fresh tubes and centrifuged at 12,000 rpm for 15 min at room temperature. Equal volumes of supernatant and 0.5% thiobarbituric acid (TBA) in 20% TCA solution were added into a new tube and boiled at 96°C for 25 min. The tubes were transferred into ice-bath and then centrifuged at 10,000 rpm for 5 min. The absorbance of the supernatant was measured at 532 nm and corrected for non-specific turbidity by subtracting the absorbance at 600 nm, 0.5% TBA in 20% TCA solution was used as the blank. MDA contents were calculated using the extinction coefficient of 155 M^{-1} cm^{-1}. Values of MDA contents were taken from measurements of three independent samples, and SD of the means were calculated.

2.2.1.2 Electrolyte leakage

Ion leakage was determined as electrical conductivity (EC%) according to Hassanein et al. [2]. Leaf samples were cut into discs of uniform size and placed in 10 ml of double-distilled water at 40°C for 30 min, and its conductivity recorded (C1) using conductivity meter (Jenway 470 portable conductivity meter). Then it was kept in a boiling water bath (100°C) for 15 min and its conductivity also recorded (C2). The percentage of electrolyte leakage was calculated according to this formula: EC (%) = (C1/C2) × 100. Where C1 and C2 are the electrolyte conductivities measured before and after boiling, respectively.

2.2.1.3 Membrane stability index

The membrane stability index (MSI) was determined by placing certain weight (200 mg) of leaf fresh material in 10 ml deionized water in test tubes. Two sets of the test tubes are prepared. One set was heated at 40°C for 30 min in a water bath and, then, the electrical conductivity (C1) was measured using a conductivity meter (ME977-C, Max Electronics, India). The second set of the tubes was boiled at 100°C in a boiling water bath for 10 min and then, the conductivity (C2) was measured again. The MSI value was calculated using the following equation as described by Premchandra et al. [21]:

$$MSI = [1 - (C1/C2)] \times 100$$

2.2.2 Determination of photosynthetic pigments

The photosynthetic pigments (chlorophyll a, chlorophyll b and carotenoids) were determined spectrophotometrically according to Metzner et al. [22]. A known fresh weight (500 mg) of leaves was homogenized in 85% aqueous acetone for 5 min. The homogenate was centrifuged and the supernatant was made up to known volume with 85% acetone and measured against a blank of pure 85% aqueous acetone at 3 wavelengths of 452.5, 644 and 663 nm using spectrophotometrically (Shimadzu, RF-5301PC, Japan). Taking into consideration the dilution made, it was possible to determine the concentrations of the pigment fractions (chlorophyll a, chlorophyll b and carotenoids) as *g* /ml using the following equations:

$$\text{Chlorophyll a} = 10.3\ E\ 663 - 0.918\ E\ 644$$
$$\text{Chlorophyll b} = 19.7\ E\ 644 - 3.87\ E\ 663$$
$$\text{Carotenoids} = 4.2\ E\ 452.5 - (0.0264\ \text{chl a} + 0.426\ \text{chl b}).$$

Pigments then were calculated on the bases of mg/g fwt.

2.2.3 Determination of soluble sugars

One gram of the dried leaf materials was homogenized with 80% ethanol then placed in a boiling water bath for 15 minutes. After cooling, the extract was filtered and the filtrate was oven dried at 60ºC then dissolved in 50 ml of distilled water to be ready for determination of soluble sugars. Soluble sugars were then determined by the anthrone sulfuric acid method described by Scott and Melvin [23]. Briefly, One ml of the extract was mixed with 9 ml of anthrone sulphuric acid reagent in a test tube and heated for 7 min at 100°C. The absorbance was read spectrophotometrically (Shimadzu, RF-5301PC, Japan) at 620 nm, against a blank containing only distilled water and anthrone reagent. All data were calculated as mg 100 g^{-1} Dwt of leaves.

2.2.4 Determination of free proline

Free proline content was determined colorimetrically in aqueous sulfosalicylic acid as described by Bates et al. [24]. Briefly, lyophilized plant material (0.1 g) was homogenized in 10 ml of 3% aqueous sulfosalicylic acid and the homogenate filtered through Whatman #2 filter paper. Two ml filtrate was reacted with 2 ml acid-ninhydrin ($C_9H_6O_4$) and 2 ml of glacial acetic acid ($C_2H_4O_2$) in a test tube for 1 h at 100°C, and the reaction terminated in an ice bath. The reaction mixture was extracted with 4 ml toluene, mixed vigorously with a test tube stirrer for 15–20 s. After 1 h, toluene was added and absorbance at 520 nm was measured by using spectrophotometer (Shimadzu, RF-5301PC, Japan). The standard curve for proline was prepared by dissolving proline in 3% sulfosalicylic acid to cover the concentration range 0.5–10 μg ml^{-1}. The proline concentration of the extract was determined from the standard curve and calculated on a dry weight basis.

2.2.5 Determination of antioxidant enzyme activities

2.2.5.1 Enzyme extraction

The samples were prepared as described by Mukherjee and Choudhuri [25]. Fresh leaf samples were submersed for 5 min in liquid nitrogen. The frozen leaves were kept at -80°C for further analyses. Enzymes were extracted by grinding weight of 500 mg fresh leaf materials, using a mortar and pestle, in 5 ml of extraction buffer solution containing 50 mM potassium phosphate buffer (pH 7.6) and 0.1 mM Na-EDTA. The homogenate was centrifuged at 15,000 *g* for 15 min and the supernatant fraction was used to assay for the various enzymes. All steps of enzyme extraction and preparation were performed at 4°C.

2.2.5.2 Superoxide dismutase (SOD)

SOD was assayed using tetrazolium reagent according to Karanlık [26]. The method depends on measuring the absorbance of the nitro-blue tertazolium (NBT) color at 560 nm wavelength. One unit of SOD activity was defined as the amount of enzyme which causes 50% inhibition of the photochemical reduction of NBT.

2.2.5.3 Catalase (CAT)

Catalase activity was determined spectrophotometry by monitoring the disappearance of H_2O_2 according to the method of Cakmak and Marschner [27].

2.2.5.4 Ascorbate peroxidase (APX)

The activity of ascorbic acid peroxidase was determined spectrophotometry by measuring the absorbance at 290 nm and estimating the

Table 1. Physical and chemical analyses of the soil used in the experiment

Physical properties		**Chemical properties**			
Particle size distribution:		**CaCO3 (%)**	**0.41**	**Soluble anions (meq/l)**	
Sand (%)	92.3	OM (%)	0.26	CO32-	0.22
Silt (%)	6.2	ES (dSm^{-1})	0.53	HCO3-	0.86
Clay (%)	1.5	Soluble cations (meq/l)		Cl-	1.83
Soil texture	(Sandy)	Ca2+	2.96	Avail. elements (mg/kg)	
		Mg2+	1.68	N	19.2
		Na+	2.04	P	8.3
pH	8.01	K+	0.21	Fe	2.4

Soil suspension (1 soil: 2.5 water)

consumption of ascorbate substrate. One unit of APX activity was defined as the amount of enzyme required to consume 1 mol ascorbate min^{-1} [27].

2.2.5.5 Glutathione reductase (GR)

GR activity was determined spectrophotometry by measuring the absorbance at wavelength 340 nm. This method depends on the oxidation of NADPH by the enzyme. One unit of GR activity was defined as the amount of enzyme that oxidized 1 mol NADPH min^{-1} [27].

2.2.6 Determination of nutrient elements

Nitrogen (N), phosphorus (P), sodium (Na), potassium (K), and calcium (Ca) analysis, dried shoot samples were ground to pass a 20-mesh sieve and digested with a mixture of H_2SO_4-$HClO_4$ using microwave energy, a modified technique of Lachica et al. [28].

In a mixture of sulfuric and perchloric acid total nitrogen was determined by the micro-Kjeldahl method [29]. For total inorganic phosphorus estimation, 0.5 g of plant material was extracted in 8 ml trichloroacetic acid (6%) and centrifuged for 15 min at 18000 X g. phosphate in supernatant was determined colorimetrically after adding 5 mol sulphuric acid, 2.5% ammonium molybdate and 0.25% 1,2,4-aminonaphtholsulphonic acid solution. After 15 min incubation at 37°C the absorbance was measured at 660 nm [30]. For Measurements of Na+ and K+ concentrations, the leaves were dried in 60°C for 48 h. Then 1 gr of leaves was powdered and burned in 560°C to obtain ash then ashes digested in 10 ml of 1N HCL. The concentration of Na+ and K+ in the digested samples was determined using a flame photometer (Model 420, Sherwood, Cambridge, UK) as described by Yousufinia et al. [31]. Calcium was measured on acid-digested samples by atomic absorption spectrophotometry in a Perkin Elmer Analyst 800 (Perkin Elmer Inc., Wellesley, MA) spectrophotometer equipped with a PE6017 lamp, and measured at 422.7 nm [32].

2.2.7 Statistical analysis

The collected data were analyzed statistically using Factorial Completely Randomized Design (FCRD) and analysis of variance according to Gomez and Gomez [33] with the aid of COSTAT computer program. Treatment means were compared using the least significant difference test (LSD) at 5% level.

3. RESULTS AND DISCUSSION

3.1 Lipid Peroxidation and Electrolyte Leakage

Data registered in (Table 2) indicated clearly that electrolyte leakage (EL), lipid peroxidation (LP) and membrane stability index (MSI) of salt stressed *Vicia faba* leaves showed different patterns of response when treated with NSi and Si. It was clear that the different concentrations of NaCl caused a significant increase in EL and LP, as determined by the amount of Malon dialdehyde (MDA) produced, comparing with the control plants. The results illustrated remarkable increase in Malon dialdehyde content and ion leakage level in response to 100 and 200 mM NaCl. Maximum values of electrolyte leakage and lipid peroxidation were registered at the 200 mM NaCl treatment. However, the application of NSi and Si caused a significant decrease in the EL and LP of the stressed plants as compared with those of the control. In contrast, the MSI of plant leaves was decreased when plants were exposed to NaCl. NSi and Si treatments, on the other side, significantly increased the MSI of salt treated faba bean plants.

Values of electrolyte leakage (EL) and membrane stability index (MSI) can be used to predict indirectly cell membrane damage caused by salt stress [34]. The stimulation effect of saline water on the value of MDA and EL % might be attributed to injury of plasma membrane. That damage caused by ROS which could induce Lipid peroxidation and consequently Electrolyte leakage [35]. Extensive membrane damage and change in membrane integrity, resulting from salinity stress, are shown as increase in lipid peroxidation and electrolyte leakage leading to decrease in MSI values.

Reduction of MDA levels and EL% in response to NSi and Si treatments might be due to induction of antioxidant responses that protect the plants from oxidative damage, increased membrane stability and tolerance of plants which in turn enhanced scavenging of harmful free radicals [36] and elevated ca uptake that protects the plant from the oxidative damage by silicon treatments [37]. On the other side, NSi and Si treatments appeared to have a positive effect on cell membranes and repair the damage mediated by salt stress to plasma membrane. This conclusion was evident from the significant increase in MSI and the significant decrease in EL of NSi and Si treated faba bean plants as compared with control plants. Similar results were obtained by Hamada [38], who found that brassinolide modifies membrane structure and stability under stress conditions.

In the present study, the decrease in lipid peroxidation, indicated by MDA content, in NSi or Si treated plants may be considered as a mechanism for the improved stability of cell membrane in salinity stressed plants in response to nanosilicon or silicon treatments. Lower lipid peroxidation, lower electrolyte leakage and higher membrane stability have also been reported in salt-tolerant genotypes of rice [39] and sugarcane [40] grown under salt stress conditions.

It is well known that, the deterioration of cell membrane is one of the physiological responses to salt stress. Oxidative stress, generated due to the presence of NaCl stress, promotes lipid peroxidation and causes a simultaneous increase in electron leakage and decrease the MSI. These physiological interferences and membrane disturbances could be mitigated by the addition of silicon and nano-silicon through the activation of antioxidant enzymes that scavenging the reactive oxygen species (ROS), as discussed later.

3.2 Photosynthetic Pigments

The present results showed that chl *a*, chl *b*, carotenoids (Table 3), total chlorophyll content (Fig. 1) and chl *a*: chl *b* ratio (Fig. 2) were all significantly decreased in faba bean leaves as salinity stress increased. The decrease in pigment concentrations of salt treated plants was more observed in the absence of NSi and Si treatments. Data in the table indicate clearly that the 50 mM salt treatment caused a slight increase in chl a, and chl b and carotenoid content as compared with salt untreated control (Table 3). This observation was true either with or without Si and NSi treatments. At this salinity level, the most observed increase in total chlorophyll (Chl a + b) content was recorded under NSi2 treatment (Fig. 1). In this regard, Chookhampaeng [41] found that chlorophyll content increased in pepper plants under low levels of salinity. One reason of that was the thicker leaves produced under salt stress. Increases in leaf thickness tended to compensate slightly for the negative effects of salinity on leaf chlorophyll [41].

Increasing salinity levels from 50 to 100 and 200 mM significantly decreased chlorophyll a, chlorophyll b and carotene contents. At 200 mM of NaCl, Chla, Chl b and carotene decreased by about 50%, 50% and 44%, respectively, as compared with control treatment. The application of Si and NSi as foliar spray alleviate the deleterious effect of salinity on the chlorophyll content. Application of 2 mM of Si or NSi produced the highest values of chlorophyll a, b and carotene as compared to the other treatments or control (Table 3). At no salt treatments, Si2 and NSi2 treatments caused an increase in Chl a by about 34% and 50%, respectively, Chl b by about 20% and 30%, respectively and carotene by about 20% and 50%, respectively, as compared with silicon untreated plants. It seems that, at any salt treatment Si and NSi treatments decrease the harmful effect of salinity stress on chlorophyll content. The reduction in leaf chlorophyll content under NaCl stress has been reported in early studies on *V. faba* [12], Zea maize [42] and green bean [43]. In this regard, Yuvakkumar et al. [42] found that total chlorophyll content of maize was increased by (13–17%) when treated with nanosilicate.

The decrease in chlorophyll content under salinity conditions is reported by Yasar et al. [43] and might have been due to salt-induced increase in the activity of the chlorophyll degrading enzyme, chlorophylase [44]. In this concern, a decrease in cholorophyll content (chl. a, b and total chl) of fennel plants under salt stress was observed [5]. The results obtained in this study are in agreement with those of Azooz et al. [12] for *V. faba*. The decrease in chlorophyll content of salt-stressed faba bean plants that associated with the decrease in enzyme activities (Table 4) and the increase in proline content (Table 5) is reliable with the suggestion that nitrogen might be redirected to the synthesis of proline instead of chlorophyll. In addition, Djanaguiraman and Ramadass [45] ascribed the decrease in pigment content of salt-stressed rice plants to the increase in the activity of chlorophyllase or to the disruption of the fine structure of the chloroplast, as well as to the instability of the chloroplast membrane and pigment-protein complex. In this regard, Sevengor et al. [46] attributed the reduction in leaf chlorophyll content under NaCl stress to the destruction of chlorophyll pigments and to the instability of the pigment-protein complex.

Table 2. Effects of nano-silicon (NSi) and silicon (Si) on electrode leakage (EL), lipid peroxidation (LP) and membrane stability index (MSI) of faba bean plants grown under different levels of salinity stress

Silica treatments	Salinity levels (mM)											
	00	50	100	200	00	50	100	200	00	50	100	200
	Electrolete leakage (%) (EL %)				Lipid peroxidation (MDA µg/g fwt)				Memberane stability index (MSI %)			
Cont	13.5	15.4	19.8	20.5	6.41	7.92	9.52	12.2	86.5	84.6	80.2	79.5
NSi1	11.6	14.2	17.3	18.6	5.23	6.23	8.53	10.2	88.4	85.8	82.7	81.4
NSi2	10.4	11.3	15.2	17.8	4.82	5.42	7.31	7.63	89.6	88.7	84.8	82.2
NSi3	13.2	14.7	14.6	18.7	5.14	5.55	7.13	8.24	86.8	85.3	85.4	81.3
Si1	11.2	13.8	16.6	18.2	5.12	6.06	8.26	9.85	88.8	86.2	83.4	81.8
Si2	10.4	11.5	14.4	17.3	4.24	5.21	7.08	7.43	89.6	88.5	85.6	82.7
Si3	10.0	11.1	13.7	16.8	4.32	5.04	6.55	7.22	90.0	88.9	86.3	83.2
LSD (5%)	1.41	2.33	2.16	1.45	0.55	0.62	1.24	1.45	2.56	2.14	2.22	1.57

Table 3. Effects of nano-silicon (NSi) and silicon (Si) on chlorophyll a (Chl a), chlorophyll a (Chl a) and carotenoids (Carot.) of faba bean plants grown under different levels of salinity stress

Silica treatments	Salinity levels (mM)											
	00	50	100	200	00	50	100	200	00	50	100	200
	Chl a (mg/g fwt)				Chl b (mg/g fwt)				Carot. (mg/g fwt)			
Cont	1.41	1.46	1.02	0.66	1.05	1.08	0.87	0.52	0.62	0.77	0.72	0.35
NSi1	1.53	1.66	1.28	1.06	1.28	1.31	1.18	1.05	0.78	0.81	0.66	0.45
NSi2	2.12	2.18	1.87	1.25	1.78	1.79	1.36	1.16	0.94	0.98	0.78	0.48
NSi3	1.66	1.72	1.46	1.01	1.25	1.28	1.05	0.88	0.55	0.58	0.52	0.38
Si1	1.45	1.48	1.15	1.04	1.12	1.14	1.02	0.77	0.71	0.79	0.73	0.41
Si2	1.89	1.92	1.53	1.11	1.25	1.28	1.07	0.82	0.75	0.82	0.75	0.45
Si3	1.52	1.61	1.36	1.09	1.16	1.12	0.97	0.78	0.65	0.72	0.68	0.37
LSD (5%)	0.22	0.18	0.17	0.11	0.12	0.14	0.16	0.14	0.11	0.15	0.10	0.11

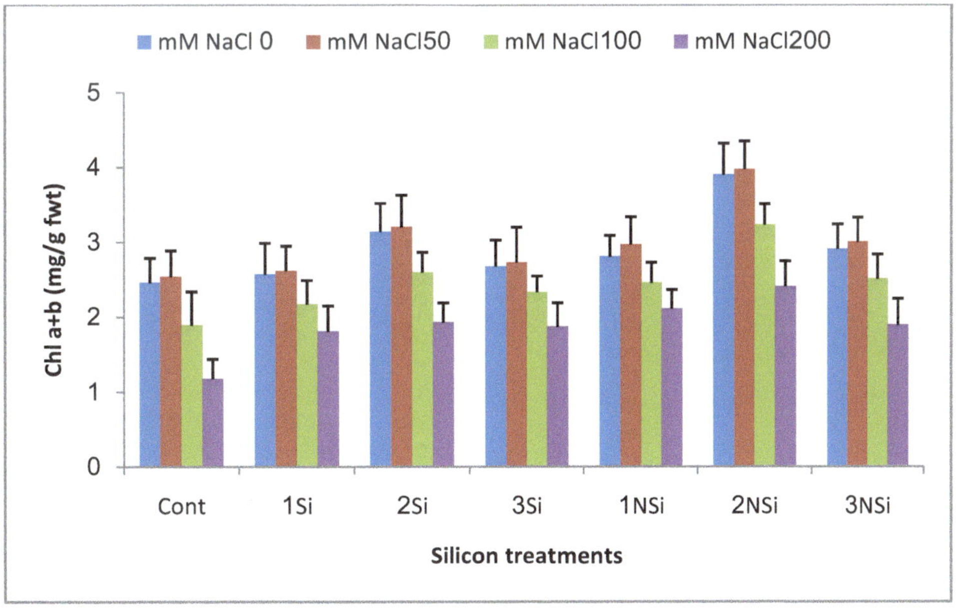

Fig. 1. Effects of nano-silicon (NSi) and silicon (Si) on total chlorophyll (Chl a +b) content of faba bean plants grown under different levels of salinity stress

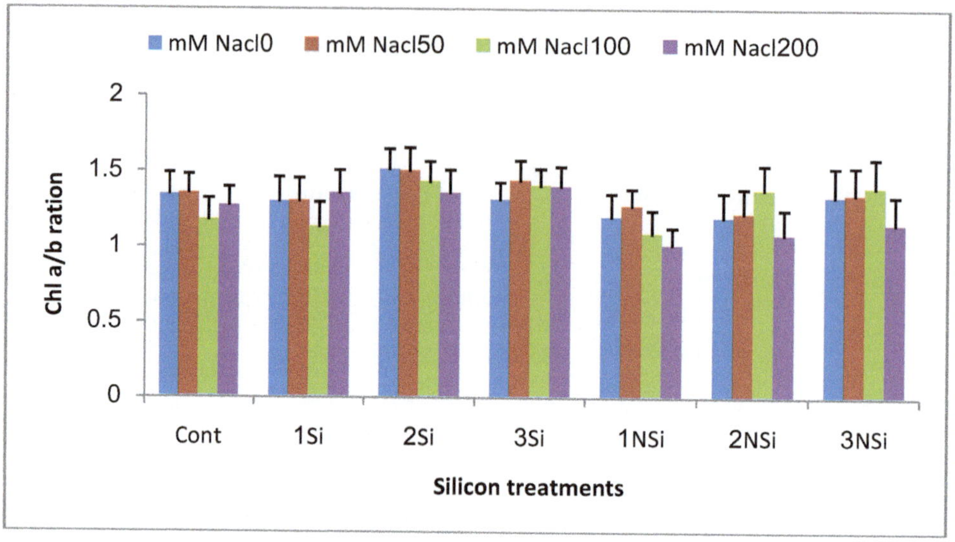

Fig. 2. Effects of nano-silicon (NSi) and silicon (Si) on Chl a/Chl b ration in faba bean plants grown under different levels of salinity stress

In the present study NSi or Si reduced the damage effects of salt stress on photosynthetic pigments through decreasing the electrolyte leakage and increasing the membrane stability index MSI (Table 2) compared with those of the control. Moreover, the chlorophyll content may be protected probably because of the high antioxidant enzyme activities that increased with NSi and Si and prevented degradation of leaf chlorophyll [46;47]. The results showed that nano Si could stabilize the integrality of chloroplast membrane and protect the chloroplasts from salt stress. Therefore with nano Si treatment, content of chlorophyll (a + b), was higher than the control (Fig. 1). In fact, NSi can improve structure of chlorophyll and can facilitate manufacture of pigments and protect chloroplasts from ageing [48]. However, bulk Si effects on pigments was not as significant as nano Si, as the grain size of nano Si is much smaller than that of bulk Si, which entered Faba bean cells more easily.

It was found that ROS produced by salt stress affect cellular membranes and cause a destruction of chroroplasts [12], while scavenging capacity during salt stress was increased with NSi and Si, exhibited as increased of antioxidant enzymes activity. Ameliorative effects on salinity stress may be attributed to silicon supplements included reductions in tissue Na content, maintenance of chloroplast cell membrane integrity, reductions in lipid peroxidation and lignification, and increases in ROS scavenging capacity. These findings are in accordance with those reported by Lei et al. [49] and Gao et al. [50] in a study on the effects of different concentrations of nano particles on the spinach traits, concluded that, chlorophyll amounts in treatment with nano particles showed significant increase, and it was 17 times of control amount. In another study, Yang and Hong [51] found that nano TiO2 increased protein and chlorophyll content of leaves more than bulk particles.

3.3 Soluble Sugar Contents

Effects of salt treatments and NSi or Si on soluble sugar contents of *V. faba* plants are shown in (Fig. 3). It was clear that total soluble sugars decreased with increasing NaCl concentration, to produce the lowest value of soluble sugars at the highest level of salt concentration, as compared with those of salt unstressed plants. In this regard the 200 mM NaCl resulted in a decrease of about 36% in the soluble sugar content as compared to unstressed plants. NSi and Si applications caused significant increases in the content of soluble sugars in faba bean plants. The maximum content of soluble sugars was estimated at moderate level of NaCl (100 mM) particularly with NSi2 and Si2 treatments compared with that determined in control plants. In this concern, NSi2 and Si2 treatments resulted in an increase of about 86% and 89%, respectively, in plants treated with 100 mM of NaCl.

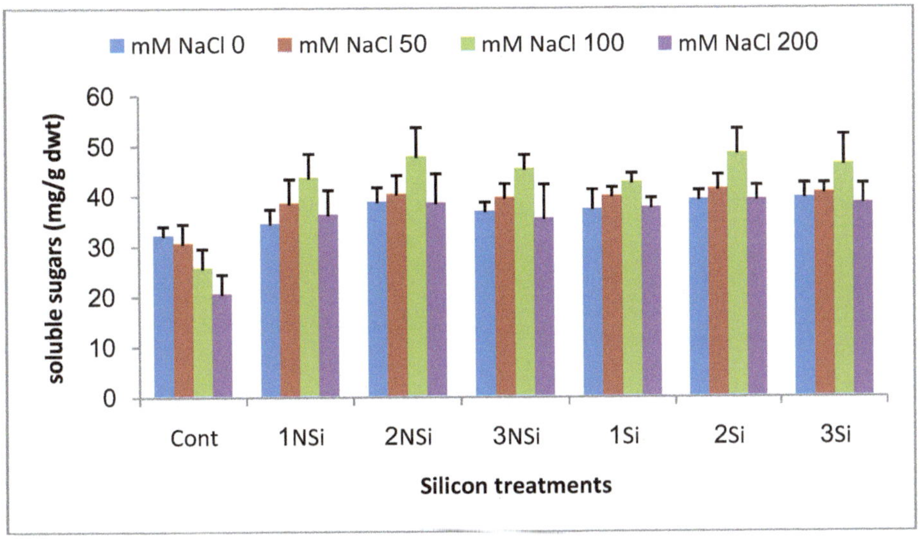

Fig. 3. Effects of nano-silicon (NSi) and silicon (Si) on total soluble sugars in faba bean plants grown under different levels of salinity stress

It was clear that soluble sugar content of plant leaves decreased significantly in salt-stressed *V. faba* plants compared to the unstressed plants. Similar results showing inhibitions in sugar formation under salinity stress were recorded in earlier studies [52,53]. The decrease in soluble sugar content and the reduction in photosynthetic pigment concentration under salt stress conditions were directly proportional to the applied concentration of NaCl in the growth media. These findings may lead to the conclusion that NaCl inhibits photosynthetic activity or increase partial utilization of carbohydrates in other metabolic pathways. Generally, NSi and Si enhanced the accumulation of sugars in salt-treated *V. faba* plants and alleviate the inhibitory effects of salt stress on sugar formation [47].

It is well known that organic solutes play a major role in the mitigation of salt stress. Accumulation of these compatible solutes reduces osmotic potential in the cytoplasm and contributes to maintaining water homeostasis among several cellular compartments. Among all organic solutes, soluble sugars and proline represent the most osmotically active organic solutes. A strong correlation between soluble sugar accumulation and salt tolerance has been widely reported [12,47]. In this regard, the enhancement effect of silicon on carbohydrate biosynthesis, especially soluble sugars, was found to be the main organic osmotica in a number of glycophytes when exposed to salt stress conditions [2]. This effect highlights another possible mechanism by which silicon plays a positive role in alleviation of the harmful effects of salt stress. Subjecting water stressed faba bean plants to silicon synergistically increased the amounts of soluble sugars than in untreated stressed ones which indicated that accumulation of these compounds by silicon plays a key role in retaining the water capacity of stressed cells which thereby can tolerate severe drought and salinity stress [36,54,55].

3.4 Antioxidant Enzyme Activities

The effect of NSi and Si treatments on the activity of antioxidant enzymes in faba bean plants, grown at different concentrations of NaCl, is presented in Tables 4 and 5. The recorded date showed a progressive increase in the activity of enzymes SOD, CAT, POD and APX with increasing salinity level in the growth media, whereas the activity of GR significantly decreased with increasing NaCl concentration as compared with those recorded for control plants. Adding Si or NSi had significant effects on the activity of all enzymes under salt stress Table 4 and 5. With increasing Si treatments a gradual increase in the activity of SOD, CAT, POD and APX was observed regardless of salinity treatment. Whereas, with NSi the highest increase in enzyme activity was observed under NSi2 treatment above which activities of the enzymes tended to decrease. The addition of Si or NSi markedly decreased the GR activity as compared with control treatment.

Table 4. Effects of nano-silicon (NSi) and silicon (Si) on SOD, CAT and POD antioxidant enzyme activities of faba bean plants grown under different levels of salinity stress

Silica treatments	Salinity levels (mM)											
	00	50	100	200	00	50	100	200	00	50	100	200
	SOD (unit g^{-1} FW)				CAT (µM H2O2 oxidized g^{-1} FW)				POD (O.D g^{-1} FW min^{-1})			
Cont	29.1	35.1	64.1	52.5	4.41	4.02	3.56	3.24	2.25	4.33	4.46	3.16
NSi1	34.3	36.6	66.5	60.6	5.12	6.82	7.11	5.64	2.88	4.75	5.75	3.65
NSi2	40.2	42.5	71.2	62.5	5.76	7.45	8.45	6.55	3.05	5.12	5.96	4.32
NSi3	28.6	30.7	60.4	48.5	4.14	5.16	7.06	5.14	2.62	4.23	4.67	3.12
Si1	30.4	36.5	67.3	53.3	4.56	6.12	7.68	5.12	2.45	4.65	5.11	3.57
Si2	34.7	38.2	69.2	58.2	4.64	6.86	7.94	5.34	2.87	4.87	5.32	3.84
Si3	36.3	40.5	69.8	55.2	4.86	7.11	8.12	6.45	3.11	4.46	5.65	4.11
LSD (5%)	2.88	2.12	2.45	3.62	NS	1.05	2.12	1.44	0.66	0.45	0.32	0.44

Table 5. Effects of nano-silicon (NSi) and silicon (Si) on APX andGR antioxidant enzyme activities and proline content of faba bean plants grown under different levels of salinity stress

Silica treatments	Salinity levels (mM)											
	00	50	100	200	00	50	100	200	00	50	100	200
	APX (mM ascorbate g^{-1} FW min^{-1})				GR (µg g^{-1} FW)				Proline (mg per 100 g DW)			
Cont	0.22	0.30	0.48	0.37	0.52	0.44	0.35	0.27	3.22	5.34	7.75	6.14
NSi1	0.37	0.41	0.49	0.38	0.47	0.39	0.33	0.25	3.11	4.01	5.66	6.01
NSi2	0.45	0.48	0.62	0.48	0.43	0.35	0.29	0.26	2.82	4.12	5.98	6.06
NSi3	0.26	0.35	0.50	0.40	0.52	0.45	0.33	0.30	3.27	4.50	6.14	6.15
Si1	0.28	0.34	0.52	0.39	0.50	0.41	0.30	0.31	3.20	3.35	5.11	6.34
Si2	0.32	0.37	0.56	0.42	0.45	0.40	0.26	0.28	2.24	3.88	5.29	6.84
Si3	0.36	0.43	0.58	0.46	0.40	0.38	0.24	0.24	2.11	3.96	5.88	6.98
LSD (5%)	0.12	0.24	0.16	0.04	0.05	0.03	0.05	0.03	0.03	0.46	0.34	0.02

It is well known that salinity stress causes generation of excessive reactive oxygen species (ROS), which leads to cell toxicity, membrane dysfunction and cell death [41]. Plants have developed enzymatic and nonenzymatic mechanism to scavenge ROS [2]. Among the active oxygen species superoxide is converted by SOD enzyme to H2O2, which is further scavenged by CAT and APX. Over expression of the APX gene in plants has showed improvement in protection against oxidative stress [43]. In the present study we found that the activities of antioxidant enzymes, SOD, POD and APX, in *V. Faba* plant leaves were enhanced when plants subjected to salinity stress condition. These results are in agreement with those of Hassanein *et al.* [53], who observed that salt stress increased the activities of antioxidant enzymes in leaves of *Zea mays* plants. In addition, Farag [56] reported that in pea (*Pisum sativum*), high concentrations of NaCl (110–130 mM) enhanced the activities of cytosolic and chloroplastic SOD. Increased activity of these antioxidant enzymes is considered to be a salt-tolerance mechanism in most plants [57].

The present results showed that salt stress caused a decrease in GR activity. GR deactivation by salt stress may be a result of prevention of new enzyme synthesis [6]. GR activates the glutathione-ascorbate cycle and converts GSSG to reduced glutathione (GSH) [58]. In addition, GR regulates GSH/GSSG ratio and supplies GSH for GPX and DHAR, which convert H_2O_2 to H_2O and reduce oxidized ascorbate, respectively. The changes in GR activity (Table 5) and lipid peroxidation (LP), as pointed out by the increase in MDA concentration (Table 2), in plants exposed to different levels of salinity stress and treated with NSi and Si were recorded. The GR activity gradually decreased with increasing NaCl concentration, whereas lipid peroxidation showed a gradual increase with increasing salt concentration as compared with control plants. In this regard, the maximum reduction in GR activity that reached 43% of that of the controls plants, and the maximum increase in MDA (LP) content which reached 77.7% of that of the control plants, was detected in faba bean plants treated with 200 mM NaCl (Table 2).

It seems that nanosilicon could alter the activity of antioxidative enzymes in plant organs to improve the salt tolerance. Results in Tables 4 and 5 showed that application of Si or NSi caused an increase in the activity of SOD, POD,

APX and CAT in *V. Faba* plants and this could ameliorate the effect of salinity. High activity of CAT in Si and NSi treated plants under salt stress suggests that the treated plants possess a better scavenging ability. The present study was consistent with the results reported by Mazorra et al. [58]. The effect of Si on the antioxidant enzymes activity under salt stress has been reported by Liang et al. [59] who described an increase in SOD activity in salt-stressed barley leaves and increases in SOD, GPX, CAT and GR activity in salt-stressed barley roots. In the present study, applying Si to Faba bean under NaCl stress significantly increased SOD, POD, APX and CAT activity. In another study, it was found that applying Si and nSi to plants under salt stress could increase antioxidative enzymes activity which played great role to counterbalance salinity damages [47] and [60]. In this regard Helaly et al. [61] reported that SOD, CAT and POX activities were increased significantly in banana plants when the nanoparticles doses increased. Moreover, Bao-shan et al. [62] tested nanoparticle silicon dioxide (NSi) on growth of Changbai larch (*Larix olgensis*) and soybean plants and observed that NSi showed the highest activity of nitrate reductase, superoxide dismutase, catalase and peroxidase. In contrast to our results, Rubinowska et al. [36] reported that at interaction of both salinity and NSi the rate of GR reduction was high. It was indicated that NSi might have toxic effects or it can stimulate salinity effects. Although there was a report on toxicity of $NSiO_2$ on *Arabidopsis thaliana*, but these toxicity was not stronger as other nanoparticles such as N-ZnO and N-Fe_3O_4 [63].

3.5 Proline Content

Analysis of the proline content of *V. Faba* showed that NaCl application significantly increased free proline contents while under Si treatment, the free proline contents of plant leaves markedly decreased but NSi treatment enhanced the accumulation of proline (Fig. 4). Proline content increased with increasing salinity level up to 100 mM of NaCl then, tended to decrease at 200 mM of NaCl (Fig. 4). At 100 mM of NaCl proline content increased in silicon untreated plants by about 140% as compared to 0 mM NaCl control treatment (Table 5). Si and NSi treatment decreased proline accumulation particularly when plants were not under salt stress. While at salt stress Si and NSi seemed to enhance proline accumulation.

Our results showed that the accumulation of proline in *V. faba* plants was linked with the increase in salinity stress level. These findings were in agreement with those obtained by Rahimi et al. [5] who reported that proline accumulation, in response to environmental stress, such as salinity stress and water stress, protected plant cells via osmotic adjustment effect by balancing the osmotic pressure of the cytosol with that of the vacuole and the external environment [5]. In this regard, proline accumulation could be a protective response, because of the osmoregulation role of proline that prevents water deficit stress under high salinity, and also because of its capability to scavenge the active radicals, produced at salt stress conditions, and stabilize the protein [15]. In addition, proline accumulation was reported to serve as a nitrogen storage compound and thus protects the cellular structure [5].

The amount of free proline in sole Si treated plants were however similar to that of control treatments. The free proline contents significantly increased under NaCl stress, while Si caused a reduction in free proline contents. Various earlier studies reported that proline contents significantly increased in many species such as common bean [64] and soybean [65] under salt stress condition. Current results also suggest that proline contents decreased with Si application, which show the favorable role of Si in mitigating the adverse effects of salt stress on plants. It was thus concluded that addition of Si and nSi is beneficial in hydroponically grown plants as it significantly improves growth attributes and effectively mitigate the adverse effects of NaCl induced salt stress. However, further studies are needed for understanding the mechanism of physiological or biochemical roles of Si in higher plants. Crusciol et al. [54] found that silicon increased proline (a key solute in osmotic adjustment) content in stressed plant tissue. In this regartd, Helaly et al. [61] found that total proline was increased significantly in banana plants when the nanoparticles doses increased.

3.6 Nutrient Elements Content

Data in Table 6 indicated clearly that the low values of N, P and Ca content of plant shoots were recorded by plants growing under salinity when compared with control or plants grown under low saline soil. A gradual reduction in N and Ca concentrations were recorded with increasing salinity level. The reduction in element concentration was more pronounced at the highest salinity level. In this regard, the 200 ppm of NaCl reduced the N% and Ca% by about 25% and 19%, respectively, as compared with control plants. On the other side, P% and K% in plant shoots increased with increasing salinity level up to 100 ppm then decreased at 200 ppm of NaCl, while Na% increased progressively with increasing salt level in the growth medium to reach its maximum at 200 ppm of NaCl. Therefore, K/Na ration decreased at high salt concentration as compared to non salinized plants. The negative effect of salinity stress on nutrient elements of bean plants was reported by Matijević et al. [66].

It is obvious from the data in Table 6 that all Si treatments caused an observed reduction in N% and Na% in plant shoots. While P, Ca, K, and Na concentrations increased with Si application. On the other side, low and medium levels of NSi (NSi1 and NSi2) caused a significant increase in N, P and K contents. The increase in K/Na ratio with the application of NSi may be attributed to the enhancing effect of nanosilica on the uptake of K or to the reduction in the absorbtion of Na. These results were agreed with those reported by Hanafy Ahmed et al. [67] on wheat. Phosphorus (P) tended to accumulate in the faba bean plants grown under low and moderate saline conditions but was reduced at strongly saline compared to control values. It might be due to an adaptation mechanism developed by the plants to overcome osmotic stress caused by salinity while further decrease in P might be related to the antagonistic relation between salt ions and P [68]. Potassium content was lower in plant shoots grown under high salinity levels than those grown in non saline conditions, while K accumulated at low levels of salinity.

As for the interaction effect of salinity stress and silicon treatments on nutrient composition of faba bean shoots, it is clear from the data recorded in Table 7 that Si treatments did not cause significant changes on N% of shoots under salt treatments, while NSi treatments showed an observed increase in N% of shoots under salinity stress particularly at 50 ppm of NaCl. Moreover, P and Ca concentrations in shoots of salt treated faba bean plants were also enhanced by the application of Si and NSi treatments. The most effective treatments in this regard were Si2 and NSi2, which produced the highest percent increase in P and Ca of salt stressed shoots (Table 7).

In addition, K concentration in salt stressed shoots was increased significantly by the application of Si and NSi as compared with salt stressed shoots without Si or NSi treatments (Table 8). Again, the most pronounced effect was recorded at Si2 and NSi2 treatments. At which K% in the 50 ppm NaCl treated shoots was increased by about 10% and 11% at Si2 and NSi2 treatment, respectively, as compared with the values obtained at 50 ppm NaCl in the absence of Si or NSi. The analogous increase in K content at 100 ppm NaCl were about 37% and 47%, respectively. While the corresponding increases at 200 ppm NaCl were about 6% and 28%, respectively. Contrary, Na concentration in salt stressed shoots was significantly decresed by Si and NSi applications. A gradual decrease in Na, of salt stressed shoots, with increasing Si or NSi concentrations was observed. The most reduction in Na was recorded at Si3 and NSi3 treatments, at which the Na% was decreased by about 37% and 32%, respectively at 50 ppm NaCl treatment, and by about 37% and 34%, respectively, at 100 ppm treatment and by about 36% and 48%, respectively, at 200 ppm treatment as compared with salt treated plants without Si or NSi application. In this regard, Liang et al., (2007) [59] reported a significant increase in K uptake and decrease in Na uptake under salt stress when Si was included because of increased activity of plasma membrane H-ATPase. K/Na ratio was significantly lower under salinity stress when Si was not applied. Silicon application enhanced K/Na selectivity ratio in Faba bean shoots thus enhancing pod and shoot yield. Application of Si in the salinity soil increased the K/Na. The exclusion of Na+ ions and a higher K/Na ratio in bean plants grown under saline conditions have been confirmed as important selection criteria for salt tolerance [69].

Application of Si significantly increased the contents of P and K, and the K/Na ratio and decreased Na ion contents of salt-affected plants. Therefore, the results shown in Table 6 agree with experimentations of Faba bean by Abdelhamid et al. [69] which indicate that salt tolerance is associated with an enhanced K/Na discrimination trait. The ability of plant to limit Na transport into the shoot is critically importance for the maintenance of high growth rates and protection of the metabolic processes in elongation cells from the toxic effects of Na. Sodium content was higher in plants grown under saline soil condition; however Si application significantly reduced Na content in shoot of Faba bean plants. Increased K content and reduced Na in shoot may be one of the possible mechanisms of increased salinity tolerance by Si application in Faba bean plants

[70]. In this regard, X-ray analysis revealed silicon deposits in unique silica filled cells of leaf margins located mainly near hydathodes and around trichome bases and along the leaf margins [71]. This leaf silica deposition may function in leaf strengthen, reduction in transpiration, and increasing biotic and abiotic resistance [72].

Table 6. Effects of salt stress and silicon treatments on mineral contents of faba bean plants

	PPM	N%	P%	Ca%		PPM	K%	Na%	K/Na
Salt treatments	00	2.39	0.26	1.37	Salt treatments	00	2.95	2.61	1.14
	50	2.17	0.28	1.27		50	3.23	2.83	1.16
	100	2.03	0.31	1.11		100	3.52	2.92	1.24
	200	1.79	0.22	0.86		200	2.46	2.97	0.83
LSD5%		0.11	0.04	0.16	LSD5%		0.46	0.26	0.28
		N%	P%	Ca%			K%	Na%	K/Na
Silicon treatments	Con	2.16	0.22	1.19	Silicon treatments	Con	2.72	3.67	0.76
	NSi1	2.21	0.29	1.19		NSi1	3.27	2.82	1.26
	NSi2	2.22	0.32	1.17		NSi2	3.38	2.69	1.25
	NSi3	2.05	0.25	1.02		NSi3	2.80	2.44	1.14
	Si1	2.03	0.25	1.21		Si1	3.09	2.85	1.08
	Si2	1.98	0.29	1.23		Si2	3.17	2.76	1.15
	Si3	2.04	0.28	1.08		Si3	2.69	2.43	1.20
LSD5%		0.05	0.05	0.06	LSD5%		0.15	0.23	0.14

Table 7. Effects of silicon (Si) and nano-silicon (NSi) on N, P and Ca percentages in faba bean plants under different levels of salinity stress

	N (%)				P (%)				Ca (%)			
	00	50	100	200	00	50	100	200	00	50	100	200
Con	2.45	2.22	2.11	1.85	0.22	0.23	0.24	0.18	1.52	1.27	1.11	0.85
NSi1	2.43	2.32	2.15	1.94	0.26	0.29	0.33	0.25	1.46	1.28	1.12	0.90
NSi2	2.46	2.38	2.10	1.88	0.30	0.32	0.38	0.28	1.36	1.30	1.17	0.86
NSi3	2.30	2.26	2.11	1.52	0.25	0.28	0.30	0.19	1.20	1.15	1.04	0.65
Si1	2.42	2.00	1.91	1.80	0.24	0.26	0.28	0.20	1.44	1.40	1.12	0.91
Si2	2.32	1.99	1.85	1.79	0.27	0.32	0.34	0.22	1.41	1.36	1.22	0.95
Si3	2.38	2.03	1.96	1.80	0.28	0.30	0.32	0.23	1.26	1.15	1.02	0.87
LSD5%	0.02	0.07	0.06	0.04	0.03	0.05	0.03	0.03	0.06	0.04	0.05	0.04

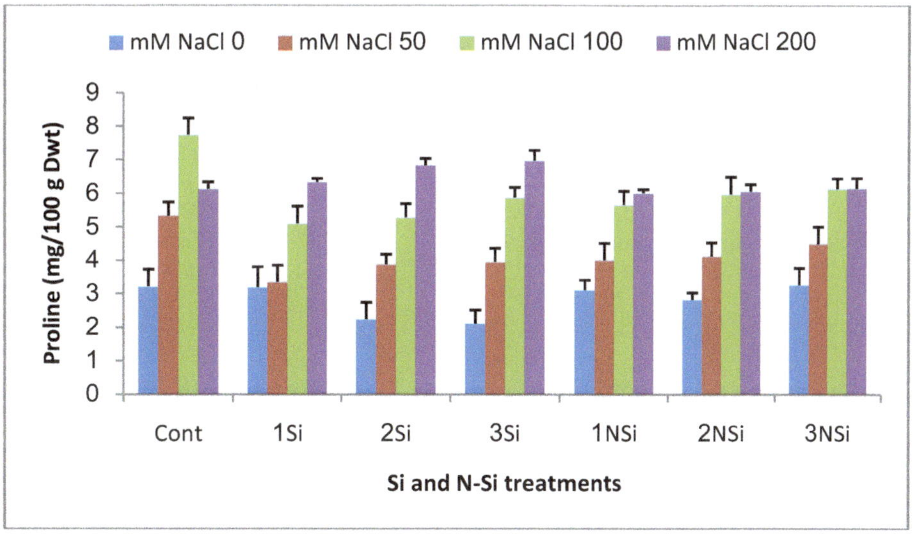

Fig. 4. Effects of nano-silicon (NSi) and silicon (Si) on free proline content of faba bean plants grown under different levels of salinity stress

Table 8. Effects of silicon (Si) and nano-silicon (NSi) on K, Na and K/Na ration in Faba bean plants under different levels of salinity stress

	K (%)				Na (%)				K/Na			
	00	50	100	200	00	50	100	200	00	50	100	200
Con	2.86	3.11	2.66	2.24	2.88	3.74	3.95	4.11	0.99	0.83	0.67	0.55
NSi1	2.88	3.30	3.88	3.02	2.75	2.88	2.89	2.77	1.05	1.14	1.34	1.09
NSi2	3.32	3.45	3.90	2.87	2.60	2.72	2.88	2.58	1.28	1.27	1.36	1.11
NSi3	2.67	3.15	3.26	2.12	2.45	2.55	2.62	2.14	1.09	1.24	1.24	0.99
Si1	2.95	3.21	3.89	2.30	2.76	2.82	2.86	2.95	1.07	1.14	1.36	0.78
Si2	3.21	3.42	3.65	2.38	2.60	2.77	2.80	2.90	1.23	1.23	1.30	0.82
Si3	2.82	3.02	3.44	2.28	2.24	2.37	2.45	2.65	1.26	1.27	1.42	0.86
LSD5%	0.44	0.12	0.25	0.11	0.15	0.25	0.18	0.26	0.16	0.11	0.14	0.23

4. CONCLUSION

The present study revealed that when faba bean plants were subjected to salt stress conditions the plants might not be able to overcome the oxidative damage. Meanwhile, nanosilicon and silicon treatments enhanced the activity of antioxidant enzymes, thus reduced the oxidative damage and decreased the absorption of Na ions in NaCl-salt stress conditions. The increase in the degree of salt tolerance induced by NSi and Si was also reflected by the improvement in the contents of photosynthetic pigments and carbohydrates in the presence of salinity. Therefore, data of this study provide an additional evidence for the stimulatory effects of nanosilicon to induce salt tolerance in faba bean plants. The results also highlight the role of NSi in regulating salinity responses, and indicate that nanosilicon could protect faba bean plants against the hazardous effect of salinity. There were no meaningful differences between the silicon (Si) and nanosilicon (NSi) applications; both forms of silicon were effective at lower levels in most measured characteristics.

COMPETING INTERESTS

Author has declared that no competing interests exist.

REFERENCES

1. Kopke U, Nemecek T. Ecological services of faba bean. Field Crops Res. 2010;115: 217–233.
2. Hassanein RA, Hashem HA, Khalil RR. Stigmasterol treatment increases salt stress tolerance of faba bean plants by enhancing antioxidant systems. Plant Osmics J. 2012;5:476-485.
3. Hungria M, Vargas MA. Environmental factors affecting nitrogen fixation in grain legumes in the topics, with an emphasis on Brazil. Field Crops Res. 2000; 65: 151-164.
4. Ashraf M, Foolad MA. Improving plant abiotic-stress resistance by exogenous application of osmoprotectants glycine betaine and proline. Environ Exp Bot. 2007; 59:206–216.
5. Rahimi R, Mohammakhani A, Roohi V, Armand N. Effects of salt stress and silicon nutrition on cholorophyll content, yield and yield components in fennel (*Foeniculum vulgar* Mill.). Intl J Agric Crop Sci. 2012;4: 1591-1595.
6. Liang Y, Zhang W, Qin C, Youliang L, Ruixing D. Effect of exogenous silicon (Si) on H+-ATPase activity, phospholipids and fluidity of plasma membrane in leaves of salt-stressed barley (*Hordeum vulgare* L.). Environ Exp Bot. 2006;57:212–219.
7. Prakash D, Upadyay G, Pushpangadan P. Antioxidant potential of some under – utilized fruits. Indo-Global J Pharm Sci. 2011;1:25-32.
8. Sabra A, Daayf F, Renault S. Differential physiological and biochemical responses of three *Echinacea* species to salinity stress. Scientia Hort. 2012;135:23-31.
9. Eyidogan F, Öz MT. Effect of salinity on antioxidant responses of chickpea seedlings. Acta Physiol Plant. 2007;29: 485-493.
10. Hernandez JA, Jimenez A, Mullineaux P, Sevilla I. Tolerance of pea plants (*Pisum sativum*) to long term salt stress is associated with induction of antioxidant defenses. Plant cell Enviorn. 2000;23:853-862.
11. Heidari M, Golpayegani A. Effects of water stress and inoculation with plant growth promoting rhizobacteria (PGPR) on antioxidant status and photosynthetic pigments in basil (*Ocimum basilicum* L.). J Saudi Society Agricult Sci. 2012;11:57-61.

12. Azooz MM, Alzahrani AM, Youssef MM. The potential role of seed priming with ascorbic acid and nicotinamide and their interactions to enhance salt tolerance in broad bean (*Vicia faba* L.). Aust J Crop Sci. 2013;7:2091-2100.
13. Ruiz-Carrasco K, Antognoni F, Coulibaly AK, Lizardi S, Covarrubias A, Martinez EA, Molina-Montenegro MA, Biondi S, Zurita-Silva A. Variation in salinity tolerance of four lowland genotypes of quinoa (*Chenopodium quino*) as assessed by growth, physiological traits, and sodium transporter gene expression. Plant Physiol Biochem. 2011;49:1333-1341.
14. Szabados L, Savouré A. Proline: a multifunctional amino acid. Trends Plant Sci. 2010;15:89-97.
15. Ben Ahmed C, Ben Rouina B, Sensoy S, Boukhriss S, Abdullah F. Exogenous proline effects on photosynthetic performance and antioxidant defense system of young olive tree. J Agricult Food Chem. 2010;58:416–422.
16. Suriyaprabha R, Karunakaran G, Yuvakkumar R, Rajendran V, Kannan N. Silica nanoparticles for increased silica availability in maize (*Zea mays* L) seeds under hydroponic conditions. Curr NanoSci. 2012;8:1–7.
17. Debnath N, Das S, Seth D, Chandra R, Bhattacharya SC, Goswami A. Entomotoxic effect of silica nanoparticles against *Sitophilus oryzae* (L.). J Pest Sci. 2011;84: 99–105.
18. Kulikova ZL, Lux A. Silicon influence on maize (*Zea mays* L.) hybrids exposed to cadmium treatment. Bull Environ Contam Toxicol. 2010;85:243–250.
19. Abdul Qados, AM. Influence of silicon and nano-silicon on germination, growth and yield of faba bean (*Vicia faba* L.) under salt stress. Amer J Exp Agric. 2015;5:509-524.
20. Unyayar S, Celik A, Cekic FO, Gozel A. Cadmium-induced genotoxicity, cytotoxicity and lipid peroxidation in *Allium sativum* and *Vicia faba*. Mutagenesis. 2006;21:77-81.
21. Premchandra GS, Saneoha H, Ogata S. Cell membrane stability, an indicator of drought tolerance as affected by applied nitrogen in soybean. J Agr Sci Camb. 1990; 115:63-66.
22. Metzner H, Rau H, Senger H. Untersuchungen Zur Synchronisier barkeit einzelner Pigmentmangel-Mutanten Von Chlorella. Planta. 1965;65:186-194.
23. Scott TA, Melvin EH. Anthrone colorimetric method. In: Whistler RL, Walfrom ML (ed.) Methods in Carbohydrate Chemistry. Academic Press, New York, London, 1956; 384.
24. Bates LS, Wladren RP, Tear LD. Rapid determination of free proline for water-stress studies. Plant Soil. 1973;39:205-207.
25. Mukherjee SP, Choudhury MA. Implications of water stress induced changes in the levels of endogenous ascorbic acid and hydrogen peroxide in *Vigna* seedlings. Physiol Plant. 1983;58: 166-170.
26. Karanlik S. Resistance to salinity in different wheat genotypes and physiological mechanisms involved in salt resistance. Ph.D. thesis, institute of natural and applied science, university of cukurova, Adana. 2001;122.
27. Cakmak I, Marschner H. Magnesium deficiency and high light intensity enhance activities of superoxide dismutase, ascorbate peroxidase and glutathione reductase in bean leaves. Plant Physiol. 1992;98:1222–1226.
28. Lachica M, Aguilar A, Yanez J. An´alisis foliar.M´etodos utilizados en la Estaci´on Experimental del Zaid´ın C.S.I.C. (II). Anna Edafol Agrobiol. 1973;32:1033–1047.
29. Bremner JM. Nitrogen. In Methods of Soil Analysis, Part 3: Chemical Methods; Sparks, D.L. (ed.); Soil Science Society of America: Madison, Wisconsin. 1996;1085–1121.
30. Sacala E, Demczuc A, Grazys E, Spiak Z. Effect of salt and water stress on growth, nitrogen and phosphorus metabolism in *Cucumis sativa* L. seedlings. Acta Soc Bot Pol. 2008;77:23-28.
31. Yousufinia M, Ghasemian A, Safalian O, Asadi A. The effect of NaCl on the growth and Na+ and K+ content of barley (*Hordeum vulgare*, L.) cultivares. Ann Biol Res. 2013;4:80-85.
32. Tejera NA, Campos IR, Sanjuan J, Lluch C. Effect of Sodium Chloride on Growth, Nutrient Accumulation, and Nitrogen Fixation of Common Bean Plants in Symbiosis with Isogenic Strains. Journal of Plant Nutrition. 2005;28:1907–1921.
33. Gomez KA, Gomez AA. Statistical procedures for agricultural research. John Wiley & Sons Inc., Singapore. 1984;680.
34. Ali B, Hayat S, Hasan SA, Hayat Q, Yadav S, Fariduddin Q, Ahmad A. A role for brassinosteroids in the amelioration of

aluminium stress through antioxidant system in mungbean (*Vigna radita* L. Wilczek). Environ Exp Bot. 2008;62:153-159.
35. Kassab OM, Orabi SA, Abo Ellil AA. Physiological response to potassium application in fodder beet plant grown under water stress. Australian Journal of Basic and Applied Sci. 2012;6:566-574.
36. Rubinowska K, Pogroszewska E, Laskowska H, Szot P, Zdybel A, Stasiak D, Kozak D. The subsequent effct of silicon on physiological and biochemical parameters of *Polygonatum multiflorum* (L.) 'Variegatum' cut shoots. Acta Sci. Pol., Hortorum Cultus. 2014;13:167-178.
37. Salwa A, Orabi B, Mekki B, Sharara FA. Alleviation of Adverse Effects of Salt Stress on Faba Bean (*Vicia faba* L.) Plants by Exogenous Application of Salicylic Acid. World Applied Sciences Journal. 2013;27: 418-427.
38. Hamada K. Brassinolide in crop cultivation. In: Macgregor P (ed) Plant Growth Regulators in Agriculture. Food Fertility Technology, Central Asia Pacific Region. 1986;190-196.
39. Tijen D, Ismail T. Comparative lipid peroxidation, antioxidant defense systems and proline content in roots of two rice cultivars differing in salt tolerance. Environ Exp Bot. 2005;53:247-257.
40. Gomathi R, Rakkiyapan P. Comparative lipid peroxidation, leaf membrane thermostability, and antioxidant system in four sugarcane genotypes differing in salt tolerance. Inter J Plant Physiol Biochem. 2011;3:67-74.
41. Chookhampaeng S. The effect of salt stress on growth, chlorophyll content proline content and antioxidative enzymes of pepper (*Capsicum annuum* L.) seedling. Eur J Sci Res. 2011;49:103-109.
42. Yuvakkumar R, Elango V, Rajendran V, Kannan NS, Prabu P. Influence of Nanosilica Powder on the Growth of Maize Crop (*Zea mays* L.). Inter J Green Nanotechn. 2011;3(3):180-190.
43. Yasar F, Ellialtioglu S, Yildiz K. Effect of salt stress on antioxidant defense systems, lipid peroxidation, and chlorophyll content in green bean. Russian J. Plant Physiol. 2008;55:782-786.
44. Nazarbeygi E, Yazdi HL, Naseri R, Soleimani R. The effects of different levels of salinity on proline and A-, B- chlorophylls in canola. Amer-Eurasian J. Agric. Environ. Sci. 2011;10:70-74.
45. Djanaguiraman M, Ramadass R. Effect of salinity on chlorophyll content of rice genotypes. Agricult Sci Digest. 2004;24: 178-181.
46. Sevengor S, Yasar F, Kusvuran S, Ellialtioglu S. The effect of salt stress on growth, chlorophyll content, lipid peroxidation and antioxidative enzymes of pumpkin seedling. African J Agric Res. 2011;6:4920-4924.
47. Siddiqui MH, Al-Whaibi MH. Role of nano-SiO_2 in germination of tomato (*Lycopersicum esculentum* Mill.) seeds. Saudi J Biol Sci Jan. 2014;21:13–17.
48. Morteza E, Moaveni P, Farahani HA, Kiyani M. Study of photosynthetic pigments changes of maize (*Zea mays* L.) under nano Tio2 spraying at various growth stages. Springer Plus. 2013;2:247-252.
49. Lei Z, Mingyu S, Xiao W, Chao L, Chunxiang Q, Liang C. Antioxidant stress is promoted by nano-anatase in spinach chloroplasts under UV-B radiation. Biol Trace Elem Res. 2008;121:69-79.
50. Gao X, Zou C, Wang L, Zhang F. Silicon decreases transpiration rate and conductance from stomata of maize plants. J Plant Nutr. 2006;29:1637–1647.
51. Yang F, Hong FS. Influence of nano-anatase Tio2 on the nitrogen metabolism of growing spinach. Biol Trace Element Res. 2006;110:179–190.
52. Younis MF, Hasaneen MNA, Ahmed AK, El-Bialy DMA. Plant growth, metabolism and adaptation in relation to stress condition XXI. Salinity and nitrogen rate effects on growth, photosynthesis, carbohydrate content and activities of antioxidant enzymes in *Lactuca sativa* transplents. Aust J Crop Sci. 2008;2:83-95.
53. Hassanein RA, Hassanein AA, Haider AS, Hashem HA. Improving salt tolerance of *Zea mays* L. plant by presoaking their grains in glycine betaine. Aust J Basic Appl Sci. 2009;3:928-942.
54. Crusciol CAC, Pulz AL, Lemos LB, Soratto RP, Lima GPP. Effects of silicon and drought stress on tuber yield and leaf biochemical characteristics in potato. Crop Sci. 2009;49:949–954.
55. Abdalla MM. Beneficial effects of diatomite on the growth, the biochemical contents and polymorphic DNA in *Lupinus albus* plants grown under water stress. Agric Biol J N Am. 2011;2:207-220.

56. Farag AA. Increasing tolerance of *Vigna sinensis* L. to salt stress using an organic acid and a polyamine. M.Sc. thesis. Ain Shams University, Cairo; 2009.
57. Hu L, Li H, Pang H, Fu J. Responses of antioxidant gene, protein and enzymes to salinity stress in two genotypes of perennial ryegrass (*Lolium perenne*) differing in salt tolerance. J plant physiol. 2012;169:146-156.
58. Mazorra LM, Nunez M, Hechavarria M, Coll F, Sanchez-Blanco MJ. Influence of brassinosteroids on antioxidant enzymes activity in tomato under different temperatures. Biol Plant. 2002;45:593-596.
59. Liang Y, Sun W, Zhu YG, Christie P. Mechanisms of silicon mediated alleviation of abiotic stresses in higher plants: a review. Environ Poll. 2007;147:422–428.
60. Wang X, Wei Z, Liu D, Zhao G. Effects of NaCl and silicon on activities of antioxidative enzymes in roots, shoots and leaves of alfalfa. Afric J Biotech. 2011;10: 545-549.
61. Helaly MN, El-Metwally MA, El-Hoseiny H, Omar SA, El-Sheery NI. Effect of nanoparticles on biological contamination of in vitro cultures and organogenic regeneration of banana. Aust J Crop Sci. 2014;8:612-624.
62. Bao-shan L, Shao-qi D, Chun-hui L, Li-jun F, Shu-chun Q, Min Y. Effects of TMS (nanostructured silicon dioxide) on growth of *Changbai larch* seedlings. J Forest Res. 2004;15:138-140.
63. Lee SK, Sohn EY, Hamayun M, Yoon LY, Lee IJ. Effects of silicon on growth and salinity stress of soybean plant grown under hydroponic system. Agroforest Syst. 2010;80:333-430.
64. Khadri M, Tejera NA, Lluch C. Alleviation of salt stress in common bean by exogenous abscisic acid supply. J Plant Growth Reg. 2006;25:110–119.
65. Chon SU, Park JH, Choi WY, Jung SY. Differential physiological responses of soybean. Korea Soybean Dig. 2003; 20: 17–27.
66. Matijević L, Romić D, Maurović N, Romić M. Saline irrigation water affects element uptake by bean plant. Eur. Chem. Bull. 2012;1:498-502.
67. Hanafy Ahmed AH, Harb EM, Higazy MA, Morgan ShH. Effect of Silicon and Boron Foliar Applications on Wheat Plants Grown under Saline Soil Conditions. Inter J Agric Res. 2008;3:1-26.
68. Bulut F, Akıncı S. The effect of salinity on growth and nutrient composition in broad bean (*Vicia faba* L.) seedlings. Fresenius Environmental Bulletin, PSP. 2010;19:12.
69. Abdelhamid MT, Shokr M, Bekheta MA. Growth, root characteristics, and leaf nutrients accumulation of four faba bean (*Vicia faba* L.) cultivars differing in their broomrape tolerance and the soil properties in relation to salinity. Comm Soil Sci Plant Anal. 2010;41:2713-2728.
70. Hellal FA, Abdelhameid M, Abo-Basha DM, Zewainy RM. Alleviation of the adverse effects of soil salinity stress by foliar application of silicon on faba bean (*Vica faba* L.). J App Sci Res. 2012;8:4428-4433.
71. Frantz JM, Locke JC, Datnoff L, Omer M, Widrig A, Sturtz D, Horst L, Krause CR. Detection, distribution, and quantification of silicon in floricultural crops utilizing three distinct analytical methods. Comm. Soil Sci. Plant Analysis. 2008;39:2734-2751.
72. Postek MT. The occurrence of silica in the leaves of *Magnolia grandiflora* L. Botanical Gazette. 1981;142:124–134.

20

Effect of *Albizia zygia* and NPK Fertilizer on the Improvement of Soil Fertility on a Humid Alfisol in Southwestern Nigeria

M. R. Olojugba[1*] and D. O. Oke[2]

[1]*Department of Soil Science and Management, Joseph Ayo Babalola University, Ikeji-Arakeji, Nigeria.*
[2]*Department of Forestry and Wood Technology, Federal University of technology, Akure, Nigeria.*

Authors' contributions

This work was carried out in collaboration between both authors. Both authors read and approved the final manuscript.

Editor(s):
(1) Peter A. Roussos, Agricultural University of Athens Lab. Pomology Iera Odos 75, Greece.
Reviewers:
(1) Samia Osman Yagoub Ali, Biology, Shagra University, KSA.
(2) Anonymous, Brazil.
(3) Stephen Yeboah, CSIR-Crops Research Institute, Ghana.

ABSTRACT

The effect of integrating *Albizia zygia* and NPK 15-15-15 on soil fertility and maize yield were studied to investigate the effect of *Albizia zygia* and NPK 15-15-15 on soil properties, growth and yield of maize. The experiment was laid out as a Randomized Complete Block Design (RCBD) with five treatments and three replicates. The experiment was conducted from 2009 to 2011 at the Teaching and Research Farm, Joseph Ayo Babalola University, Ikeji-Arakeji, Nigeria. The treatments were: control (no treatment), *Abizia zygia* at 5 tons/ha, *Albizia zygia* + NPK 15-15-15 at 2.5 tons/ha and 125kg/ha respectively, NPK 15-15-15 at 250kg/ha and 0PK at 250 kg/ha of single superphosphate (SSP) and 90 kg/ha of muriate of potash (KCl). Soil was sampled before planting for determination of soil nitrogen conten, phosphorous, potassium, organic matter, soil pH, total soil porosity, Water holding capacity and soil bulk density were determined. Maize yield was determined at end of growing seasons as well as profitability of the treatments were also evaluated. Soil physical properties such as total porosity, water holding capacity, soil organic matter, total exchangeable bases, cations exchange capacity and soil pH were significantly improved by applying *Albizia* zygia biomass. Maize yields (1.33, 1.93 and 1.33 ton/ha in 2009, 2010 and 2011 planting seasons) were found to have increased significantly in the plot treated with *Albizia* + NPK

**Corresponding author: Email: michaelolojugba@gmail.com;*

15-15-15 over other treatments. In the same vein, the same treatment (*Albizia* + NPK 15-15-15) was found to have returned more money (N37, 459 and N76,204 in 2009 and 2010 cropping seasons respectively) to the farmer over other treatments.

Keywords: Fertility; soil nutrients; Albizia zygia; NPK 15-15-15; integrated nutrient management; green manure; mulching.

1. INTRODUCTION

Soil degradation which is defined as the decline in the physical, chemical, biological as well as hydrological properties of soil is widely recognized as a global problem associated with desertification in dry lands [1]. In sub-Saharan Africa, it is influenced with the depletion in soil nutrient potential (mainly nitrogen, phosphorus and carbon content), this has been identified as a major threat to food security [2]. This problem is exacerbated by wind, water and stream bank erosion [3], poor rainfall distribution [4], low and unbalance rates of mineral fertilizers application. The region is also characterized by climatic conditions that influenced the depletion of soil organic matter [5] which affects the water holding capacity of the soils and also lead to deficiency in soil nitrogen and phosphorus.

In Nigeria, like many other Sub-Saharan African countries, soil fertility and hence productivity are declining at an alarming rate because areas of high agricultural potential are densely populated, for example, Nigerian population is increasing at the rate of 0.3% annually. In most cases, farms holdings are less than 1ha. Farmers practice intensive continuous cropping with limited or no replacement of nutrients through fertilizer application due to the high cost and non availability of inorganic fertilizers as well as perceived negative effect of inorganic fertilizer on weakly buffered West African soils which may lead to soil degradation through a rapid decrease in the soil organic matter, exchangeable bases and pH, hence cannot sustain crop productivity in the long term [6].

Organic inputs, including composts, animal manure, crop residues and green manure are good methods of enhancing physical, chemical and biological properties of soils and crop performance [7]. It contributes to improving soil structure and aggregation and decrease soil bulk densities and increased the percentages of pore spaces [8]. Due to this, soil water infiltration and water holding capacity increase [9]. Organic inputs are also source of energy and slowly available carbon to support soil organism's activity which are the primary agent for decomposition in the soil.

The role of agroforestry systems becomes important in the marginalized area in Nigeria with low capital and technological inputs. The conservation benefits, sustainable yield potential, biological and socio-economic characteristics make it more desirable in soil fertility management. One of such trees is *Albizia zygia*, a leguminous plant, deciduous, up to 25m high in forest, usually a shrubby weed in bush re-growth. One of the practices of agroforestry for soil fertility in Nigeria is biomass transfer which is defined as the transfer and application of cut and fallen mulch or tree litter from the source to the field on which crops are grown or to be grown [10]. In Nigeria, inorganic fertilizer addition is the major method of replenishing soil fertility and boosting crop performance. However, inorganic fertilizer use in Nigeria is estimated at 13 kg/ha in 2009 by the Federal Ministry of Agriculture and Rural Development (FMARD), is far lower than 200 kg/ha recommended by the FAO as well as the 104 kg/ha in South Asia and 142 kg/ha in Southeast Asia [11]. The high cost and the gap between net benefit from harvested products and fertilizer purchase make the use of inorganic fertilizer use unpopular. On the other hand, significant investments in labour, land availability, bulkiness and unbalanced nutrient content [12] as well as increase in soil heavy metal as a result of continuous application of compost limit the fullest adoption of organic inputs in Nigeria. In view of the limitation of both organic and inorganic fertilizers, a combined application of the two is worth researching. The abundance of *Albizia zygia* coupe with its nitrogen fixing ability in the study area makes it an excellent agroforestry species of choice.

This work is aimed at determining the potential of *Abizia zygia* leaf biomass as source of fertilizer and the combined effect of both *Albizia zygia* and NPK fertilizer for soil fertility improvement in a degraded humid Alfisol in Nigeria.

2. MATERIALS AND METHODS

2.1 Site Description

The experiment was conducted on the teaching and research field plot of the college of agricultural science of Joseph Ayo Babalola University, Ikeji Arakeji, Nigeria which lies between latitude 07°16' and 07° 18' N and longitude 05° 09° and 05° 1' E. The study area is situated in the humid tropical forest zone of Nigeria. It has an annual average rainfall of between 1500-1800mm and relative humidity of between 80-85% annually Fig. 1. It has a gentle undulating elevation of about 1150m-1250m above sea level.

2.1.1 Layout of experiment

The experiment was laid out as a Randomized Complete Block Design (RCBD) with five treatments and three replicates. A 20m x 35m plot was demarcated at the Teaching and Research Farm of Joseph Ayo Babalola University, Ikeji–Arakeji, Nigeria from 2009 to 2011. This was partitioned into three blocks of 5m x 35m each. Each block was partitioned into five 5m x 5m plots. Each 5m x 5m plot was separated by buffer of 2.5m wide. The treatments were:

a. Control (no treatment).
b. *Abizia zygia* at 5 tons/ha
c. N: P: K 15-15-15 at 250 kg/ha
d. *Albizia* zygia at 2.5 tons/ha +NPK 15-15-15 at 150 kg/ha respectively
e. 0PK (N at zero level) at 250 kg/ha of single superphosphate (SSP) and 90 kg/ha of murate of potash (KCl)

All the treatments were allocated to the plots in each block at random. *Albizia* leaves (not senescent) after curing were incorporated with hoe in the top of 15cm-20cm at 12.5 kg/plot or 5 tons/ha [6] for two consecutive cropping seasons (2009 and 2010), while NPK was applied at 18g/plant or at 250 kg/ha at two weeks, 10 cm from the maize plant. This was followed by another growing season (2011) without treatment application to study the residual effect of the different treatments. Maize seeds variety TZPBSR-N (IITA) obtained from International Institute of Tropical Agriculture, Ibadan, Nigeria (IITA), Ibadan were planted at 60 x 30 cm to give a plant population of 55, 000 stands per hectare, and 133 stands per plot. Two seeds were sown per hole and thinned to one after germination. Weeding was done as appropriate.

2.1.2 Nutrient concentration in the *Albizia zygia* leaves

This was carried out to determine the nutrient concentration in *Albizia zygia* leaves and other chemical parameters influencing their leaf quality and suitability for soil fertility maintenance.

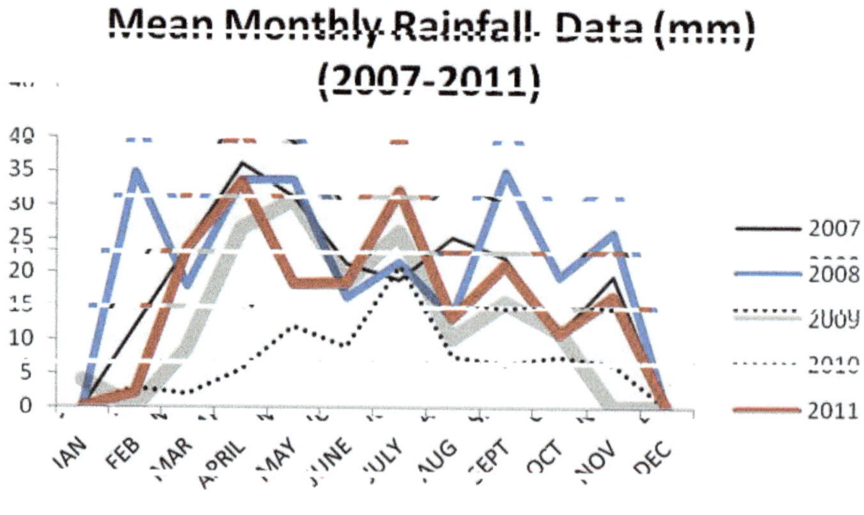

Fig. 1. Mean monthly rainfall in the study area (2007-2011)
Source: Meteorology department, federal university technology, akure

Matured but not senescent *Abizia zygia* leaves were collected from existing plants along roadsides and on abandoned farmlands very close to the campus of Joseph Ayo Babalola University, Ikeji-Arakeji. The leaves were bulked, air dried for five days and composite samples were taken and analyzed for some nutrients (Nitrogen, Phosphorus, Potassium, Calcium and Magnesium) content, total carbon content, total soluble polyphenol and lignin. Total nutrients and total carbon were determined by the method described by [13] while total soluble polyphenols and lignin contents were determined using the method described by [14].

2.1.3 Plant samples collection and analysis

Maize yield determination

At maturity, four plants were selected from each plot, harvested and weighed and calculated on tone per ha based on this formula: Pp= (B + b) (L + l) / Lb × N [15] and [16], where Pp = Plant population, L = Length of the field, B = Breadth of the field, l = Length of spacing of the field, b= Breadth of spacing of the field and N = Number of cobs per field stand. At maturity (90 DAP), maize cobs were harvested from a plot size of 3m x 3m from the central rows of each plot. The ears were removed and cob length was measured from base to the top. The diameter of each cob was measured at the middle point using a veneer caliper. Maize cob were dried at 100°C to reduce the spoilage and to get a constant weight, they were shelled to separate the rachis and grains, the grains were oven dried again at 65°C for about 10 minutes to 14% moisture content and weigh to determine the grain weight. The 14% grain moisture content was confirmed with the use of grain moisture tester.

2.1.4 Soil samples collection and analysis

2.1.4.1 Physical analysis

Soil samples were collected before planting at40, 60 & 90 days after planting (DAP) using a 3.5 cm diameter soil auger. There were four auger points in each plot and samples were collected at 0-20cm depth. Samples from each plot were bulked and composite soil samples were collected and taken to the laboratory for analysis. The soil samples were air dried for few days, sieved to pass through 2mm mesh and chemically analyzed. Particle size analysis was done by hydrometer method [17] using sodium hexametaphoshate (Calgon) as dispersing agent, water holding capacity was determined using a pressure plate apparatus at 0.1kPa (field capacity), bulk density was determined on a core cylinder of dried soil and calculated as: Bulk density = Oven dry weight of soil (g)/Volume of the cylinder (cm3) and total Porosity was calculated as: Total Porosity = 100 (1-Db/Dp; Where Db = bulk density, Dp = particle density (2.65g/cm^3)

2.1.4.2 Chemical analysis

The pH was determined using a pH meter in slurries from a ratio of 1g soil to 2.5 cm^3 water. The organic carbon content of the soil was determined according to [18] dichromate oxidation method. The percentage organic matter content in the samples was calculated by multiplying the values of organic carbon by the conventional Van Bam meller factor of 1.724 based on the assumption that soil organic matter contains 58% carbon [19]. Total soil nitrogen was determined by macro kjeldahl methods [20]. Available phosphorus was extracted using Bray II method [21] and determined by spectrophotometer.

Exchangeable Na, K, Ca and Mg were extracted with $BaCl_2$ 0.1m [22] and analysed by atomic absorption. The exchangeable acidity was determined from 0.1N KCl extracts and titrated with 1.0N HCl.

Cation Exchangeable Capacity (CEC) was determined by summing up total exchangeable bases (TEBS) and total exchangeable acidity (TEA).

2.1.5 Basic economic data collection

Production cost (maize seeds, fertilizers, organic matter, weeding and fertilizer application) for maize production.

Economic analysis

a) Economic analysis was done using the prevailing market prices for inputs at planting and for outputs at the time the crop was harvested. All costs and benefits were calculated on hectare basis in Naira (N). The following concepts used in the partial budget analysis are defined as Mean grain yield is the average yield (t/ha) of each treatment.

b) The total production cost (TPC) is the sum of field cost of fertilizer plus the cost of organic manure and the costs of their applications.
c) The value of yield is the product of mean grain yield and the price of ton of maize grain.
d) The net benefit (NB)/ha for each treatment is different from the yield value and the total production cost.

2.1.6 Statistical analysis

Data were subjected to analysis of variance using the general linear model procedure (GLM) for randomized complete Block design (RCBD) in SAS [23]. Analysis of variance was computed to determine the significance of treatments. Mean separation was done using Duncan New Multiple Range Test.

3. RESULTS AND DISCUSSION

3.1 Result

3.1.1 Pre-planting physico-chemical properties

The pre-planting physico-chemical properties of soil of the experimental site showed the soil is slightly acidic with a value of 5.9. The organic matter has the value of 0.6%. The values of total soil nitrogen, phosphorous, potassium, calcium, magnesium, sodium, exchangeable acidity and cations exchange capacity were 0.067%, 3.44mg /kg, 0.15 cmol/kg of soil, 0.80cmol/kg of soil, 0.35cmol/kg of soil, 0.17 cmol/kg of soil, 1.36 cmol/kg of soil and 2.83 cmol/kg of soil respectively. The texture of the area is sandy clay loam.

3.1.2 Nutrient composition of *Albizia* leaves used for the experiment

Table 1 shows the nutrient composition of *Albizia* leaves used for the experiment.

The result show that nitrogen content in 2010 with the values of 4.71% was higher than of 2009 with the value of 4.11%, phosphorus content was also higher with value of 0.2% in 2010, while the value in 2009 was 0.19%, calcium content was higher in 2009 than that of 2010 with values of 0.27% and 0.25% respectively. Magnesium content was higher in 2010 than 2009 with value of 0.25% and 0.28% respectively. Carbon content was higher in 2009 with a value of 65.6% while in 2010, it was the value of 57.21%. The lignin content was higher in 2010 with a value of 34.24% while in 2009, the value was 33.4%. The polyphenols content value was higher in 2010 with a value of 7.2% and 6.5% in 2009 planting season.

3.1.3 Effect of organic and Inorganic fertilization on soil porosity, water holding capacity and soil bulk density

In 2009 planting season, the value of total porosity ranged from 35.99% in 0PK (plot without N and with the application of full doses of P and K) plot to 43.92% in the plot mulched with *Albizia zygia* leaves. The values of water holding capacity (WHC) ranged from 29.29% in the control plot to 35.67% in the plot mulched with *Albizia* leaves. The values of bulk density ranged from 1.67g/cm^3 in NPK plot to 1.49g/cm^3 in the plot mulched with *Albizia* leaves Table 2. There were no significant differences at $P<0.05$ among *Albizia* as well as in NPK and control plots.

In 2010, the values of soil total porosity ranged from 33.93% in the control plot to 44.15% in the plot mulched with *Albizia zygia* leaves. The values of soil water holding capacity ranged from 29.12% in the plot amended with NPK to 35.55% in the plot mulched with *Albizia* leaves. While the values of soil bulk density ranged from 1.47g/cm^3 with the plot mulched with *Albizia* leaves to 1.67g/cm3 in the 0PK plot. There were no significant differences among *Albizia* and *Albizia* + NPK plot.

Table 1. Nutrient composition of the *Albizia zygia* leaves used for the experiment

	2009	2010
Nitrogen (%)	4.11	4.71
Phosphorus (%)	0.19	0.2
Calcium (%)	0.27	0.25
Magnesium (%)	0.25	0.28
Potassium (%)	1.93	1.91
Carbon (%)	65.6	57.21
Lignin (%)	33.4	34.24
Polyphenols (%)	6.5	7.2

In 2011 planting season, the values of soil porosity ranged from 33.11% in NPK plot to 42.26% in the plot mulched with *Albizia zygia* leaves. The values of soil water holding capacity ranged from 26.18% in the plot amended with NPK to 32.64% in the plot mulched with *Albizia* leaves. The values of bulk density ranged from 1.52g/cm^3 in the plot mulched with *Albizia* leaves to 1.77g/cm3 in the plot amended with NPK.

3.1.4 Effect of organic and inorganic fertilization on soil pH, organic matter and soil phosphorous

In 2009 planting season, the values of soil pH ranged from 5.35 in the plots amended with *Albizia* + NPK to 5.93 in the plot applied with *Albizia*. The values of organic matter ranged from 1.05% to 1.74% in the control and *Albizia* respectively. The values of soil phosphorus ranged from 3.20mg/kg to 15.02mg/kg in the control and *Albizia* + NPK respectively. In 2010, the plot values ranged from 5.31 in OPK and *Albizia* + NPK plots to 5.92 in the plot mulched with *Albizia* leaves. While the values of Organic matter ranged from 1.07% to 1.79% in the control and *Albizia* plot respectively. The soil phosphorus values ranged from 3.29 mg/kg in the control plot to 12.7mg/kg in the plot mulched with *Albizia* leaves. In 2011, pH values ranged from 5.23 to 5.04 in NPK and *Albizia* plots respectively.

The organic matter values ranged from 0.69% in the control plot to 1.67% in the *Albizia* plot. The values of phosphorus ranged from 3.14mg/kg in the control plot to 12.16mg/kg in the plot amended with *Albizia* + NPK Table 3. Soil pH, organic matter and available phosphorous were all significant in all the years at $p \geq 0.05$

3.1.5 Effect organic and Inorganic fertilization total exchangeable bases and cation exchangeable capacity

Table 4 shows the effect of organic and Inorganic fertilizer on total exchangeable bases and cation exchangeable capacity. In 2009, total exchangeable bases (TEBs) values ranged from 1.34cmol/kg of soil in the control plot to 2.23 cmol/kg of soil in the plot mulched with *Albizia* leaves. The values of cation exchangeable capacity (CEC) ranged from 2.70 cmol/kg of soil to 3.75cmol/kg of soil in the control and *Albizia* mulched plots respectively.

In 2010 planting season, the values of total exchangeable bases ranged from 1.32 cmol/kg of soil to 2.25cmol/kg of soil in the control and *Albizia* plots respectively. The values of cation exchangeable capacity (CEC) ranged from 2.68 cmol/kg of soil to 3.77 cmol/kg of soil in the control and *Albizia* plots respectively.

In 2011, planting season the values of total exchangeable bases ranged from 0.81 cmol/kg of soil in the control plot to 1.48 cmol/kg of soil in the plot mulched with *Albizia* leaves. Also, the values of cation exchangeable capacity were low and ranged from 2.20 cmol/kg of soil to 3.04cmol/kg of soil in the control and *Albizia* mulched plots respectively.

3.1.6 Effect of organic and Inorganic fertilization on the total soil nitrogen at different sampling time

In 2009 planting season, 40 days after planting (DAP), the values of total nitrogen ranged from 0.067% to 0.074% in the control and *Albizia* + NPK plots respectively and showed significant difference at $p \leq 0.05$ among all the treatments.

Table 2. Effects of organic and inorganic fertilization on soil total porosity, water holding capacity (%) and bulk density (g/cm^3) for 2009, 2010 and 2011 planting seasons

Treatments	2009			2010			2011		
	TP	WHC	BD	TP	WHC	BD	TP	WHC	BD
ALBI ZIA	43.92^a	35.67^a	1.49^d	44.15^a	35.55^a	1.47^b	42.26^a	30.64^a	1.52^c
ALB +NPK	43.35^a	34.09^b	1.49^d	43.77^a	33.87^b	1.46^b	40.11^b	31.83^a	1.60^b
NPK	36.25^b	30.12^c	1.67^a	36.47^b	29.12^c	1.66^a	33.11^c	26.18^b	1.77^a
OPK	35.99^c	34.21^b	1.55^c	36.98^b	30.48^c	1.67^a	33.35^c	27.37^b	1.76^a
CONTROL	36.90^b	29.29^c	1.66^b	33.93^c	29.89^c	1.66^a	33.97^c	27.59^b	1.74^a

Means on the same column followed by the same letter are not significantly different at $P \leq 0.05$
TP = soil total porosity, WHC = water holding capacity, BD = soil bulk density
ALB = Albizia. statistical analysis: analysis of variance was computed to determine the significance of treatments. mean separation was done using Duncan new multiple range test

Table 3. Effect of organic and inorganic fertilization on soil pH, organic matter (%) and soil phosphorous

Treatments	2009			2010			2011		
	pH (H_2O)	OM (%)	P (mg/kg)	pH (H_2O)	OM (%)	P (mg/kg)	pH (H_2O)	OM (%)	P (mg/kg)
ALBIZIA	5.93^a	1.74^a	11.99^c	5.92^a	1.79^a	12.07^e	5.64^a	1.67^a	10.25^b
ALB +NPK	5.35^d	1.57^b	15.02^a	5.37^d	1.65^{ab}	16.02^b	5.37^c	1.67^a	12.16^a
NPK	5.64^c	1.27^c	14.54^b	5.65^c	1.28^{cd}	15.33^c	5.23^d	1.00^b	12.06^a
OPK	5.36^d	1.22^d	9.50^d	5.37^d	1.24	9.90^c	5.41^b	1.17^b	9.41^c
Control	5.85^b	1.05^e	3.20^e	5.84^b	1.07^{de}	3.29^e	5.24^d	0.69^c	3.14^d

Means in the same column followed by the same letter are not significantly different at $P \leq 0.05$. statistical analysis: analysis of variance was computed to determine the significance of treatments. mean separation was done using Duncan new multiple range test. OM = soil organic matter, N = soil total nitrogen, P = available phosphorus

Table 4. Effect of organic and inorganic fertilization on soil total exchange bases & cation exchange capacity

Treatments	2009		2010		2011	
	TEB	CEC	TEB	ECEC	TEB	ECEC
	(cmol/kg of soil)					
ALBIZIA	2.23^a	3.75^a	2.25^a	3.77^a	1.48^b	3.04^a
ALB +NPK	1.93^a	3.21^a	1.98^b	3.27^b	1.89^a	3.21^a
NPK	1.82^b	2.87^b	1.82^c	3.01^c	1.11^c	2.27^b
OPK`	1.64^b	2.76^b	1.63^d	2.90^d	0.95^c	2.32^b
Control	1.34^c	2.70^b	1.32^e	2.68^e	0.81^d	2.20^c

Means on the same column followed by the same letter are not significantly different at $P \leq 0.05$. ALB = Albizia. statistical analysis: analysis of variance was computed to determine the significance of treatments. mean separation was done using Duncan new multiple range test

At 60 days after planting the values of total nitrogen ranged from 0.066% in both control and OPK plots to 0.073% in the *Albizia* + NPK plot and there were significant differences among the treatments at $p \leq 0.05$. At 90 days after planting the values ranged from 0.064% in the control to 0.071% in both *Albizia* and *Albizia* + NPK plots (Table 5).

In 2010 planting season, at 40 days after planting, the values of total soil nitrogen ranged from 0.064% in both control and OPK plot to 0.075% in the *Albizia* + NPK plot. At 60 days after planting (DAP). It ranged from 0.063% in control and OPK plot to 0.075% in *Albizia* + NPK plots. At 90 days after planting the values ranged from 0.062% and *Albizia* + NPK plots. However, there were no significant differences between the control and OPK at 60 & 90 days after planting as well as between *Albizia* and *Albizia* + NPK at 90 days after planting at $p \leq 0.05$ (Table 5).

In 2011 planting season, at 40 days after planting (DAP), the values ranged between 0.059% to 0.073% in the control and *Albizia* + NPK respectively and significantly different at $p \leq 0.05$. At 60 days after planting the values ranged from 0.058% in the control plot to 0.070% in the *Albizia* plot. At 90 days after planting, the values ranged from 0.055% in the control plot to 0.072% in the plot mulched with *Albizia* and statistically different among the treatments at $p \leq 0.05$ (Table 5).

3.1.7 Effect of organic and Inorganic fertilization on maize cob length (cm) cob diameter (cm) and grain weight (t/ha)

In 2009 planting season, the values of cob length ranged from 18.35 cm in the control plot to 22.58cm in the plot treated with *Albizia* + NPK. However, there was no statistical difference between cob length in NPK and *Albizia* + NPK plots at $P \leq 0.05$ Table 6. The values of cob diameter ranged from 3.14cm to 5.60cm in the control and *Albizia* + NPK plots respectively. In the same vein, there was no statistical difference between the plots amended with NPK and *Albizia* + NPK at $P \leq 0.05$ Table 6. Grain weight values ranged from 0.74 t/ha in the plot without amendment to 1.33 t/ha in the plot amendment with *Albizia* + NPK.

Table 5. Effects of organic and inorganic fertilization on total soil nitrogen (%) at different sampling time

Treatments	2009			2010			2011		
	40*	60*	90*	40*	60*	90*	40*	60*	90*
ALBIZIA	0.073[b]	0.072[b]	0.071[a]	0.074[a]	0.072[a]	0.072[a]	0.072[a]	0.070[a]	0.0720[a]
ALB +NPK	0.074[a]	0.073[a]	0.071[a]	0.075[a]	0.073[a]	0.072[a]	0.073[a]	0.069[a]	0.0710[a]
NPK	0.070[c]	0.069[c]	0.067[b]	0.068[b]	0.066[b]	0.067[b]	0.065[b]	0.060[b]	0.0590[b]
OPK	0.068[d]	0.066[d]	0.065[c]	0.064[c]	0.063[c]	0.062[c]	0.060[c]	0.060[b]	0.0570[c]
Control	0.067[d]	0.063[d]	0.064[c]	0.064[c]	0.063[c]	0.062[c]	0.059[c]	0.058[c]	0.0550[d]

*Means on the same column followed by the same letter are not significantly different at P<0.05. ALB = Albizia, *=days after planting. statistical analysis: analysis of variance was computed to determine the significance of treatments. mean separation was done using Duncan new multiple range test*

Table 6. Effect of organic and inorganic fertilization on maize cob length (cm), cob diameter (cm) and grain weight (t/ha)

Treatments	Cob length			Cob diameter			Grain weight		
	2009	2010	2011	2009	2010	2011	2009	2010	2011
ALBIZIA	20.24[b]	19.83[b]	17.98[c]	4.14[b]	5.14[b]	3.11[c]	1.17[c]	1.44[b]	1.13[c]
ALB +NPK	22.58[a]	22.65[a]	20.90[a]	5.60[a]	5.24[a]	5.13[a]	1.33[a]	1.93[a]	1.33[a]
OPK	19.90[b]	18.98[b]	16.67[d]	4.05[b]	4.57[d]	3.04[d]	1.05[c]	1.15[d]	1.01[d]
NPK	22.40[a]	22.45[a]	19.35[b]	5.46[a]	5.04[c]	4.51[b]	1.26[b]	1.18[c]	1.15[b]
Control	18.35[c]	16.70[c]	15.27[e]	3.14[c]	4.05[e]	2.05[d]	0.74[d]	0.67[e]	0.33[e]

Means on the same column followed by the same letter are not significantly different at P<0.05. ALB = Albizia, statistical analysis: analysis of variance was computed to determine the significance of treatments. mean separation was done using Duncan new multiple range test

In 2010 planting season, the values of maize cob length ranged from 16.70 cm in the control plot to 22.65 cm in the plot amended with *Albizia* + NPK and showed no significant difference in the plots amended with NPK and *Albizia* + NPK at $P \leq 0.05$. The cob diameter values ranged from 4.05cm to 5.24cm in the control and *Albizia* + NPK plots respectively. On the other hand, the grain weight ranged from 0.67 t/ha in the control plots to 1.93 t/ha in the plot amended with *Albizia* + NPK.

In 2011 planting season, the cob length values ranged from 15.27cm in the control plot to 20.90 cm in the plot amended with *Albizia* + NPK. The cob diameter ranged from 2.05 cm to 5.13 in the control and *Albizia* + NPK plots respectively. The grain weight values ranged from 0.33 t/ha in the control plot to 1.33 t/ha in the plot amended with *Albizia* + NPK.

3.1.8 Profitability analysis of maize yield, organic and inorganic fertilizers for 2009 and 2010 planting seasons

In 2009 planting season, the total yield (discounted at 10%) ranged from 0.67 t/ha in the control plot to 1.2 t/ha in the plot amended with Alibizia + NPK, and there were significant among the treatments at $P \leq 0.05$. The yield value in (Nigerian Naira) ranged from N46, 900 (USD 296.84) in the control plot to N84, 000 (USD 531, 65) in the plot amended with *Albizia* and NPK. The cost of production ranged from N39, 690 (USD 251.20) in the control plot to N50, 055 (USD 316.99) in the plot mulched with *Albizia* leaves. The net profit, values ranged from N7, 210 (USD 45.63) in the control plot to N37, 459 (USD 237.08) in the plot amended with *Albizia* + NPK (Table 7).

In 2010 planting season, the yield value ranged from 0.60 t/ha to 1.3 t/ha in the control and *Albizia* + NPK plot respectively, however, there was no significant different at $P \leq 0.05$ between OPK and NPK plots. The yield values ranged between N42, 000 (USD 265.82) to N121, 800 (USD 770.89) in the control and *Albizia* + NPK plots respectively. The cost of production values ranged from N39, 218 (USD 248.22) to N48, 667 (USD 188.63) in the control plot and *Albizia* mulched plots. There was significant differences among the treatment at $P<0.05$. The net profit, values ranged from N2782 (USD 17.61) to N76204 (USD 482.30) in the control and *Albizia* + NPK plots respectively. There were significant differences among the treatments at $P \leq 0.05$. The net profit were higher in all the plots except in the control and NPK plots (Table 7).

Table 7. Profitability analysis of maize production, organic and Inorganic fertilizers for 2009 and 2010 planting seasons

Treatments	2009				2010			
	Yield	Value of yield	Cost of production	Net profit	Yield	Value of yield	Cost of production	Net profit
ALBIZIA	1.06[c]	73,500[c]	50,085[a]	23,415[b]	1.3[d]	91,000[b]	48,667[a]	42,333[b]
ALB + NPK	1.2[a]	84,000[a]	46,541[b]	37,459[a]	1.74[a]	121,800[a]	45,596[b]	76,204[a]
OPK	0.95[d]	66,500[d]	43,942[c]	22,557[b]	1.04[c]	72,800[d]	44,415[c]	28,385[d]
NPK	1.13[b]	79,100[b]	42,998[d]	36,102[a]	1.06[c]	74,200[c]	42,525[d]	31,675[c]
CONTROL	0.67[e]	46,900[e]	39,690[e]	7,210[c]	0.60[d]	42,000[e]	39,218[e]	2782[e]

Means on the same column followed by the same letter are not significantly different at P<0.05. statistical analysis: analysis of variance was computed to determine the significance of treatments. mean separation was done using Duncan new multiple range test

3.2 Discussion

3.2.1 Pre-planting physico-chemical properties of soil of the experimental site

The texture of the area was sandy clay loam. This may be attributed to the lithology of the parent material [24]. Nitrogen, phosphorous, potassium, calcium, magnesium and sodium contents were low in the study area. This may be due to the over cropping, leaching of soluble cations, soil erosion and lack of proper land management practices in the area. The acidic nature of the soil may be due to the leaching of soluble cations observed in the area as well as the distribution of exchangeable acidity [25]. Olojugba, 2002, [26] Reported that sufficient exchangeable acidity is responsible for soil acidity. Also [25] observed that leaching of Na, K, Ca and Mg were largely responsible for the development of acidity in the soil. The low value of organic matter in the area may be due to continuous cropping without the addition of organic manure.

3.2.2 Nutrient concentration in *Albizia* leaves used for the experiment

In 2009 and 2010 cropping seasons, the concentration of nitrogen in the leaves showed that *Albizia* leaves are good source of nitrogen; this reason might be due to the nitrogen fixing ability [27]. The lignin, polyphenols and carbon were higher which makes *Albizia* resistant to decomposition thereby enabling it to stay more on soil as organic humus [27]. The low values of calcium, magnesium, potassium and phosphorus suggested that inorganic source of these nutrients or plant biomass richer in these nutrients is needed to be integrated when using *Abizia* leaves as organic materials.

3.2.3 Effect of organic and inorganic fertilization on soil total porosity, water holding capacity and soil bulk density

The moderate to higher values of soil total porosity and water holding capacity in most of the *Albizia* amended plots over the low values in the control and NPK plots may be due to moderate organic matter content as this encouraged better and higher activities of soil micro-organisms which in turn enhanced better soil structure and porosity [28]. They were of the opinion that organic inputs such as compost, animal manure, crop residues and green manures are a good way of enhancing both physical, chemical and biological soil properties which includes soil porosity and available water capacity. Anikwe 2000, [15] Submitted that moderate to higher total porosity in the soil was beneficial because as it improves gaseous exchange and creates healthier environment for root elongation, expansion and nutrient uptake. The low to moderate soil bulk density in the *Albizia* amended plot over its higher values in un-amended and NPK plots may be due to the distribution of soil organic matter. Soil organic matter contributes to improving soil structure, aggregate stability, decrease soil bulk density and increase the percentage of pore spaces [8].

3.2.4 Effect of organic and inorganic fertilization on soil pH, organic matter and soil phosphorous

There were significant differences in soil pH in 2009 and 2010 planting season and only the plot mulched with *Albizia* at 5 ton/ha had increased pH while other treatments had decreased soil pH. This is consistent with the findings of Opala 2011, [28] who attributed a similar situation to the proton exchange between the soil and added *Albizia* which contains more phenolic, humic–like material. Another mechanism that has been proposed to explain the increase in soil pH by such materials as *Albizia* is the specific adsorption of humic material and/or organic acids (the products of decomposition of organic materials) onto hydrous surfaces of aluminum and iron oxides by ligand exchange with corresponding release of OH^- [29]. The moderate values of organic matter recorded in 2009 and 2010 planting season in the plots treated with *Albizia* and *Albizia* + NPK over the control and OPK plots might be due to the decomposition of the mulch materials used which gave rise to the build-up of soil organic matter in this study. Moderate organic matter was found in the plots mulched with more resistant mulch materials which could be attributed to their ability to decompose more slowly over time leading to the build-up of organic matter as it was observed in the plot mulched with *Albizia* and NPK + *Albizia*. Schmel 2007, [30] reported that litter which decomposes slowly favours the build-up of soil organic matter. They further observed that the decomposition rate of this resistant litter was enhanced by mixing it with nutrient–rich litter [30]. This type of increase is attributed to N-transfer from nutrient–rich to nutrient poor litter [31]. In 2011 planting season, the soil organic matter was low except for the plots mulched with *Albizia* and NPK + *Albizia* which are attributable to build–up of soil organic matter.

3.2.5 Effect of organic and inorganic fertilization on total nitrogen

The moderate value of total nitrogen in *Albizia* and *Albizia* + NPK plots and in all the sampling days (40, 60 and 90 DAP) and across the years may be due partly to the nitrogen fixing ability of *Albizia* Table 1, in addition to NPK treatments and in part to the soil organic matter distribution. [32] Stated that organic matter of plant origin serves as a reservoir of nutrient such as nitrogen, sulphur, phosphorous and many minor elements. In 2011, plots treated with *Albizia* had the highest nitrogen content at 90 DAP due to build-up of organic matter by *Albizia* because of its slow decomposition which might lead to build–up of nutrient [30]. Salamanca, Kaneko & Katagiri 1998, [31] submitted that the build-up of soil nutrients may be attributed to N-transfer from nutrient-rich litter to nutrient-poor litter.

3.2.6 Effect of organic and inorganic fertilization on maize cob length, cob diameter and grain weight

The maize cob length, cob diameter and grain yield were higher in the plot mulched with *Albizia* + NPK with the highest in 2009, 2010 and 2011 planting seasons. This may be due to the supplying of large quantities of needed nutrients as well as improving the soil physico-chemical properties [27]. The improvement in soil fertility exceeded expectation in an integrated system probably because of combined effect of soil conservation, nutrient enrichment, enhancement of biological activities and improvement in moisture retention capacity [33]. Adekayode and Olojugba, 2010 [34] and Awodun, Ojeniyi, Adeboye and Odedina, 2007 [35] showed that highest yield of maize grain were obtained in the plot amended with NPK and wood ash due to the improvement in the soil fertility.

3.2.7 Profitability analysis of maize production, organic and inorganic fertilization

The profitability analysis of maize production was discounted at 10% for all treatments in 2009 and 2010 cropping seasons. Generally, all the treatments had higher net benefit over the control in the two seasons. In 2009 planting season, the low net benefit recorded in the control plot may be due to low yield as a result of non amendments while the low net benefit recorded in plots amended with *Albizia* could be attributed to high labour cost associated with this practice. On the other hands, higher net benefit recorded in the plot amended with NPK + *Albizia* might be due to moderate labour cost as a result of application of the component treatments at their half dose each which invariably reduced their total variable cost.

4. CONCLUSION AND RECOMMENDATIONS

Improving soil quality and crop production on farms is a major issue for agricultural research. This study has contributed to knowledge on some of the processes governing integrated nutrient management using low income technology such as agro forestry systems in the form of biomass transfer. This work has shown that there is a potential to improve yield of maize by integrating NPK fertilizer with agroforestry species (*Albizia zygia*). It was also shown that integrating *Albizia zygia* with NPK fertilizer increased the release of soil nutrients as manifested by maize yield increase. Thus, the slowly-decomposing litter of *Albizia zygia* is useful contribution to improve soil fertility, rather than being burnt as currently practiced by the farmers. It was also established in this study that combined used of *Albizia* + NPK fertilizer resulted in yields of maize grain, nitrogen, phosphorus, pH compared to those achieved with the application of either inorganic fertilizer or *Albizia* alone. Organic matter was found to be necessary for soil health; hence, its application is advised. Application of *Albizia* had much protective function on both physical and chemical properties of the soil because of its slow decomposing pattern. *Albizia* alone has more residual effects on soil physical and chemical properties, while *Albizia* + NPK have residual effects on maize yield. Labour requirement is very high for *Albizia* which could have affected negatively its full adoption. The financial benefits to farmer were more when using NPK + *Albizia* over all other treatments which made it more viable option among the treatments.

The following recommendations are offered;

(a) Further investigations into effects (i.e. crop yield and nutrient up take) of *Albizia zygia* biomass (litter) either applied directly or used to produce compost is required before recommendations can be made about its use as a source of organic fertilizer.
(b) Labour requirement studies should be further investigated by comparing biomass transfer and green manuring methods.

COMPETING INTERESTS

Authors have declared that no competing interests exist.

REFERENCES

1. Gisladoltir G, Stocking M. Land degradation control and its global environmental benefits. Land degradation and Development. Bationo A, Kimetu J, Ikerra S, Kimani S, Mugendi D, Odendo M, Silver M, Swift MJ, Sagina N. The African network for soil biology and fertility: New challenges and opportunities in: Managing Nutrient cycles to sustain soil fertility in Sub-saharan Africa. A. Bationo (Ed). Academy Science Publishers. Nairobi. 2005;16:99-122.
2. Bationo A, Kimetu J, Ikerra S, KImani S, Mugende D, Odendo M, Silver M, Swift MJ, Sanginga N. The African network for soil biology and fertility: New challenges and opportunities In: Managing Nutrient cycles to sustain soil fertility in Sub-Saharan Africa. Bationo A, (Ed.). Academy Science Publishers Nairobi. 2004;341.
3. Zougmore R. Integrated water and nutrient management for sorghum production in semi-arid Burkina Faso. Tropical Resource Management. Wageningen University, Wageningen. The Netherlands. 2003;45.
4. Sivakumar MVK, Wallace JS. Soil water balance in the Sudan-Sahelian Zone: Need, relevance and objectives of the workshop. In: Soil water balance in the Sudan-Sahelian zone. Sivakwnar MVK, Wallance JS, Renard C, Giroux C, (Eds.). IAHS. Press, Institute of Hydology, Wallingford, UK. Proceedings of the International Workshop, Niamey. 1991;3-10.
5. Andren O, Kihara J, Bationo A, Vanlauwe B, Katterer T. Soil climate and decomposer activity in Sub-Saharan Africa estimated from standard weather station data-A simple climate index for soil carbon Balance Calculations. Ambio. 2007;36: 379-386.
6. Bationo A, Buekert A. Soil organic carbon management for sustainable land use in Sudano-Sahelian West Africa. Nutrient cycling in Agro ecosystems. 2001;61:131-142.
7. Harris F. Review paper: Management of manure in farming systems in semi-arid West Africa. Experimental Agriculture. 2002;38:131-148.
8. Sylvia MD, Fuhrmann JJ, Hartel PG, Zuberer DA. Principles and applications of soils microbiology. Practice hall inc. upper saddle river New Jersey, USA. 1999;550.
9. Webber J, Karczewska A, Drozd J, Licznar M, Licznar S, Jamroz E, Kocowicz A. Agricultural and ecological aspects of a sandy soil as affected by the application of municipal solid waste compost. Soil Biology and Biochemistry. 2007;39:1294-1302.
10. Sanchez D. Improved fallows come of age in the tropics. Agroforestry systems. 1999;47:3-12.
11. Saweda LT, Babatunde O, Shehu S, James S. Nigeria strategy support program (NSSP) working paper: A review of fertilizer policy issues in Nigeria, International food policy research institute publication. 2010;5.

Available: www.ifpri.org

12. Palm CA, Myers RJK, Nanda SM. Combined use of organic and inorganic nutrient sources for soil fertility maintenance and replenishment in: Replenishing soil fertility in Africa. SSSA special publication. SSSA. Madison WI, Buresh RJ, Sanchez PA, (Ed). 1997;51.
13. Anderson JM, Ingram SJ. Tropical biology and fertility: A handbook of methods. 1993;587.
14. Constantinides M, Fownes JH. Nitrogen mineralization from leaves and litter of Tropical plants, Relationship to nitrogen, lignin and polyhenols concentrations. Soil Biochemistry. 1999;26:49-55.
15. Anikwe MAN. Use of numerical index methods for quantifying the productivity of some soils in Southeastern Nigeria. University of Nigeria, Nsukka. PhD Thesis (Unpublished). 2000;142.
16. Adeboye OC, Ajadi SO, Fagbohun AB. An accurate mathematical formula for estimating Plant population in four dimensional fields of sole crop. Journal of agronomy. 2006;289-292.
17. Gee GW, Bauder JW. Particle-size analysis. In Methods of Soil Analysis Part 1. Klute A, (edi.). Soil Science Society of America Book Series 5, Madison, Wisconsin, USA. 1986;383-411.
18. Walkley A, Black A. Method for determining soil organic matter and a proposed modification of the chromic acid titration method. Soil Science. 1934;37:29-38.
19. Allison FE. Soil organic matter and its roles in crop production. Elsevier scientific publication Co. Amsterdam. 1973;637.
20. Bremner JM. Nitrogen method of soil analysis, Part 1 In: Black CC, (ed.). American Society of Agro forestry. Madison. Wisconsin, USA. 1965;1149-1178.
21. Bray RH, Kurtz LT. Determination of total organic carbon and available form of phosphorous in soil. Journal of Soil Sci. 1945;59:39-45.
22. Hendershot WH, Lalande H, Duquette M. Soil reaction and exchangeable acidity. In: Soil Sampling and Methods of Analysis Carter MR, (eds). Lewis publishers, USA. 1993;141-145.
23. Institute SAS. SAS/Start User's Guide, 2 version 9.2. SAS Inst. Cary N.C, USA; 2005.
24. Olojugba MR. Characterization, classification and fertility evaluation of the forest soil of basement complex in the south western Nigeria. Evidence from forestry plantations of the Federal University of Technology, Akure, Nigeria. International Journal of Agriculture and Food Science. 2010;1:208-222.
25. Enwezor WO, Udo EJ, Subulo RA. Fertility status and productivity of "Acid-sands" in Acid sand of Southern Nigeria. SSSN Special publication, Mon. 1981;1:56-73.
26. Olojugba MR. Soil characterization for the forestry plantation site of the Federal University of Technology, Akure, Nigeria, MTech. Thesis (Unpublished). 2002;74.
27. Gnankambary Z. Compost and fertilizer mineralization, effects on soil and Harvest in parkland systems in the south-Sudanese zone of Burkina Faso, Swedish University of Agricultural Sciences, Umea. PhD thesis (Unpublished). 2007;44.
28. Harris F. Review paper: Management of manure in farming systems in semiarid West Africa. Experimental Agriculture. 2002;38:131-148.
29. Opala PA. Comparative soil chemical properties materials on selected soil chemical properties and nutrient uptake by maize in an Acid soil. Arch. Applied sci. Rec. 2011;3(1):96-107.
30. Schmel JP, Hettenschwiler S. Nitrogen Transfer between decomposing leaves of different N status. Soil Biology and Biochemistry. 2007;39:1428-1436.
31. Salamanca EF, Kaneko N, Katagiri S. Effects of leaf litter mixtures on the decomposition of *Quercus serrata* and *Pinus densiflora* using field and laboratory microcosm methods. Ecological Engineering. 1998;10:53-73.
32. Schlecht E, Buerkert A, Tielkes E, Bationo AA. Critical analysis of challenges and opportunities for soil fertility restoration in Sudano-Sahelian West Africa. Nutrient cycling in Agro ecosystems. 2006;36:109-136.
33. Upawansa GK. New Kekulam rice cultivation: A practical and scientific ecological approach. Rebuilding lost soil fertility. LEISA, ILEIA Newsletter. Leusden, Netherlands. 1997;13(3):20-21.

34. Adekayode FO, Olojugba MR. The utilization of wood ash as manure to reduce the use of mineral fertilizer for improved performance of maize (*Zea mays* L.) as measured in the chlorophyll content and grain yield. Journal of Soil Science and Environmental Management. 2010;1(3):40-45.

35. Awodun MR, Ojeniyi SO, Adeboye A, Odedina SA. Effect of oil palm bunch refuse ash on soil and plant nutrient composition and yield of maize. American-Eurasian Journal of Sustainable Agriculture. 2007;1(1):50-54.

21

Yielding Reliability of Legumes Grown as Stubble Catch Crop

Edward Wilczewski[1*] and Zbigniew Skinder[1]

[1]Department of Agrotechnology, Faculty of Agriculture and Biotechnology, UTP University of Science and Technology, 85-225 Bydgoszcz, Poland.

Authors' contributions

This work was carried out in collaboration between both authors. Author EW performed field study, developed the statistical analysis and wrote the first draft of the manuscript. Author ZS managed the field study. Both authors designed the field study, read and approved the final manuscript.

Editor(s):
(1) Marco Aurelio Cristancho, National Center for Coffee Research, CENICAFÉ, Colombia.
Reviewers:
(1) Palmiro Poltronieri, National Research Council of Italy, Agriffod Dept. CNR-ISPA, Italy.
(2) Anonymous, Nigeria.

ABSTRACT

Aims: The aim of the study was to determine the yielding reliability of serradella, field pea and yellow lupine, sown as stubble catch crop in August on a typical *Alfisol*.
Study Design: The randomized complete block design with four replications.
Place and Duration of Study: Field study was conducted in 2005–2009, at the Research Station in Mochełek near Bydgoszcz (17°51′ E; 53°13′ N) in Midwest Poland.
Methodology: The objects of the study were three leguminous plants: Serradella 'Igela', field pea 'Grapis' and yellow lupine 'Legat', grown as stubble catch crop on *Alfisols* formed of a sandy loam. The main studied features were yield of green matter and post harvest residues of plants. Moreover yield reliability in years for particular plants were calculated.
Results: The total yield of dry biomass of plants grown as stubble catch crop ranged from 1.76 $Mg \cdot ha^{-1}$ in 2005, with shortage of rainfall, to 3.65 $Mg \cdot ha^{-1}$ in 2006, which was characterized by a high rainfall in August and high temperatures in September. The total dry matter yield of field pea and yellow lupine was significantly higher than from serradella. The highest aboveground dry matter yield was obtained from pea, and the lowest from serradella. Proportion of post-harvest residues in the total dry weight yield usually amounted to 30-40%. Yellow lupine produced a significantly higher average from 5 years yield of post-harvest residues than field pea and

**Corresponding author: E-mail:* edward@utp.edu.pl;

serradella.
Conclusion: Serradella, field pea and yellow lupine sown as stubble catch crop within 3-12 August are characterized by a high reliability of dry matter yield, from approximately 80.5% (yellow lupine) to 83.8% (field pea). Serradella matched the other plants in terms of the yield reliability, but the yield of green mass of the plants was significantly lower than that of pea and lupine.

Keywords: Catch crop; field pea; reliability; serradella; yellow lupin; yielding.

1. INTRODUCTION

The value of legumes in shaping the physical, biological and chemical properties of soil is well-known [1-3]. These plants used as green manure contribute to the improvement of the enzymatic activity of the soil, increase the content of soil available forms of nitrogen and improve the yield of plants grown in the succeeding year [2,4]. Despite this, the proportion of legumes in crop structure in Poland is very low [5]. This is due to the relatively low demand for green fodder for animals. At the current stock of cattle, nutritional requirements of this group of animals are covered mainly using feeds produced on permanent grassland. In addition, farmers are reluctant to grow these plants as catch crops intended for green manure. This is due to the conviction of not reliable yielding of these plants cultivated as catch crops sown in August.

To achieve high yields, these plants are recommended as stubble catch crop sown in July [6]. Currently, such sowing time of stubble catch crop in Central and Eastern Europe is mostly impossible because of the late harvest time of cereals, which are followed by these crops.

However, studies published in the past 20 years indicate that some legumes can produce satisfactory yields even at sowing approximately August the 10th - 20th [7]. The condition of the crop to go is mainly adequate supply of water in the soil during the planting and good thermal conditions in September, enabling fast growth of biomass [8-10]. In the study by Rinnofner et al. [9] dry matter yield of legumes mixture (field pea + common vetch + chickling vetch) was 3.46 Mg·ha^{-1} in the year with favorable rainfall patterns and only 1.0 Mg·ha^{-1} in the year characterized by a substantial deficiency of rainfall throughout the growth period of plants grown as the catch crop. In a study by Zaniewicz-Bajkowska et al. [11], dry matter yield of faba bean and serradella sown as the catch crop on July the 21st amounted to 5.4 Mg·ha^{-1} and 2.9 Mg·ha^{-1} respectively. In this yield 135 and 79 kg·ha^{-1} N, respectively, was accumulated. The reduction in dry matter yield of serradella and faba bean resulting from sowing delay from July the 21st to August the 18th amounted to 27.6 and 55.6%. The cited studies indicate that the sowing time is an important factor for the yield of legumes grown as the catch crop, but its delay until mid-August does not exclude the possibility of obtaining the yield of the importance as fertilizer for plants grown in the following year. More important is the amount of rainfall during the planting, conditioning the proper course of seed germination.

The aim of the study was to determine the yielding reliability of serradella, field pea and yellow lupine, sown in stubble catch crop in August on a typical *Alfisol.*

2. MATERIALS AND METHODS

2.1 Experimental Design and Conditions

The field study was conducted in 2005 – 2009, at the Research Station in Mochełek near Bydgoszcz (17º51′ E; 53º13′ N) in Midwest Poland. It was a continuation of the study carried out in 2002-2004, published in the paper "Value of selected papilionaceous crops grown in stubble intercrop on light soil. Part I. Biomass yield and plant health status" [12]. The objects of the study were three leguminous plants: Serradella 'Igela', field pea 'Grapis' and yellow lupine 'Legat'. The randomized complete block design with four replications was used. The area of plots for harvest was 27 m^2. Previous crop for plants grown as catch crops was spring barley. The field experiment was performed in *Alfisols* [13], formed of sandy loam, of IVa quality class. The soil had slightly acidic reaction (pH in 1M KCl 6.2) and contained very high amount of available phosphorus and potassium (107 and 325 mg of P and K in kg of dry soil, respectively). Content of total nitrogen and organic carbon in the soil amounted to 660 and 7200 mg in kg of dry soil respectively. Field water capacity of soil in the study area ranged from 215 to 270 mm at the depth from 0 to 100 cm [14].

The pluviotermic conditions during the period of study was determined on the basis of Sielianinov hydrothermal coefficient [15]:

$$\text{Sielianinov coefficient (k)} = \frac{P\ 10}{\sum t} \qquad (1)$$

Where P is total precipitation in the given period [mm], Σt – total of average daily air temperatures from that period [°C].

2.2 Cultivation Conditions and the Course of Experiment

After harvest of spring barley mineral fertilizers were spread (26 kg·ha^{-1} P and 66 kg·ha^{-1} K) and soil disking was performed to mix the stubble and fertilizers with the soil. Subsequently plowing to the depth of about 12 cm was performed. Before sowing the soil was cultivated using a cultivator with a crumbler roller.

The sowing parameters were as follows:

- Sowing time: 3-12 August (Table 1);
- Row-space: 12.5 cm;
- Sowing depth: Serradella – 2-3 cm; field pea – 4-5cm; yellow lupine – 3-4cm;
- Sowing rates [kg·ha^{-1}]: serradella – 60; field pea – 150; yellow lupine – 140.

The number of plant per m^2 was determined on each experimental plot 3 weeks after sowing. In the second part of October, after 70-79 days after sowing, the yield of green matter of plants was determined.

Table 1. Parameters concerning catch crops sowing and harvesting

Year	Sowing time	Harvest time	Growth day
2005	04.08	20.10	78
2006	11.08	19.10	70
2007	03.08	17.10	76
2008	12.08	21.10	71
2009	11.08	28.10	79

Dry weight of aboveground biomass was determined based on 1kg of green mass samples taken from each plot. In order to determine the yield of fresh and dry weight of the post-harvest residues, soil monoliths measuring 25·25·25 cm were collected from each plot and screened through a sieve with a mesh size of 4 mm. Subsequently, the samples were rinsed with tap water. Then after initial drying the samples on absorbent paper they were weighed (fresh matter yield), dried in an oven at 50°C and reweighed (dry matter yield).

2.3 Statistical Analysis

The results were statistically analyzed using analysis of variance, which was made with a computer program AWAR, developed in the Department of Agrometeorology and Applied Informatics, Institute of Soil Science and Plant Cultivation in Pulawy [16]. The analysis was conducted for the randomized block design. The significance of differences was determined using Tukey's half-interval at the significance level α = 0.05. To estimate the relation between the total precipitation in July and August and dry matter yield of plants regression analysis was carried out using Statistica for Windows version 10 software.

Yield reliability coefficients (Rc) were calculated using the formula by Rudnicki and Wasilewski [17]:

$$\mathrm{Rc} = \frac{\bar{x}}{\bar{x}+s} \cdot 100 \qquad (2)$$

Where $\bar{x}$ is the average yield of the years of research, while the s is standard deviation of yield in those years.

3. RESULTS AND DISCUSSION

3.1 Weather Conditions in the Study Region

Weather conditions in the area of research were very different in particular years. Most adverse conditions occurred in 2005, in which precipitation totals in all months of the catch crop growth period were significantly lower than the long-term average total for this region (Table 2). Also in the month preceding the sowing of seeds (July) precipitation accounted for only half of the total typical of this month. Due to the shortage of water available for plants also high temperatures in September were undesirable factor, which enhanced the negative effects of rainfall deficit.

In 2006, the supply of plants in rainwater was better suited to the growing of plants grown in stubble catch crop. Very high rainfall in August and average in September favored dynamic growth of plants. It was also possible due to the relatively high air temperatures in this month.

In subsequent years, the plants were well supplied with rainwater in July (2007 and 2009) or in July and August (2008). However, relatively low air temperatures were observed in September (2007 and 2008) and October (2007 and 2009).

Evaluation of pluviotermic conditions made based on the Sielianinov hydrothermal coefficient [15] showed that throughout the five-year study period there was not a single season free of drought during the catch crop growth period (Table 3).

However, as seen from the other studies conducted in this area, for yielding of plants grown as stubble catch crop most important is to supply the plant with water during July-August. With total rainfall in July and/or August of at least 100 mm high yields of dry matter can be achieved, even in drought conditions during September and October [18].

Thus, the conditions occurring in the period 2006-2009 can be considered favorable for the cultivation of stubble catch crop. In each of these years, at least in one of those months Sielianinov coefficient was above 1.

3.2 Plant Density after Emergence and Yield of Dry Matter

Plant density after emergence varied between years (Table 4). It was the lowest in 2005, in which a drought occurred both in July and in August. The highest plant density after emergence was found in 2008, in which the rainfall in July was average and in August very high. Particular species grown as stubble catch crop unevenly responded to environmental conditions during germination and emergence. Coefficient of variation of plant density after emergence in the years of the study was the lowest for the pea and the highest for serradella. The different response of individual species may be due to different depths of sowing. The seeds of pea were sown deepest, therefore they can make better use of water accumulated in the soil than the other plants.

Table 2. Mean air temperature and total rainfall, in period from July to October

Years	July	August	September	October	Mean/total VII X
		Mean air temperature [°C]			
2005	19.4	16.3	14.8	8.7	14.8
2006	22.4	16.6	15.2	9.6	16.0
2007	18.0	17.8	12.4	6.9	13.8
2008	19.2	17.8	12.4	8.4	14.5
2009	18.6	18.2	13.7	6.3	14.2
Mean 2005-2009	19.5	17.3	13.7	8.0	14.7
Mean 1949-2009	18.0	17.4	13.2	8.2	14.2
		Total monthly rainfall [mm]			
2005	33.6	43.4	17.8	15.1	109.9
2006	24.2	129.0	40.6	12.1	205.9
2007	104.7	42.1	37.6	19.9	204.3
2008	58.7	95.5	20.2	80.0	254.4
2009	118.0	17.6	34.4	66.2	236.2
Mean 2005-2009	67.8	65.5	30.1	38.7	202.1
Mean 1949-2009	70.8	52.0	41.0	32.7	196.5

Table 3. Ploviotermic conditions in period from July to October

Year	July	August	September	October
2005	Very dry*	Dry	Extremely dry	Very dry
2006	Extremely dry	Very wet	Dry	Very dry
2007	Rather wet	Dry	Rather dry	Dry
2008	Dry	Rather wet	Very dry	Extremely wet
2009	Wet	Extremely dry	Dry	Extremely wet
Mean 2005-2009	Rather dry	Rather dry	Dry	Rather wet

**Ploviotermic conditions based on Sielianinov coefficient (k): extremely dry – k≤0.4; very dry – 0.4>k≤0.7; dry – 0.7>k≤1.0; rather dry 1.0>k≤1.3; optimal – 1.3>k≤1.6; rather wet – 1.6>k≤2.0); wet – 2.0>k≤2.5; very wet – 2.5>k≤3.0; extremely wet – k>3.0 [15]*

Table 4. Plant density after emergence [$no \cdot m^{-2}$]

Year	Serradella	Field pea	Yellow lupine	Mean
2005	121	78	48	82
2006	177	96	100	124
2007	269	76	73	139
2008	328	123	81	177
2009	284	80	55	140
Mean	236	91	71	133
Variation coefficient [%]	35.8	21.8	29.1	25.8

Yields of individual species grown as stubble catch crop was varied in the years of the study (Table 5). The yield of green matter of plants grown as stubble catch crop ranged from 0.66 - 0.87 $Mg \cdot ha^{-1}$ dry mass in 2005, characterized by the shortage of rainfall in the whole growth period, to 1.92 - 2.60 $Mg \cdot ha^{-1}$ in 2006, characterized by high rainfall in August and relatively high temperatures in September. The highest yields of green matter were obtained from pea, and the lowest of serradella. Green matter yield of yellow lupine was in 2 of the 5 years higher than in serradella and only in 1 out of 5 years it was significantly lower than in pea. The average of 5 years of research post-harvest crop residues of lupine was significantly higher than pea and serradella which in 4 out of 5 years of research produced a similar mass of post-harvest residues. The total dry matter yield of pea and yellow lupine was significantly higher than that obtained from serradella. The proportion of post-harvest residues in total dry weight yield was most commonly 30-40%. Extremely high proportion of these parts of plants in the yield was obtained in the dry year 2005, when it amounted to 57.4%. This was due to the very low yield of aboveground biomass in this year.

The total dry matter yield of field pea was 16.4 and 30.0% lower compared to the tansy phacelia and oilseed radish grown under the same conditions [10]. In the case of yellow lupine the difference amounted to 19.3 and 32.5% respectively.

In the eight-year study period, a high yielding reliability of legumes grown as stubble catch crop was stated (Table 6). The differences between the studied species were 3.2 - 3.6 percentage points for the aboveground dry matter yield and only 1.3 - 3.3 percentage points for total dry matter.

Table 5. Dry matter yield of plants grown as catch crops [$Mg \cdot ha^{-1}$]

Year	Plant			Means
	Serradella	Field pea	Yellow lupine	
	Green matter			
2005	0.66 b	0.87 a	0.70 ab	0.74
2006	1.92 b	2.60 a	2.58 a	2.37
2007	1.49 b	2.44 a	1.73 b	1.89
2008	1.37 b	2.58 a	1.79 ab	1.91
2009	1.03 b	2.41 a	2.18 a	1.88
2005-2009	1.30 c	2.18 a	1.80 b	1.76
	Post-harvest residue			
2005	0.89 a	1.06 a	1.09 a	1.01
2006	1.21 ab	1.13 b	1.52 a	1.29
2007	1.04 a	0.97 a	1.19 a	1.07
2008	1.31 a	1.14 b	1.40 a	1.29
2009	0.62 b	0.63 b	1.12 a	0.79
2005-2009	1.01 b	0.99 b	1.26 a	1.09
	Total dry matter			
2005	1.56 a	1.93 a	1.79 a	1.76
2006	3.12 b	3.73 ab	4.10 a	3.65
2007	2.52 a	3.41 a	2.91 a	2.95
2008	2.68 b	3.72 a	3.19 ab	3.20
2009	1.65 b	3.04 a	3.31 a	2.67
2005-2009	2.31 b	3.17 a	3.06 a	2.85

 # – means marked with different letters in lines differ significantly at α = 0.05

Previous research on this subject indicates that the studied plants, especially serradella have unreliable yielding due to the delayed sowing [6]. In our study, the reliability coefficient of serradella yield was as high as for the peas. However, this high reliability referred to yield lower by 27% compared to the peas.

Yielding of crops grown as stubble catch crop depends mainly on rainfall in July and August as well as on the air temperature in September and October [8,10,19]. Regression analysis shows that the optimum amount of precipitation in July and August for field pea in typical Alfisols soil condition is 154 mm. It allows the dry biomass yield of 3.36 $Mg \cdot ha^{-1}$ (including 2.44 $Mg \cdot ha^{-1}$ green matter) (Table 7). The highest green matter yield of yellow lupine can be achieved with total rainfall of 142 mm during this period. The analysis showed that field pea and yellow lupine sown in stubble catch crop within 3-12 of August are characterized by a high reliability and satisfactory yielding. Serradella definitely did not match these plants in terms of the yield level.

Table 6. Yielding reliability of dry matter yield in years of the study 2002-2009 [%]

Part of plant	Plant		
	Serradella	Field pea	Yellow lupine
Green matter	77.1	77.5	73.9
Post-harvest residue	77.8	79.5	78.9
Total dry matter	81.8	83.8	80.5

Table 7. Regression equations for dry matter yield of catch crops depending on rainfall totals in the period from 1 July to 31 August (n=8)

Catch crop		Regression equation[#]	R^2	p	Optimal rainfall [mm]	Estimated yield at optimal rainfall [$Mg \cdot ha^{-1}$]
Green matter	Serradella		0.54	n.s.*		
	Field pea	$y=-0.2632x^2+80.9x-3781$	0.71	0.044	154	2.44
	Yellow lupine	$y=-0.3249x^2+92.3x-4548$	0.73	0.037	142	2.01
Total dry mass	Serradella		0.50	n.s.		
	Field pea	$y=-0.253x^2+78x-2646$	0.74	0.036	154	3.36
	Yellow lupine		0.60	n.s.		

*[#] - regression equation calculated based on data from years 2002-2009, * - n.s. - not significant*

5. CONCLUSION

Serradella, field pea and yellow lupine sown as stubble catch crop within 3-12 August showed a high reliability of dry matter yield, ranging from 80.5% (yellow lupine) to 83.8% (field pea). Serradella matched the other plants in terms of the yield reliability, but the yield of green matter of the plants was significantly lower than that of pea and lupine.

The study showed a high yield potential of pea and yellow lupine. The best conditions to obtain a high yield of dry weight of these plants occur in the years in which the rainfall in July and/or August amounts to 142 mm for yellow lupine and 154 mm for field pea.

COMPETING INTERESTS

Authors have declared that no competing interests exist.

REFERENCES

1. Thorup-Kristensen K, Dresbøll DB. Incorporation time of nitrogen catch crops influences the N effect for the succeeding crop. Soil Use Manag. 2010;26(1):27-35.
2. Piotrowska A, Wilczewski E. Effects of catch crops cultivated for green manure and mineral nitrogen fertilization on soil enzyme activities and chemical properties. Geoderma. 2012;189–190:72-80.
3. Doltra J, Olesen JE. The role of catch crops in the ecological intensification of spring cereals in organic farming under Nordic climate. Eur. J. Agron. 2013;44:98-108.
4. Wilczewski E, Piotrowska-Długosz A, Lemańczyk G. Influence of catch crop on soil properties and yield of spring barley. International Journal of Plant Production. 2014;8(3):391-408.
5. Statistical yearbook of the republic of Poland. Central Statistical Office, Warsaw, Poland; 2013.
6. Gromadziński A. Effect of the sowing date and nitrogen fertilization on the yielding of plants cultivated as stubble catch crop. Pam. Puł. [in Polish]. 1976;66:155-165.
7. Jaskulski D, Tomalak S, Rudnicki F. Regeneration of the stand after winter wheat for spring barley with catch crop plants. Zesz. Probl. Post. Nauk Rol. [In Polish]. 2000;470:49-57.
8. Sypniewski J, Skinder Z, Kluczek JP. Yielding of fodder crops as stubble catch crop fertilized with cattle slurry. Zesz. Probl. Post. Nauk Roln. [In Polish]. 1994;414:131-144.
9. Rinnofner T, Friedel JK, De Kruijff R, Pietsch G, Freyer B. Effect of catch crops on N dynamics and following crops in organic farming. Agronomy for Sustainable Development, Springer Verlag (Germany). 2008;28(4):551-558.
10. Wilczewski E, Skinder Z, Szczepanek M. Effect of weather conditions on yield of tansy phacelia and common sunflower grown as stubble catch crop. Pol. J. Environ. Stud. 2012;21:1053-1060.
11. Zaniewicz-Bajkowska A, Rosa R, Kosterna E, Franczuk J. Catch crops for green

manure: Biomass yield and macroelement content depending on the sowing date. Acta Sci. Pol., Agricultura. 2013;12(1):65-79.
12. Skinder Z, Lemańczyk G, Wilczewski E. Value of selected papilionaceous crops grown in stubble intercrop on light soil. Part I. Biomass yield and plant health status. Acta Sci. Pol. Seria Agricultura, [In Polish]. 2007;6(1):23-33.
13. Soil Survey Staff. Keys to Soil Taxonomy, 11th ed., Washington, DC: USDA Natural Resources Conservation Service; 2010.
14. Rolbiecki St, Rolbiecki R, Rzekanowski C, Żarski J. The influence of sprinkler irrigation on yields of some vegetable crops in the region of Bydgoszcz, Poland. Acta Horticulturae. 2000;537:871-877.
15. Skowera B, Puła J. Pluviometric extreme conditions in spring season in Poland in the years 1971-2000. Acta Agrophysica. 2004;3(1):171 [In Polish].
16. Filipiak K, Wilkos S. Statistical calculations. Description of system AWAR. IUNG Puławy. 1995;R324:1-52 [In Polish].
17. Rudnicki F, Wasilewski P. Importance of cereal mixtures and cereal-legumes in reducing the negative effects of the high proportion of cereals in crop rotation. Zesz. Probl. Post. Nauk Roln. 2000;470:127-135 [In Polish].
18. Wilczewski E. Environmental determinants of stubble intercrop cultivation in Kuyavian-Pomeranian Province. Part in: Understanding the Requirements for Development of Agricultural Production and of Rural Areas in the Kuyavian-Pomeranian Province as a Result of Scientific Research. Śliwińska E, Spychaj-Fabisiak E. (red.). UTP Bydgoszcz. 2009;279-288.
19. Demidowicz G, Gonet Z. Assessment of climate of Poland for the cultivation of stubble after crops. Pam. Puł. 1976;66:203-204 [In Polish].

22

Influence of Integrated Nutrient Management on Growth and Yield of Sweet Corn (*Zea mays* L. *saccharata*) under Temperate Conditions of Kashmir Valley

Shahid Rasool[1*], R. H. Kanth[2], Shabana Hamid[3], W. Raja[2], B. A. Alie[2] and Z. A. Dar[4]

[1]*CSIR- Indian Institute of Integrative Medicine, Srinagar, Kashmir, India.*
[2]*Division of Agronomy, Sher-e-Kashmir University of Agricultural Sciences and Technology of Kashmir, India.*
[3]*Department of Agriculture, J&K Government, Kashmir, India.*
[4]*Division of Plant Breeding and Genetics, Sher-e-Kashmir University of Agricultural Sciences and Technology of Kashmir, India.*

Authors' contributions

This research work was carried out in collaboration between all authors. Authors SR and RHK designed the study, wrote the protocol and wrote the first draft of the manuscript. Author BAA reviewed the experimental design and all drafts of the manuscript. Authors SR and SH managed the analyses of the experiment. Authors ZAD and WR performed the statistical analysis. All authors read and approved the final manuscript.

Editor(s):
(1) Juan Yan, Sichuan Agricultural University, China.
Reviewers:
(1) Anonymous, Brazil.
(2) Anonymous, Pakistan.

ABSTRACT

The growth and yield response of sweet maize (*Zea mays* (L.) saccharata) to varying levels of organic and inorganic fertilizers during the growing seasons of *kharif* 2010 and 2011 was studied under temperate conditions of Kashmir Valley. Twelve treatments comprising of sole and combination of organic and inorganic fertilizers were laid in a randomized block design with three replications. The results revealed that application of T_{10} [75% (NPK) + FYM (4.5 t/ha) + Biofertilizer (*Azotobacter* + Phosphate solubilizing bacteria (PSB))] significantly increased the number of days

**Corresponding author: E-mail: msrasool@iiim.ac.in;*

taken to tasseling, silking and milky stages and various other growth characters viz., plant height, leaf area index and dry matter accumulation at 15 days interval from sowing up to harvest and crop growth rate and relative growth rate at 7 days interval from 15 DAS up to harvest whereas, the lowest values of these parameters were recorded in unfertilized control. The treatment T_{10}[75 % (NPK) + FYM (4.5 t/ha) + Biofertilizer (*Azotobacter* + Phosphate solubilizing bacteria (PSB))] proved to be significantly superior to rest of the treatments including unfertilized control in increasing cob yield with and without husk, fodder yield and green biomass yield during both years of experimentation, however, ratio of cob to fodder yield during 2011 and 2012 were recorded highest in treatment T_3 [FYM (18 t ha^{-1})] and T_2 [Recommended NPK kg ha^{-1} (90:60:40)], respectively, whereas unfertilized control recorded the lowest ratio of cob to fodder yield.

Keywords: Sweet maize; organic fertilizer; inorganic fertilizers; growth; yield.

1. INTRODUCTION

Sweet corn is one of the most popular vegetables in the USA, Canada and Australia. It is becoming popular in India and other Asian countries. Sweet corn differs from other corns (field maize, pop corn and ornamental) because the kernels have a high sugar content in the milk on early dough stage. It is consumed in the immature stage of the crop. The kernels of sweet corn taste much sweeter than normal corn, especially at 25-30% maturity. The sweet corn industry is expanding because of increasing domestic consumption, export development and import replacement. It is an attractive crop for producers to grow because the plant grows quickly and is considered a valuable rotational crop and farming operation can be mechanized. Planting usually commences in spring when soil temperature reaches above 12°C. In warmer regions with longer growing season allows two crops to be planted each year, however taking the major constraints of shorter growing period (April to September) together with cold stress at early and late stages of crop growth restricts it to mono-cropping under temperate conditions.

The potential of the sweet maize crop is not being exploited satisfactorily due to many constraints among which inappropriate nutrient supply ranks first. Others are pest problems at maturity, low fertility status of the soils and the high cost of the scarce inorganic fertilizers with their potential polluting effects on the environment following continuous usage. Soils of the agro-ecology are generally low in organic matter as a result of the rapid mineralization and the fact that very little organic matter is added to the soil during and after cropping [1]. The need to use renewable forms of energy has rekindled interest in the use of organic manures such as cow dungs, poultry droppings and crop residues as alternatives for inorganic fertilizer worldwide. Application of organic manures plays a direct role in plant growth as a source of all the necessary major and minor nutrients in available forms during mineralization which improves both the physical and biological properties of the soil [2]. Nutrients contained in organic manures are released more slowly and are stored for longer periods in the soil, thereby ensuring a long residual effect [3]. To meet crop's nutrient needs, organic manures are however, required in rather large quantities which now make for a strong advocacy for fortifying these manures with inorganic fertilizers. A positive interaction has been shown to exist between the combination of organic manures and urea as N source [4,5]. [5] reported that the combined use of poultry manure and urea gave the best performance for sweet maize amongst all treatments. Manures application has also been reported to increase the N and exchangeable cations levels in the soil [6]. Keeping this in view the present study was undertaken with an objective to evaluate the performance of sweet maize under the combined use of organic and inorganic fertilizers.

2. MATERIALS AND METHODS

A field experiment was conducted for two consecutive years (*kharif* 2011 and 2012) at the Experimental Farm of the Division of Agronomy, Sher-e-Kashmir University of Agricultural Sciences and Technology of Kashmir that lies between 34° 0.8 • N latitude and 74° 83 • E longitude at an altitude of 1587 meters above mean sea level. The experimental site was well drained and had uniform topography. The experiment was laid out in a Randomized Block Design with three replications and comprising of 12 treatments/ the treatment details are given in Table 1.

The chemical fertilizers Urea, DAP and MOP were used as source of nitrogen, phosphorus and potassium, respectively, FYM and vermicompost were used as source of nitrogen, phosphorus, potassium and micro-elements also in the form of organic manure. The microbial culture of *Azotobacter* and PSB were used as bio-fertilizers. Farmyard manure (FYM) and vermicompost were applied to the respective plots, as per the layout plan, and mixed thoroughly with the soil. Half of the nitrogen was applied as basal at the time of sowing and the remaining half of nitrogen in two splits each at knee high and tassel emergence stage in the respective plots at the rates as per layout plan. Phosphorus and potassium were applied, at the rates as per layout plan to the respective plots at the time of sowing. *Azotobacter* and PSB were used as a seed treatment (50 g kg^{-1} seed) to the respective plots.

The data were recorded from five randomly selected plants which were tagged from penultimate rows of each plot and the average for every parameter was worked out. Days taken to different physiological stages were recorded at various growth stages i.e. knee high, tasseling, silking and milking stages. Plant height (cm), leaf area index and dry matter accumulation (q ha^{-1}) were recorded at 15 days interval from the date of sowing. Green cob yield with and without husk (q ha^{-1}), green fodder yield (q ha^{-1}), biomass yield (q ha^{-1}) and ratio of cob to fodder yield were recorded from 5 cobs taken randomly from each net plot and then converted to q ha^{-1}. Crop growth rate (g $g^{-1}day^{-1}$) and relative growth rate (g $g^{-1}day^{-1}$) were calculated at 7 days interval according to the formula of [7,8], respectively.

$$\text{Crop growth rate (g g}^{-1}\text{day}^{-1}) = \frac{W_2 - W_1}{t_2 - t_1}$$

W_1 and W_2 = Dry matter production per plant (g) at time t_1 and t_2, respectively.

$$\text{Relative growth rate (g g}^{-1}\text{day}^{-1}) = \frac{\text{Log}_e W_2 - \text{Log}_e W_1}{t_2 - t_1}$$

W_1 = Dry weight of plant at time t_1, W_2 = Dry weight of plant at time t_2

The data obtained in respect of various observations were statistically analyzed by the method described by [9]. The significance of "F" and "t" was tested at 5 % level of significance. The critical difference was determined when "F" test was significant.

3. RESULTS AND DISCUSSION

The present investigation indicated that application of T_{10} significantly increased the number of days taken to tasseling, silking and milky stages than other treatments but remained at par with T_2, T_8, T_9, T_{11} and T_{12} during 2011 and 2012, however, significantly lowest number of days for the crop to reach these stages was recorded in unfertilized control (Fig. 1 and 2). These results are in accordance with the results obtained by [10]. With the release and availability of sufficient nitrogen supply from the combined nutrition of inorganic and organic nature the advancement of the crop enhanced because of mere fact that nitrogen is a part of chlorophyll and is involved in cell division and enlongation. The nitrogen is also known to delay the reproductive period of the crop which is evident from the present findings i.e. increased number of days to tasseling, silking and milky stage.

Plant height is an important growth index to study the accumulation of dry matter by the plant and is very important to monitor the overall canopy architecture and also govern the orientation of the leaves that further govern the photosynthetic efficiency of a plant to utilize the natural resources. It was found that the periodic plant height of the crop went on increasing upto harvest and the magnitude of increase was more than double from 15-30 and 30-45 DAS irrespective of treatments. The results revealed that the plant height recorded with T_{10} from 60 DAS, 75 DAS and harvest was significantly higher than unfertilized control, T_3, T_4 and T_5 inoculation but at par with rest of the treatments during 2011 and 2012 (Figs. 3 and 4), however, at 15 DAS T_{10} recorded significantly higher plant height than other treatments. Significantly lowest periodic plant height was recorded under the treatment of unfertilized control. Similar results were also obtained by [11-13]. Significant effect on the increase on the plant height in sweet corn with the application of NPK may be attributed to the fact that nitrogen being an essential constituent of plant tissue favours rapid cell division and its enlargement, which together with the adequate quantity of phosphorus and potassium helps in the rapid cell division and better development of the cell size. Further, the beneficial effect of FYM may be attributed to the fact that it supplied available plant nutrients and also had solubilizing effect on fixed forms of nutrients especially phosphorus in soil [14].

Leaf area index is of paramount importance in all the crop plants, because optimum leaf area is required for maximum light interception which results for higher photosynthesis. Periodic leaf area index also increased significantly with application of T_{10} over other treatment tested during two years of study (Figs. 5 and 6). [11,15,16] also reported increased leaf area index by combined application of organic and inorganic fertilizers. Nitrogen is an essential constituent of proteins, enzymes and chlorophyll and has been observed to influence the leaf growth and its expansion, resulting in increased leaf area index. Availability of adequate phosphorus in plant results in proper leaf expansion, increase in leaf surface area and number of leaves and results in better efficiency of chlorophyll during photosynthesis and this overall improvement gets translocated into better growth of the plant. Availability of adequate quantity of potassium is essential for the better crop growth and improves the source sink relationship that helps in harvesting higher crop yield. Potassium is a cofactor of numerous enzymes and basically helps in the translocation mechanism and it improves the mobility and utilization of other elements. Besides application of FYM and biofertilizer, apart from improving soil physico-chemical and biological properties of soil releases adequate quantities of nitrogen and phosphosrus to boost up the growth of the crop thereby increasing leaf area index. Moreover, leaf area index values were highest at 75 DAS, the period coinciding silking stage and thereafter declined upto harvest. The decline in leaf area index after silking could be attributed to leaf senescence due to shading of lower leaves.

Dry matter accumulation is another important character to express the growth and metabolic efficiency of the plant, which ultimately influence the yield. Perusal of the data revealed that irrespective of treatments dry matter production went on increasing with the advancement in the age of crop upto harvest and the magnitude of increase was more than double from 15 to 30 DAS and 30 to 45 DAS. At 15, 30, 45 and 60 DAS the dry matter production recorded under treatment T(10) was significantly higher than other treatments but was at par with T_2, T_{11} and T_{12} during 2011 and 2012 (Figs. 7 and 8). However, at 75 DAS and harvest the dry matter production recorded with treatment T_{10} was significantly higher than unfertilized control and at par with rest of the treatments during two years of experimentation. The study also revealed that significantly lowest periodic dry matter was observed under unfertilized control. Similar findings were also been reported by [17-19]. The increase in periodic dry matter accumulation with application of integrated fertilizer management may be attributed to increase in plant height and leaf area index resulting thereby in better light interception by crop which accumulated more photosynthates and thus produced more dry matter. Further, FYM and biofertilizer supplied the nutrients in balanced proportion and improved the physical characters, which might have increased the availability of nutrients particularly nitrogen and phosphorus.

The results obtained in Figs. 9 and 10 revealed that application of T_{10} recorded significantly higher crop growth rate than various treatments and being generally at par with T_2, T_{11} and T_{12}, however, trend of significance recorded with T_{10} over other treatment didn't show similar pattern at all intervals of measurement, but varied from one interval to another. Similar trend was also noticed for relative growth rate at different growth intervals (Figs. 11 and 12). [20] reported that application of NPK along with FYM significantly increased crop growth rate and NAR of maize crop over control. These results are in agreement with those of [21]. The crop growth rate and relative growth rate, in general, indicate the amount of increase in dry matter over existing dry matter per unit time. The higher CGR and RGR obtained during the study under the above treatment is the reflection of accumulation of dry matter at the respective periods.

Yield is the net result of various agronomic inputs influencing growth and yield attributing characters during the life cycle of the crop. The efficiency of different factors is judged mainly by their contribution towards economic yield. The treatment T_{10} proved to be significantly superior to rest of the treatments including unfertilized control in increasing the cob yield with and without husk during both years of experimentation and pooled data over years (Table 2). The yield superiority of T_{10} over unfertilized control with regard to cob yield with husk was to the tune of 107.64, 129.96 and 119.18 per cent and with regard to cob yield without husk the increase in yield was to the tune of 99.86, 121.63 and 111.10 per cent during 2011 and 2012, respectively. Further, significantly higher green fodder and biomass yield of sweet corn were also obtained with application of T_{10} during 2011 and 2012 (Table 3). The treatments T_3 and T_2 recorded the highest cob to fodder ratio of 1:1.50 and 1:1.32 recorded the highest cob to fodder ratio during 2011 and 2012, respectively. Among the 12 treatments the lowest ratio of cob to fodder yield was observed in unfertilized control during both years of study as well as pooled data over years. Higher yields of sweet corn obtained with the application of integrated nutrients was mainly due to their positive effect on various yield contributing characters like cob length, number of cobs per plant, number of grains per cob, length and diameter of cob etc. Earlier [15,22-24 and 25] also found significant and consistent increase in cob and fodder yield with combined application of organic and mineral fertilizers. Concomitant release of nitrogen at most critical stages of their need is a key to ensure higher yields. In the present study, application of FYM and biofertilizer alongwith 75% NPK resulted in maximum yield that establishes the fact of synchrony between availability of nitrogen at critical stages of crop as well as other benefits derived from FYM. Further, application of FYM adds and exploits the fixed nutrients of soil in available form and regulates its supply to the crop through mineralization and prevents them from leaching and other losses [26]. Therefore, this study clearly revealed the importance of both inorganic and organic nutrients for enhancing the productivity of sweet corn. Increase in fodder yield with application of T_{10} could be attributed to significant improvement in plant height and dry matter accumulation of sweet corn. Increase in fodder yield with integrated nutrient management has also been reported by [20,27,28]. The results of many trails including those of long term manorial experiments conducted at different locations in India have also revealed that combined application of mineral fertilizers and organic manures is an appropriate method to achieve yields. The biomass yield is a function of cob and fodder yield, representing vegetative and reproductive growth of the crop. The results are in agreement with those of [29,30].

Table 1. Treatment details

Treatments		Details
T_1	:	Control
T_2	:	Recommended NPK kg ha^{-1} (90:60:40)
T_3	:	FYM (18 t ha^{-1})
T_4	:	Vermicompost (3.6 t ha^{-1})
T_5	:	Biofertilizer (*Azotobacter* + Phosphate solubilizing bacteria (PSB))
T_6	:	75 % (NPK) + FYM (4.5 t ha^{-1})
T_7	:	75 % (NPK) + Vermicompost (0.9 t ha^{-1})
T_8	:	75 % (NPK) + Biofertilizer
T_9	:	75 % (NPK) + FYM (2.25 t ha^{-1}) + Vermicompost (0.45 t ha^{-1})
T_{10}	:	75 % (NPK) + FYM (4.5 t/ha) + Biofertilizer (*Azotobacter* + Phosphate solubilizing bacteria (PSB))
T_{11}	:	75 % (NPK) + Vermicompost (0.9 t/ha) + Biofertilizer (*Azotobacter* + Phosphate solubilizing bacteria (PSB))
T_{12}	:	75 % (NPK) + FYM (2.25 t/ha) + Vermicompost (0.45 t/ha) + Biofertilizer (*Azotobacter* + Phosphate solubilizing bacteria (PSB))

Table 2. Effect of integrated nutrient management on green cob yield with and without husk (q ha^{-1}) of sweet corn

Treatments	Cob yield with husk			Cob yield without husk		
	2011	2012	Mean	2011	2012	Mean
T_1	167.42	161.74	164.58	126.17	121.38	123.77
T_2	261.94	286.08	273.89	188.80	206.42	197.52
T_3	249.47	314.91	281.43	185.45	228.66	206.57
T_4	260.68	288.88	275.55	191.20	210.26	201.30
T_5	210.60	226.07	219.08	159.01	172.56	166.34
T_6	277.67	295.08	287.24	202.42	214.46	209.08
T_7	276.10	281.33	278.72	200.10	204.45	202.28
T_8	250.56	277.95	264.12	182.43	201.80	192.02
T_9	273.93	301.45	287.55	199.50	219.97	209.63
T_{10}	347.63	371.94	360.72	252.17	269.02	261.28
T_{11}	287.25	320.82	303.81	210.88	234.52	222.55
T_{12}	301.65	333.48	317.38	221.57	244.53	232.92
SEm±	3.187	3.455	3.343	3.766	4.075	4.017
CD ($p \le 0.05$)	9.41	10.20	9.87	11.12	12.03	11.86

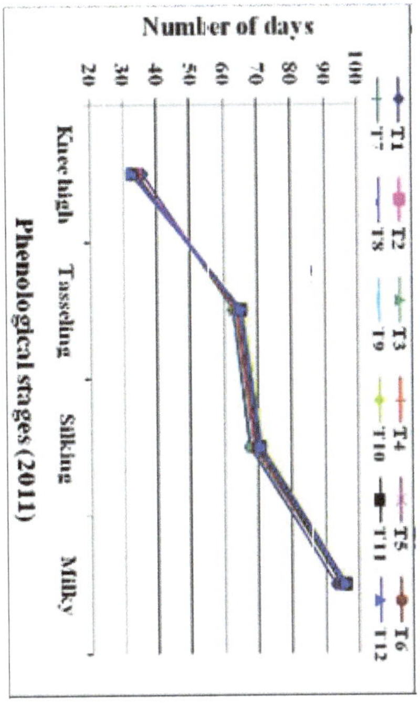

Fig. 1

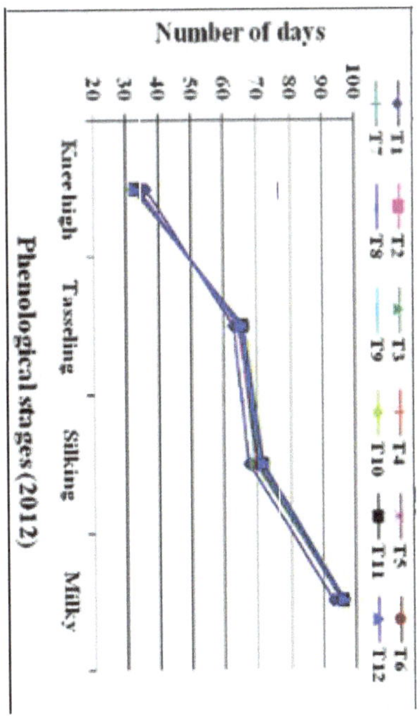

Fig. 2

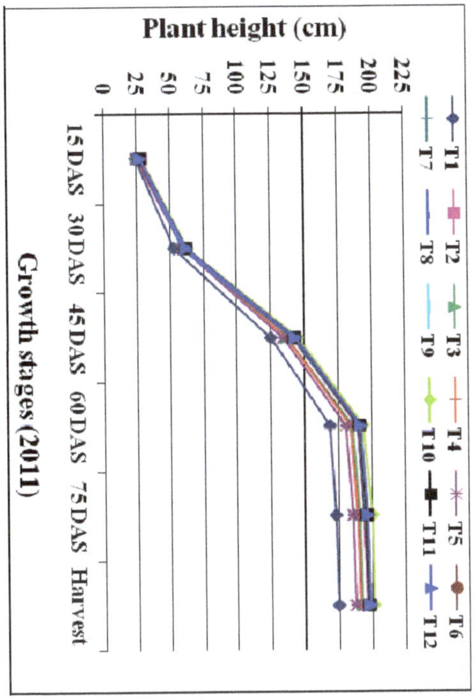

Fig. 3

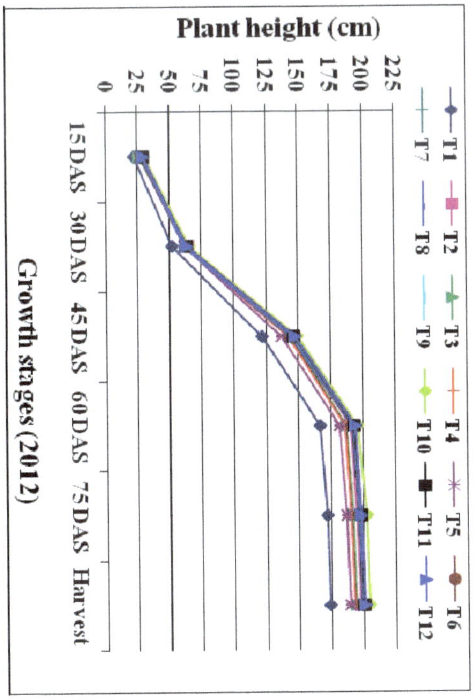

Fig. 4

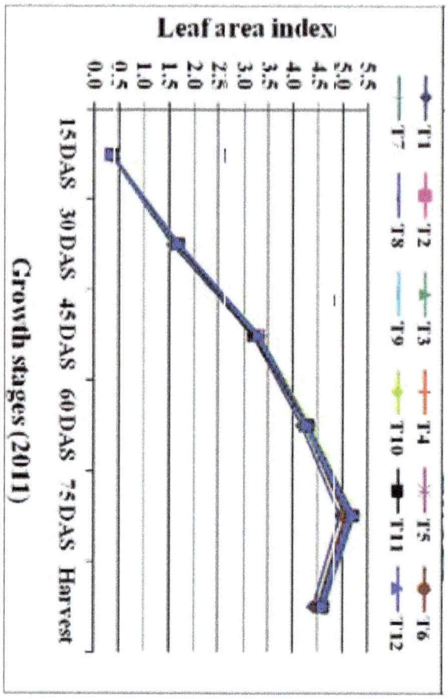

Fig. 5

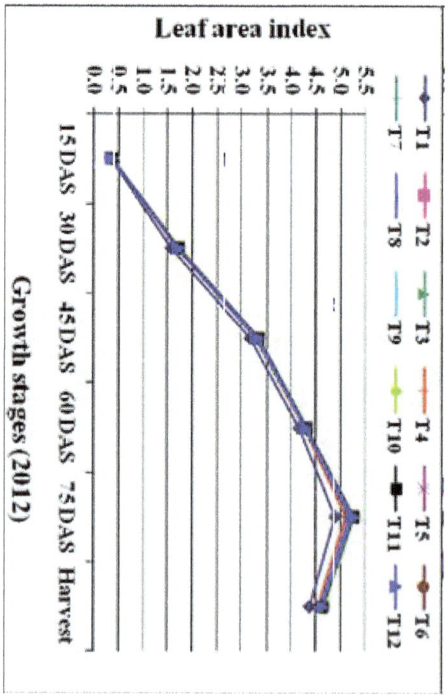

Fig. 6

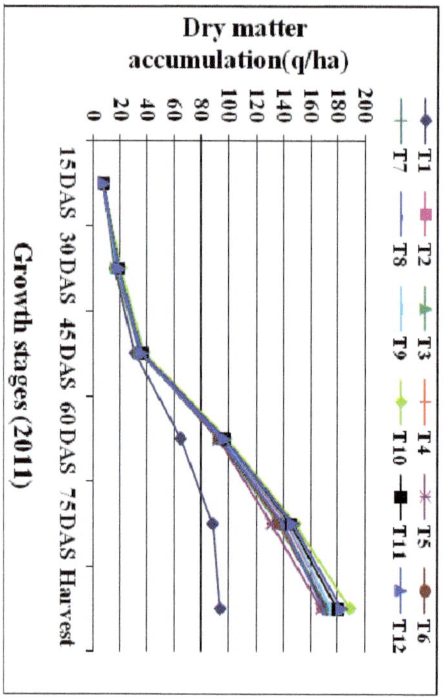

Fig. 7

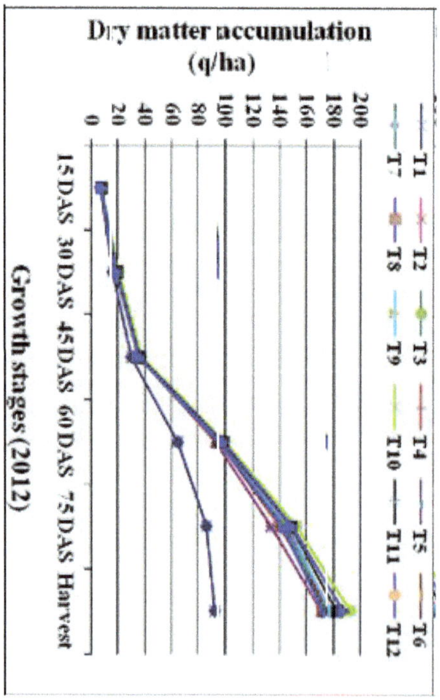

Fig. 8

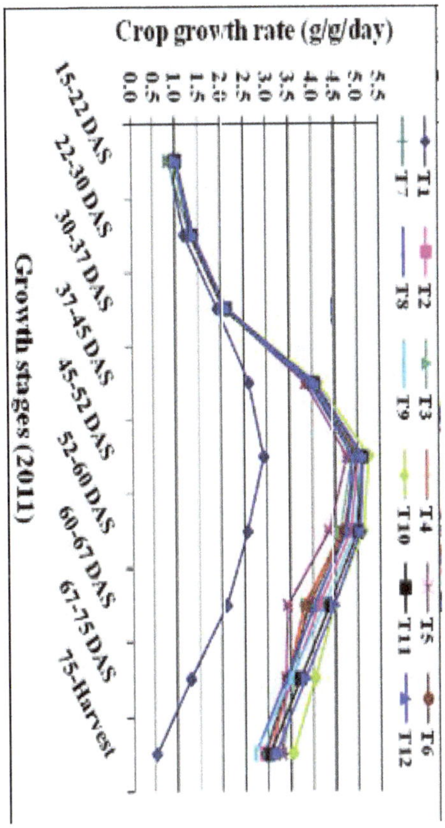

Fig. 9

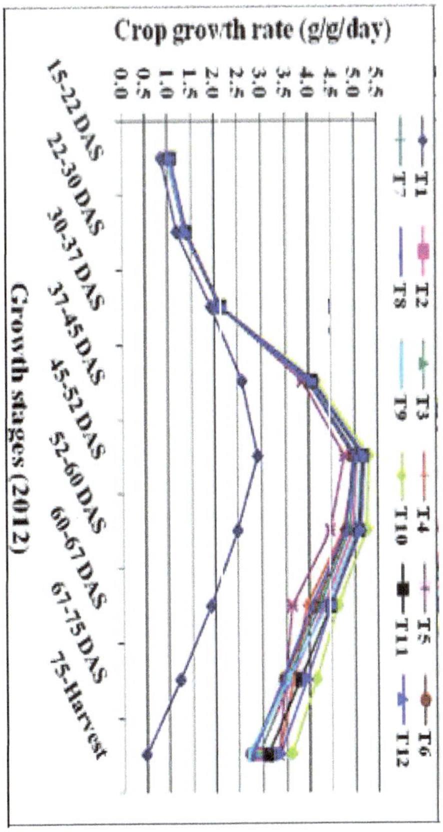

Fig. 10

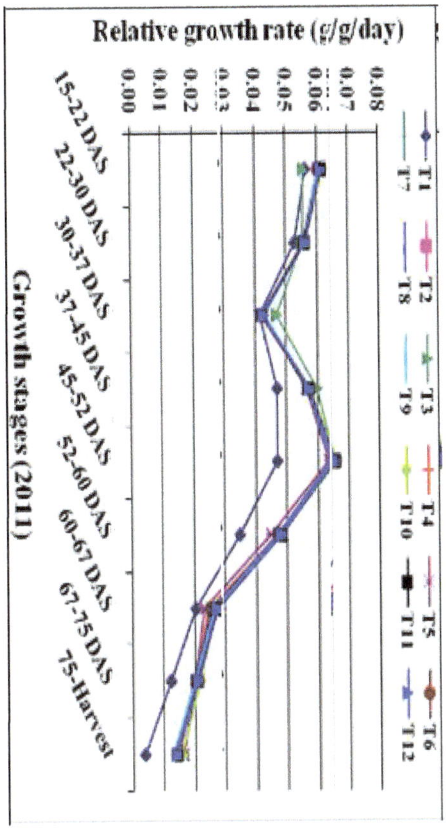

Fig. 11

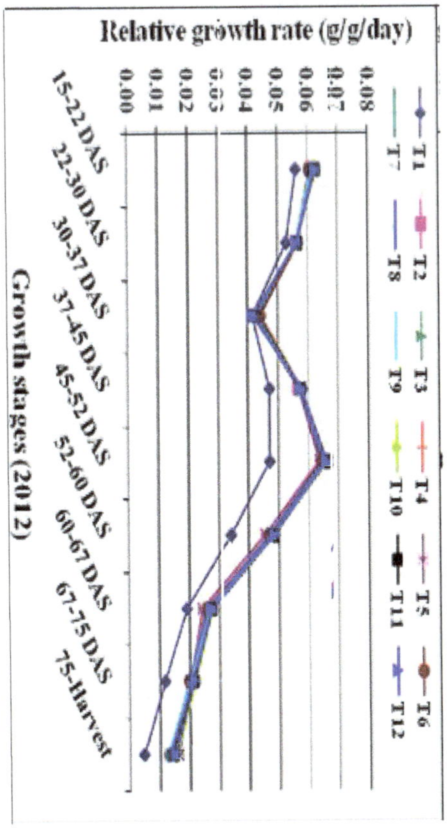

Fig. 12

Fig. 1-12. Growth parameters of sweet corn (*Zea mays* var. *saccharata*) as affected by integrated nutrient management

Table 3. Effect of integrated nutrient management on fodder yield (q ha^{-1}), green biomass yield (q ha^{-1}) and ratio of green cob and biomass yield of sweet corn

Treatments	Fodder yield			Green biomass yield			Ratio of cob to fodder yield		
	2011	2012	Mean	2011	2012	Mean	2011	2012	Mean
T_1	187.98	182.52	185.25	355.40	344.26	349.83	1:1.12	1:1.13	1:1.13
T_2	367.70	378.09	372.90	629.64	664.17	646.79	1:1.40	1:1.32	1:1.36
T_3	374.87	389.69	382.28	624.34	704.60	663.71	1:1.50	1:1.24	1:1.36
T_4	354.03	363.65	358.84	614.71	652.53	634.39	1:1.36	1:1.26	1:1.30
T_5	273.53	295.45	284.49	484.13	521.52	503.57	1:1.30	1:1.31	1:1.30
T_6	351.40	357.79	354.60	629.07	652.87	641.84	1:1.27	1:1.21	1:1.23
T_7	358.46	354.15	356.30	634.56	635.48	635.02	1:1.30	1:1.26	1:1.28
T_8	339.75	357.02	348.39	590.31	634.97	612.51	1:1.36	1:1.28	1:1.32
T_9	382.05	390.98	386.51	655.98	692.43	674.06	1:1.39	1:1.30	1:1.34
T_{10}	406.01	410.27	408.14	753.64	782.21	768.86	1:1.17	1:1.10	1:1.13
T_{11}	387.99	390.96	389.47	675.24	711.78	693.28	1:1.35	1:1.22	1:1.28
T_{12}	388.65	391.76	390.20	690.30	725.24	707.58	1:1.29	1;1.17	1:1.23
SEm±	5.179	5.259	5.215	12.281	13.924	13.385	-	-	-
CD ($p \leq 0.05$)	15.29	15.52	15.43	36.26	41.11	39.52	-	-	-

4. CONCLUSION

Two year investigations on "Effect of integrated nutrient management on growth and yield of sweet corn (*Zea mays* var. *saccharata*) under temperate conditions of Kashmir" revealed that application of T_{10} being at par with T_9 and T_{11} proved significantly superior in terms of cob yield with and without husk, yield attributes and growth parameters. In view of this, it may be concluded that for obtaining maximum cob and fodder yield in sweet corn, it needs to be fertilized with T_{10}. However, such studies require more critical testing at various locations over a longer period before final recommendations are made.

COMPETING INTERESTS

Authors have declared that no competing interests exist.

REFERENCES

1. Lal R, Greenland DJ. Effects of organic amendments in soil physical properties and crop production in the tropics. Lal R, Greenland DJ. (eds). J. Wiley and Sons New York; 1979.
2. Abou El-Magd MA, El-Bassiong M, Fawzy ZF. Effect of organic manure with or without chemical fertilizers on growth, yield and quality of some varieties of broccoli plants. Journal of Applied Science Research. 2006;2(10):791-798.
3. Sharma AR, Mittra BN. Effect of different rates of application of organic and nitrogen fertilizer in a rice-based cropping system. Journal of Agricultural Science (Camb), 1991;117:313-318.
4. Bocchi S, Tano F. Effect of cattle manure and components of pig slurry on maize growth and production. European Journal of Agronomy. 1994;3:235-41.
5. Khaliq T, Mahmmod T, Kamel J, Masood A. Effectiveness of farmyard manure, poultry manure and nitrogen for corn (*Zea mays* L.) productivity. International Journal of Agriculture and Biology. 2004;6(2):260-263.
6. Boateng SA, Zickermann J, Kornaharens M. Effect of poultry manure on growth and yield of maize. West Africa Journal of Applied Ecology. 2006;9:1-11.
7. Redford PT. Growth analysis formulae-Their use and abuse. Crop Science. 1967;7:171-175.
8. Blackman VH. The compound interest law and plant growth. Annals of Botany. 1919;33:353-360.

9. Cochran WG, Cox GM. Experimental Designs. 2nd Ed. Johan Wiley and Sons Inc. New York; 1967.
10. Sharma MP, Gupta JP. Effect of organic materials on grain yield and soil properties of maize-wheat cropping system. Indian Journal of Agricultural Sciences. 1988;68(11):715-717.
11. Pathak SK, Singh SB, Singh SN. Effect of integrated nutrient management on growth, yield and economics in maize-wheat cropping system. Indian Journal of Agronomy. 2002;47(3):325-332.
12. Paradar VK. Development of integrated nutrient management system through organic sources to enhance productivity of maize based cropping under rainfed ecosystem at Chhindwara. 48th Annual Progress Report, All India Coordinated Maize Improvement Project, Directorate of Maize Research, Pusa Campus, New Delhi. 2005;48:77.
13. Singh H, Singh P, Sumeriya HK. Effect of fertility level on fodder yield and HCN content of sorghum (*Sorghum bicolor* L. Moench) genotypes. International Journal of Tropical Agriculture. 2008;26:417-420.
14. Ganai BA, Singh CM. Effect of FYM applied to rice-wheat rotation on physiochemical properties of soil. Indian Journal of Agronomy. 1988;33(3):327-329.
15. Afifi MH, Manal FM, Gomaa AM. Effect of biofertilizer under different levels of chemical fertilizers on maize (*Zea mays* L). Annals of Agricultural Science Moshtonor. 2003;41:1411–1420.
16. Kumar A, Thakur KS, Sharma S. Integrated nutrient management in maize (*Zea mays*) - gobhi sarson (*Brassica napus* ssp *oleifera* var. *annua*) cropping system under rainfed condition. Indian Journal of Agronomy. 2005;50:274-277.
17. Amujoyegbe BJ, Opabode JT, Olayinka A. Effect of organic and inorganic fertilizer on yield and chlorophyll content of maize (*Zea mays* L.) and sorghum (*Sorghum bicolour* (L.) Moench). African Journal of Biotechnology. 2007;16:1869-1873.
18. Singh RK, Singh SK, Singh LB. Integrated nitrogen management in wheat (*Triticum aestivum*). Indian Journal of Agronomy. 2007;53:124-126.
19. Kumar A, Dhar S. Evaluation of organic and inorganic sources of nutrients in maize (*Zea mays* L.) and their residual effect on wheat (*Triticum aestivum*) under different fertility levels. Indian Journal of Agricultural Sciences. 2010;80:364-371.
20. Singh G, Marwaha TS, Kumar D. Effect of resource conserving techniques on soil microbiological parameters under long term maize (*Zea mays* L.) – wheat (*Triticum aestivum* L.) crop rotation. Indian Journal of Agricultural Sciences. 2009;79:94-100.
21. Verma A, Nepalia V, Kanthaliya PC. Effect of nutrient supply on growth, yield and nutrient uptake by maize (*Zea mays* L.) - wheat (*Triticum aestivum* L.) cropping system. Indian Journal of Agronomy. 2006;51:3-6.
22. Kumar A, Thakur SK. Effect of integrated nutrient management on promising composite maize varieties under rainfed mid-hill conditions of Himachal Pradesh. Indian Journal of Agricultural Sciences. 2004;74(1):40-42.
23. Anonymous. 47thAnnual Progress Report, All India Coordinated Maize Improvement Project, Directorate of Maize Research, Pusa Campus, New Delhi. 2004;47:75.
24. Anonymous. 49th Annual Progress Report, 2005-2006. All India Coordinated Maize Improvement Project, Directorate of Maize Research, Pusa Campus, New Delhi. 2006;49:76.
25. Dadarwal RS, Jain NK, Singh D. Integrated nutrient management in baby corn (*Zea mays* L.). Indian Journal of Agricultural Sciences. 2009;79:1023-1025.
26. Singh GR, Choudhary KK, Chaure NK, Pandya KS. Effect of seed bacterization and nitrogen level on soil properties, yield parameters and economics of sunflower. Indian Journal of Agricultural Sciences. 1996;66(4):250-252.
27. Das A, Lenka NK, Sudhishri S, Patnaik US. Influence of integrated nutrient management on production, economics and soil properties in tomato (*Lycopersicon esculentum*) under on-farm condition in Eastern Ghats of Orissa. Indian Journal of Agricultural Sciences. 2008;78:40-43.
28. Balai ML, Verma A, Nepalia V, Kanthaliya PC. Productivity and quality of maize (*Zea mays* L.) as influenced by integrated nutrient management under continuous cropping and fertilization. Indian Journal of Agricultural Sciences. 2011;81:374-376.

29. Mahala HL, Shaktawat MS, Shivran RK. Direct and residual effect of sources and levels of phosphorus and farmyard manure in maize (*Zea mays* L.) -mustard (*Brassica juncea*) cropping sequence. Indian Journal of Agronomy. 2006;51:10-13.

30. Panwar AS. Effect of integrated nutrient management in maize (*Zea mays* L.) – mustard (*Brassica campestris* var. *toria*) cropping system in mid hills altitude. Indian Journal of Agricultural Sciences. 2008;78:27-31.

23

Standard Ear Yield and Some Agronomic Characteristics of Baby Corn var. ksc 403 su under Influence of Planting Date and Plant Density

Atena Rahmani[1*], Majid Nasrolah Alhossini[1] and Seyed Mohsen Nabavi Kalat[2]

[1]Department of Agronomy and Plant Breeding, Young Researchers and Elite Club, Mashhad Branch, Islamic Azad University, Mashhad, Iran.
[2]Department of Agronomy and Plant Breeding, Mashhad Branch, Islamic Azad University, Mashhad, Iran.

Authors' contributions

This work was carried out in collaboration between all authors. All authors read and approved the final manuscript.

Editor(s):
(1) Sławomir Borek, Department of Plant Physiology, Adam Mickiewicz University, Poland.
(2) Daniele De Wrachien, State University of Milan, Italy.
Reviewers:
(1) Mrityunjoy Biswas, Department of Agronomy & Haor Agriculture, Sylhet Agricultural University, Sylhet, Bangladesh.
(2) Anonymous, Ain Shams University, Cairo, Egypt.

ABSTRACT

In order to investigate of standard ear yield and some agronomic characteristics of baby corn under influence of planting date and plant density an experiment was conducted at Khorasan Razavi Agricultural and Natural Resources Research center, Mashhad, Iran in 2010. In this research planting date (14 June, 3 July and 24 July) and plant densities (65000, 85000 and 105000 plant ha^{-1}) were arranged in main and sub plots, respectively. This experiment was laid out using a split plot design with four replications. The results indicated that different planting dates had significant effects on agronomic traits such as plant height, ear height, number of leaves above ear, stem diameter, ear length, and ear diameter, dehusked and husked baby corn yield, standard and sub-standard dehusked ear yield, standard and sub-standard ear percentage. The highest husked ear yield with 13240 kg ha^{-1} was belonged to 24 July planting date. Although, delay in planting date, led to increase of growth period prevail suitable weather (low temperatures) at anthesis, therefore, ear yield increased but caused to the reduction of ear length and increased ear diameter. Also, ear

**Corresponding author: E-mail: atenarahmani@mshdiau.ac.ir;*

marketing obviously decreased, not only for fresh consumption but also for industry. The highest dehusked standard ear yield (722 kg ha^{-1}) was obtained from the highest plant density. Interaction effect of planting date × plant density had significant on standard ear percentage at a 5% level of probability. Finally, planting date of 14 June and the highest planting density of 105000 plant ha^{-1}, is recommended in order to maximize and produce high quality standard ear that can be presentable in market.

Keywords: Baby corn; planting date; plant density; standard ear yield.

1. INTRODUCTION

Baby corn has had a special place among fresh and conserved vegetables as a product with high nutritional value in recent years. This product is the very green ear, which is harvested and consumed in pollination stage. From decades ago, baby corn has been consumed freshly in the east west of Asian countries. In fact, it is a long-term that Chinese, Thai, and Taiwanese farmers have used dehusked ears. It is also used as a new nutritional substance in some foods such as salads, soups and pickles [1]. Baby corn contains vitamin B and C, fiber, carotenoids with high nutritional value [2].

Production and processing of baby Corn as an exportable vegetable in countries such as Taiwan, Sri Lanka, Indonesia, India, Zambia, Zimbabwe, South Africa, Guatemala and Nicaragua become increasingly popular and has been adapted [3]. The most effective factor in producing baby corn is Suitable variety. Accordingly, the most important criteria for the selection of suitable varieties are: early maturity, prolificacy (more number of cobs/plant), and synchronized ear emergence [1].

Investigations indicate that there are no researches related to the baby corn in Iran and most of the researches have focused on sweet corn. Fundamentally, planting date is one of the important components of cropping systems. Selecting appropriate planting date leads to produce high yield of corns. In a research being done in Arlington Mexico, the effect of planting date in improving the products and ear yield has been studied and indicates that late planting date is the cause of yield reduction [4]. The results of experiment on a variety of sweet corn (*Zea mays L.* Saccharata sturt) in eight sowing dates (26 April, 11 May, 26 May, 11 June, 26 June, 11 July, 26 July and 11 August) conducted at Agricultural Research Center, Haran University of Turkey determined that plantings dated there was no such planting dates as mentioned earlier were significant in terms of grain yield, grain size and single grain weight [5]. Noldin and Mundstock reported that different planting dates (11 October, 2 November) had significant effect on three varieties of sweet corn and it was determined that late planting date was the cause of yield reduction which was about 9% and the number of ear per plant were not affected by planting date [6]. The investigation associated with effect of planting date and plant density on corn yield (cultivar SC 403) indicated that maximum yield being about 2250 plant ha^{-1} was obtained on 29 April [7]. In this research, corn production rate, ear height and the harvest index were affected by plant density. However, with increasing density, ear length decreased. Accordingly, spring sowing sweet corn with density of 65000-55000 plant ha^{-1} for market was recommended.

Also, result of the other researcher on four different planting dates (25 April, 10 May, 25 May, 9 June) of sweet corn (SC404 variety) revealed that different planting dates had significant differences with each other in terms of ear height and ear diameter. Planting date had significant effect on yield in a way that maximum yield (11.2 ton ha^{-1}) was obtained with 65000 density plant ha^{-1} on 10 May [8]. Determination of an appropriate plant density may lead to optimum higher yield [9,10]. Genter et al. found that with increase in plant density, the number of plants having no ear increases [11]. In the investigation associated with yield and yield components of 35 corn genotypes with two plant densities of 55000 and 110000 plant $ha-^{1}$ have reported that levels of plant density have significant effect on all investigated traits except ear height and weight of husked ear [12]. Hybrids 27A x 29B with plant density of 110000 plant ha^{-1} had priority over the other studied genotypes in terms of plant height, ear height, number of ear plot-1, height and diameter of husked ear and dehusked ear. In an experiment associated with 35 genotypes of single gross yield of baby corn with two levels of density (55000, 110000 plant $ha-^{1}$) it was revealed that hybrid (27A x 31B) had the highest yield of husked and dehusked ear with plant density of 55000 plant ha-^{1}. Hybrid (27A x 29B) had the better yield with plant density of 110000 plant ha^{-1} [12]. Considering the nutritional value, baby corn needs to be added to the individuals' daily consumption. So far, there has not been any arable investigation associated with baby corn at Khorasan Razavi province. Determining planting date and plant density is a necessary requirement for planting baby corn. The objectives of this study were to investigate the effect of plant density and planting date and their interaction effects on baby corn in order to achieve the highest yield of standard baby corn. The purpose of this study was to investigate the effects of plant density, planting date, and Their Interaction of Baby corn in order to produce high-performance standard Baby corn.

2. MATERIALS AND METHODS

The present study was conducted at field experiment of the Khorasan Razavi Agricultural and Natural Resources Research Center (Latitude: 36° 13́ N., Longitude: 59° 40́E. and 985 meters above sea level), Mashhad, Iran in 2010. Soil texture at the experimental area was silt loam (having pH 8 and EC =1.67 mmhos cm^{-1}). In order to investigation summer planting and short growing season of Baby Corn, This experiment was carried out on split plot Based on a Randomize Completely Block design. The experiment, was consisted, three planting dates (14 June, 3 July, 24 July, the main factor) and three planting density (65, 85 and 105 thousand plants ha^{-1}, as subplots) with four replications. Each plot was considered with dimensions of 12 m^2 Consists of four rows with a pacing of 75 cm. Three seeds were planted into a hile in the 4 to 6 leaf stage seedlings were reduced to the one. To ensure germination of seeds, Irrigation was conducted with of 4 days interval. After emergence and seedling establishment irrigation was performed by based on plant demined. Fertilizer was used based on soil test. At sowing add 200 kg ha^{-1} Ammonium phosphate ($NH_4H_2PO_4$), 200 kg ha^{-1} potassium sulfate (K_2SO_4) and 300 kg ha^{-1} urea (CH_4N_2O) were used for the experiment and this was followed by 200 kg ha^{-1} N as urea at the six leaf stage. Morphological characteristics were measured and recorded at during the crop growing season such as plant height, ear height (distance from the ear to the ground); number of ears per plant, stem diameter (between the second and third nodes), the total number of leaves and number of leaves above the ear from 10 competitive randomly plants from each plot. At the harvesting time (1 to 2 cm silking length or 2 to 3 days after silking) in each plot two lateral rows Considered and removed as margin and the rest of plants were counted first, then harvested by hand. Also traits such as ear height, ear diameter, and husked and dehusked ear yield and harvest index of 10 ears in each plot were observed and recorded, on random basis. Also 10 to 12 length (cm) and 1 to 2 diameters (cm) are scale of diagnosis for standard and non-standard ear. Therefore, ears with this size, are standard and ears out of this, in order to processing are nonstandard. It should be noted that the reason for choosing ksc 403 su as sweet corn varieties in this investigation are as follows: Prolificacy, early maturity and ear emergence is simultaneously, whereas many of the other improved varieties are lacking this feature. Excel and SAS (Ver9.2) statistical software's recorded and analyzed Data related to measured characteristics, respectively. Mean comparisons were conducted by Duncan's multiple range test at a 1% level of probability.

3. RESULTS AND DISCUSSION

3.1 Plant and Ear Height

Results of variance analysis indicated that planting date had significant effect on plant height and ear height at a 1% level of probability. However, plant density was only significant in plant height ($p<.01$) (Table 1). Evaluating the effect of planting date indicated that the tallest plant and maximum ear height were obtained on 24 July with mean of 162.6 and 67.46 cm, respectively. Also, the shortest plant minimum ear height was obtained on 14 June with mean of 106 and 45.82 cm, respectively (Table 3). In addition, the highest plant height was from the plant densities of 85000 ha^{-1} with mean of 140.9 cm (Table 5). The main reason for increasing these traits in 3 July were suitability of climatic conditions and also coincide being time of pollination with desired temperature. Finally, with decreasing intensity of radiation entered into the canopy, internodes length will be increased and consequently plant height will be increased with increasing density and it could be due to plant competition, for light and nutrients. If the plant density excessively, plant in addition to light would be compete on other environmental sources and the plant height will be decreased [13,14].

3.2 Number of Leaves per Plant and above Ear

Planting date and plant density have no significant effect on number of leaves per plant. However, planting date had significant effect on number of leaves above ear at 1% level of probability (Table 1). The highest number of leaves above ear was obtained with average of 6.3 leaves on planting date of 24 July (Table 3). Number of leaves per plant is a genetic trait and environmental factors such as the plant density do not have significant effect on it. In this regard, Charles and Arnold indicated that although the number of leaves per plant is a genetic trait, it can be affected by temperature, planting date, soil conditions and agronomical cultivation [15].

3.3 Stem Diameter

Stem diameter was affected by plant density and planting date (Table 1). Accordingly, the highest and the lowest stem diameter were observed on planting date of 3 July and 14 June with amount of 23.9 and 19.8 mm, respectively (Table 3). Also, The highest and lowest stem diameter were belong to 65 and 105 thousand plants ha-^{1} with 22.19 and 21.27 mm, respectively (Table 5) These results were in agreement with those obtained by Tamadon Rastagari [8].

3.4 Ear Length and Ear Diameter

Analysis of variance of ear length and ear diameter data revealed that planting date had significant effect on these traits; while plant density did not have affect (Table 1). The highest and lowest ear length were obtained on 14 June and 24 July planting dates which were about 9.3 and 7.5 cm, respectively. Regarding the ear diameter the highest and the lowest value was obtained on 24 July and on 3 July planting date being about 15.75 and 14.05 mm, respectively (Table 3). Ear diameter and ear length are important traits being considered in selecting the high quality products in convert industry factories. Despite the fact that the ear length decreased on 24 July planting date, its popularity in markets decreased because of the increase in ear diameter and poor quality. Therefore, the best value of ear diameter and ear length concerning to marketability were obtained on 3 July planting date being about 7.9 (cm) and 14.05 (mm), respectively. Along with late planting of sweet corn, the ear length and the corn production were decreased [16].

3.5 Number of Ears per Plant

Planting date did not have a significant effect on number of ears per plant. However, plant density had significant effect on number of ears per plant at 1% level of probability (Table 1). Mean comparison number of ears per plant decreased up to 1.6 (Table 5).

3.6 Standard and Sub-Standard Ear Percentage

Standard ear percentage is one of the most important traits in producing ear in such a way that standard ear percentage is the proportion of the number of ear to the total number of harvested ear. Planting date had significant effect on standard and sub-standard ear percentage However; trait did not affect by plant density (Table 2). Investigations demonstrated that interaction planting date × plant density had significant effect on standard ear percentage at 5% level of probability (Fig 1). Mean comparison of planting dates effect on standard ear percentage indicated that the highest percentage of standard and sub-standard ear were obtained on 14 June and 24 July planting dates which were estimated about 53.68 and 66.98 percentage (Table 4). Accordingly, the highest percentage standard ear was obtained on 14 June planting date with plant density of 65000 plants ha-^{1} being about 49.31 percentage (Fig. 1).

3.7 Husked Ear Yield

Planting date had significant effect on husked ear yield at 1% level of probability. However, plant density did not have significant effect on husked ear yield (Table 2). The highest yield obtained on 24 July (13240 kg ha^{-1}) and 14 June (12250 kg ha^{-1}). However didn't have any significant different between 24 July and 14 June, but they have significant different with 3 July (7396 kg ha^{-1}) (Table 4). The main reason for increasing crop yield on 24 July it could be depends on optimal temperature on during the pollination and suitability of climatic conditions compare the previous two planting date. Researchers reported that in an investigation of different planting date of sweet corn in the southeast of Antalya the highest and lowest production, was belonged to 25 July and 25 April with 17751 and 1824 kg ha^{-1} respectively. Such result is in harmony with those obtained by Oktem et al. [5].

Table 1. Analysis of variance for standard ear yield and agronomic characteristics of baby corn on different planting date and plant density

S.O.V	df	Mean of squares							
		1	2	3	4	5	6	7	8
Replication (R)	3	0.841n.s	15.906n.s	0.656n.s	0.182n.s	2.532n.s	0.127*	0.051n.s	0.246n.s
P da	2	11137.88**	1591.18**	2.093n.s	10.304**	53.438**	11.114**	9.147**	1.903n.s
Error (a)	6	113.63	87.19	0.685	0.140	2.054	0.029	0.150	2.230
P de	2	0.551**	29.92n.s	0.242n.s	0.134n.s	4.001*	0.132n.s	0.440n.s	0.826**
P da × P de	4	37.920n.s	35.55n.s	0.345n.s	0.090n.s	1.356n.s	0.111n.s	0.455n.s	0.110n.s
Error (b)	18	49.794	22.127	0.294	0.085	1.171	0.057	0.266	0.110
CV %		5.01	7.98	4.99	5.11	5.03	2.89	3.49	15.46

(1): Plant height, (2): Ear height, (3): Number of leaves plant^{-1}, (4): Number of leaves above ear, (5): Stem diameter, (6): Ear length, (7): Ear diameter, (8): Number of ear plant^{-1}. P da: Planting date, P de: Plant density also P da × P de: Interaction between planting date and plant density. CV: Coefficient of variance; Ns, and **: Non-significant and significant at 5% and 1% levels of probability, respectively*

Table 2. Analysis of variance for standard ear yield and agronomic characteristics of baby corn on different planting date and plant density

S.O.V		Mean of squares						
	df	1	2	3	4	5	6	7
Replication (R)	3	5.512n.s	25.194n.s	4709791.72n.s	233256.30n.s	0.005n.s	344394.77n.s	7.141n.s
S d	2	531.29*	677.73**	117354024.18 **	12340853.8**	0.536**	14733650.53**	88.11n.s
Error (a)	6	104.78	79.80	2443858.32	323312.24	0.027	288057.87	21.61
P d	2	66.006n.s	36.228n.s	11578263.62n.s	1813610.75**	0.094**	1303281.83**	23.68*
S d×P d	4	86.429*	69.215n.s	1133890.85n.s	173060.93n.s	0.013n.s	487190.58n.s	4.149n.s
Error (b)	18	29.566	32.870	3865299.40	241659.74	0.012	214909.28	5.919
CV %		13.68	9.57	17.94	21.29	17.49	29.10	11.58

(1): Standard ear %, (2): Sub-Standard ear %, (3): Husked ear yield, (4): Dehusked ear yield, (5): Dehusked standard ear yield, (6): Dehusked sub-standard ear yield, (7): Ear harvest index. S d: Planting date, P d: Plant density also Sd × Pd: Interaction between planting date and plant density. Also CV: Coefficient of variance. Ns, and **: Non – significant and significant at 5% and 1% levels of probability, respectively*

Table 3. Means comparison for standard ear yield and some agronomic characteristics of baby corn on different planting date

Planting date	Treatment						
	1	2	3	4	5	6	7
14 June	106b	45.82b	10.57a	4.642b	19.81b	9.335a	14.57b
3 July	153.7a	63.46a	10.69a	6.217a	23.89a	7.904b	14.05c
24 July	162.6a	67.46a	11.35a	6.275a	20.90b	7.505c	15.75a

(1): Plant height (cm), (2): Ear height (cm), (3): Number of leaves plant^{-1}, (4): Number of leaves above ear, (5): Stem diameter (mm), (6): Ear length (cm), (7): Ear diameter (mm). Mean followed by similar letters are not significantly different at 5% probability level

Table 4. Means comparison for standard ear yield and some agronomic characteristics of baby corn on different planting date

Planting date	Treatment					
	1	2	3	4	5	6
14 June	53.68a	53.68b	12250a	2810b	833a	1977b
3 July	39.13ab	60.13ab	7396b	755c	411c	344c
24 July	33.98b	66.98a	13240a	3082a	623b	2459a

(1): Standard ear %, (2): Sub-standard ear %, (3): Husked ear yield (kg ha^{-1}), (4): Dehusked ear yield (kg ha^{-1}), (5): Dehusked standard ear yield (kg ha^{-1}), (6): Dehusked sub-standard ear yield (kg ha^{-1}). Mean followed by similar letters are not significantly different at 5% probability level

Table 5. Means comparison for standard ear yield and some agronomic characteristics of baby corn on different plant density

Plant density (plant ha^{-1})	Treatment						
	1	2	3	4	5	6	7
65000	140.5a	22.19a	2.50a	20.42b	1879c	554b	1325b
85000	140.9a	21.13b	2. 11b	20.01b	2084b	591b	1493b
105000	140.8a	21.27c	1.65c	22.62a	2683a	722a	1961a

(1): Plant height (cm), (2): Stem diameter (mm), (3): Number of ear plant^{-1}, (4): Ear harvest index(%), (5): Dehusked ear yield (kg ha^{-1}), (6): Dehusked standard ear yield(kg ha^{-1}), (7): Dehusked sub-standard ear yield(kg ha^{-1}). Mean followed by similar letters are not significantly different at 5% probability level

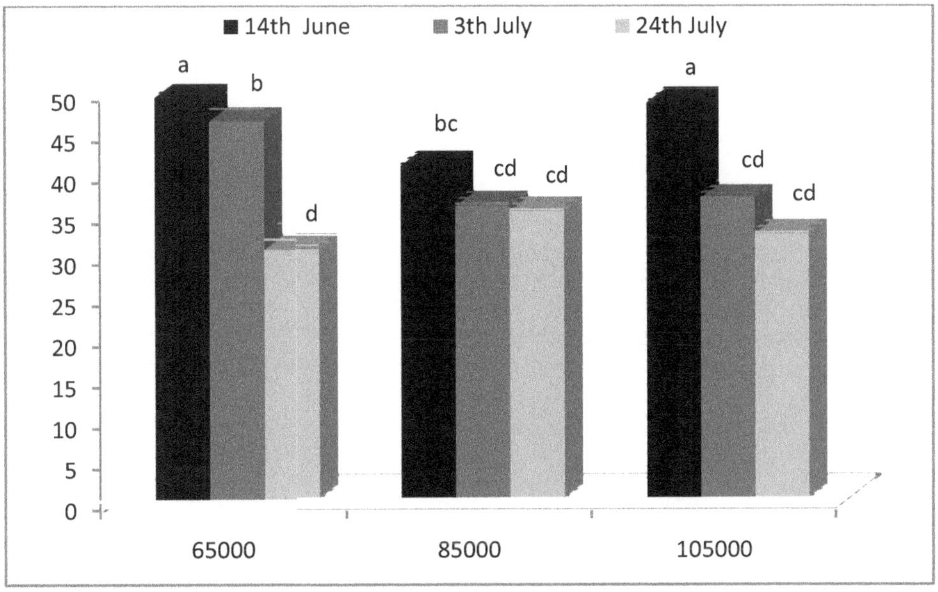

Fig. 1. The interaction effects of planting date and plant density on standard ear percentage
Mean followed by similar letters are not significantly different at 5% probability level

3.8 Dehusked Ear Yield, Dehusked Standard Ear Yield and Dehusked Sub- Standard Ear Yield

Variance analysis of dehusked ear yield, dehusked standard and sub standard ear yield indicated that planting date and plant density had significant effects on these traits (Table 2). The results revealed that the crop sown on 24 July gave the highest dehusked ear yield (3082 kg ha-^{1}) and dehusked sub-stadard ear yield (2459 kg ha-^{1}) while the crop sown on 14 June gave the highest dehusked standard ear yield (833 kg ha-^{1}) (Table 4). Also, Mean comparison of plant density indicated that the highest dehusked ear yield, dehusked standard ear yield and dehusked sub-standard ear yield were belong to plant density of 105000 plant ha-^{1} with 2683, 722 and 1961 kg ha-^{1} respectively (Table 5). It should be noted that crop sown on 3 July performed poor due to being coincide daily high temperatures with germination was. Researches indicated that prolongation of seedling emergence could lead to growth retardation and components involved in the yield [17].

3.9 Ear Harvest Index

Ear harvest index was influenced by plant densities while not by sowing dates (Table 2). Statistical mean comparison indicated the highest ear harvest index of 22.62% was observed with 105000 plants ha-^{1} (Table 5). Harvest index is the proportion of economic yield to the biological yield and it is affected by plant density, the available amount of nutrient and water and environmental temperature during the growth season [18]. This parameter is the index of grain production or transfer coefficient and distribution of photosynthetic nutrient between economic and herbal sections [19].

4. CONCLUSION

The results showed that dehusked ear yield increased on 3 August. However baby corn planting on 24 July due to early cold autumn and reduced marketability is not recommended. In this research according to the international codex to produce and supply products with high quality planting date of 14 June and density of 105,000 plants ha-^{1} recommended for producing the highest and the best standard ear. The results showed that planting date of 14 June and planting density of 105,000 plants ha-^{1} with the highest standard ear Percentage (31/49) has produced. In these circumstances, processing factories, a longer period of time worked the benefits of cultivating this plant weather in Mashhad.

COMPETING INTERESTS

Authors have declared that no competing interests exist.

REFERENCES

1. Kumar S, Kallo F. Attributes of maize genotype for baby corn production. Indian Institute of Vegetable Research; 2000. Available: http://www.Agron.missouri.edu
2. Anonymus. A flavorful appealing addition to salads pasta soups and other favorite dishes. Washington State University; 1998. Available: http://www.Agsyst.wsu.edu
3. Jason B. World market for fresh and canned baby corn. Market Information Bulletion. No 5; 1995.
4. Georg W, Dickerson, G. Speciality corn. Guide H-235. Cooperative Extension Service .College of Agriculture and Home Economics. New Mexico State University; 2005.
Available: http://www.cahe.nmsu.edu
5. Oktem A, Gulgun A, Coskum Y. Determination of sowing dates of sweet corn (*Zea mays* L. saccharta sturt) under sanlirfa conditions. Turkish Journal of Agriculture. 2004;28:83-91.
6. Noldin J, Mundstock C. Grain yield and yield components of three maize cultivars at two sowing date. Pesquisa Gropecuaria Brasileria. 1988;23:615-620.
7. Mokhtarpour H, Mosavat S, Bazi M, Saberi A. Effects of sowing date and plant density on ear yield of sweet corn (*Zea mays* var saccharata) SC 403. Iranian Journal of Crop Sciences. Persian. 2005;8:171-183.
8. Tamadon Rastegari M. Effects of sowing date and plant density yield and yield components sweet corn ksc 404. Var on Mazandaran weather conditions. Thesis of M.Sc. University of mazandaran, Persian. 2000;104.
9. Norwood C. Dryland corn in western Kansas: Effect of hybrid maturity, planting date and plant population. Agronomy Journal. 2001;93:540-547.
10. Widdicombe W, Thelen K. Row width and plant density effect on corn grain production in the northern corn belt. Agronomy Journal. 2002;94:1020-1023.
11. Genter C, Camper H. Component plant part development in maize as affected by hybrids and population density. Agronomy Journal. 1973;65:669-671.
12. Rodrigues R, Silva L, Mori E. Baby corn single-cross hybrids yield in two plant densities. Crop Breeding. 2003;3:177-184.
13. Muchow R, Sinclair T, Ennett J. Temperature and solar radiation effects on potential maize yield across location. Agronomy Journal. 1990;82:338-348.
14. Glenn F, Daynard T. Effect of genotype, planting pattern and plant density on plant-to- plant variability and grain yield of corn. Journal of Plant Science. 1973;54:323-330.
15. Charles Y, Arnold A. Environmentally induced variations of sweet corn characteristics as the relate to the time required for development. Journal of American Society for Horticultural Science. 1969;94:115-118.
16. Waligora H. The influence of sowing times on vegetative period and morphological characters of sweet corn. Prace. Z. Zakresu. Nauk. Rolniczych. 1997;83:135-140.
17. Nasrolah Alhossini M, Rahmani A, Khavari Khorasani S. Effects of salinity stress on germination stages through seed priming with chloride salts of sweet corn: growth and ion transport *(Zea Mays.* L Var saccharata*).* Journal of Crop Ecophysiology. Islamic Azad University Tabriz Branch. Persian. 2013;27:357-372.
18. Olnes A, Beneit G. Effect of planting date on time and rate nitrogen accumulation by maize .Crop Science Journal. 1990;64:42-53.
19. Donald C, Humblini J. The biological yield and harvest index of cereal as agronomic and plant breeding criteria. Advance in Agronomy. 1976;28:361-405.

Effect of Different Sowing Dates on Growth and Grain Yield of Chickpea (*Cicer arietinum* L.) Cultivars under Agro-environment of Taluka Dokri Sindh, Pakistan

Hafeez-ur-Rehman[1], Rafi Qamar[2*], Atique-ur-Rehman[3], Farhan Ahmad[4], Jamshaid Qamar[5], Muhammad Saqib[6] and Shah Nawaz[4]

[1]Agronomy Section, Agriculture Research Institute Tando Jam, Sindh, Pakistan.
[2]Agronomy Section, Central Cotton Research Institute Sakrand, Sindh, Pakistan.
[3]Department of Agronomy, Bahauddin Zakariya University, Multan, Pakistan.
[4]Entomology Section, Central Cotton Research Institute Sakrand, Sindh, Pakistan.
[5]Department of Agronomy, University College of Agriculture, University of Sargodha, Pakistan.
[6]Department of Agronomy, The Islamia University, Bahawalpur, Pakistan.

Authors' contributions

All authors contribute for designed the study, performed the statistical analysis and wrote the first draft of the manuscript. Author RQ carried out edited corrections and facilitated all e-mail correspondences. All authors read and approved the final manuscript.

Editor(s):
(1) Sławomir Borek, Department of Plant Physiology, Adam Mickiewicz University, Poland.
Reviewers:
(1) Anonymous, Tunisia.
(2) M. Monjurul Alam Mondal, Crop Physiology Division, Bangladesh Institute of Nuclear Agriculture, Bangladesh.
(3) Sevcan Oztemiz, Biological Control Research Station of Adana, Turkey.
(4) Klára Kosová, Division of Crop Genetics and Breeding, Crop Research Institute, Prague, Czech Republic.

ABSTRACT

Aim: The study was carried out to investigate the effect of different sowing dates on growth and grain yield of chickpea (*Cicer arietinum* L.) cultivars.
Study Design: A randomized complete block design (RCBD) in split plot arrangement was used.
Place and Duration of Study: Pulses Section, Rice Research Institute, Dokri, Sindh, Pakistan during Rabi season of 2012-13.
Methodology: Five sowing dates [S_I = 15th October 2012, S_2 = 30th October 2012, S_3 = 15th

**Corresponding author: E-mail: drrafi1573@gmail.com;*

November 2012, S_4 = 30th November 2012 and S_5 = 15th December 2012] in main plots and four cultivars [V_1 = DG-92, V_2 = Chhola, V_3 = DG-89 and V_4 = Sanyasi] in sub-plot.
Results: Sowing dates had significant difference ($P \leq 0.05$) on days to 50 % flowering, plant height, branches and pods per plant and 1000-seed weight while cultivars had on days to 50 % flowering and 1000-seed weight. Significant differences ($P \leq 0.05$) were observed among sowing dates and chickpea cultivars interaction for days to 90 % maturity and grain yield. The maximum days to 90 % maturity were observed for cultivar Sanyasi (169 days) when sown on 15th October while minimum for cultivar DG-92 (122 days) when sown on 15th December. Grain yield was maximized for DG-92 (3.3 Mg ha^{-1}) at sowing date of 15th November.
Conclusion: It is concluded that under agro-climatic conditions of Taluka Dokri, District Larkana, Province Sindh, Pakistan, cultivar DG-92 performed the best when sown on first fortnight of November.

Keywords: Sowing dates; cultivars; grain yield; chickpea.

1. INTRODUCTION

Chickpea is a conventional pulse crop of Pakistan and reasonable source of protein in human diet. It contains 21% protein, 61% carbohydrates and 2.2% oil [1]. It is not only the best source of protein but also a good food of high nutritive value ensuring substantial quantity of dietary fiber, vitamins C along with iron and minerals [2]. In Pakistan, chickpea is grown on an area of 975 thousands hectares with the total production of 475 thousand tons [3] and an average grain yield of 794 kg ha^{-1}, which is too low than its potential (1600 kg ha^{-1}). In Pakistan, chickpea is a broad-spectrum, reasonable and wide adaptability in all types of soil. It has ability to tolerate relatively high temperatures during the flowering and grain filling stage. Pulses have capability to fix the atmospheric nitrogen which has an important role in the profitability of farming systems by reducing cost of fertilizer application for both winter and summer cereal crops.

There are some limitations in chickpea production in which frost damage and disease are the two main limitations [4]. In sub-tropical region, the climate is temperate with winter rainfall, chickpea is conventionally seeded in spring; therefore, the crop faces high temperature and water stress towards maturity which resulted in low and variable yields. However, with new cultivars, winter seeding of chickpea in sub-tropical environments has recently been augmented, since winter sowing provides higher and more stable yield and growth [5]. To overcome the effect of cold temperature, adjustments in sowing date can be used as a strategy to increase chickpea production [6]. Time of sowing will depend on the interaction of environment and the available varietal germplasm [7]. The main reason of chickpea flower abortion has been shown when mean daily temperature of less than 15°C [8]. Flower development is a crucial stage because fluctuation in environment affects it which ultimately influence on crop production. Start of flowering in chickpeas is dependent on photo-thermal reaction, which is the main determinant [9]. The optimum sowing date results in timely initiation of flowering by minimizing threats of cold temperatures which can retard the growth of chickpea plants [4]. Early sowing of chickpea can expose it to heavy rainfall which results in lodging, diseases occurrence, and moisture deficit during grain fill stage. Late sowing can effect on plant height which may reduce vegetative cover and water use efficiency and increase the incidence of insects [10]. Sowing time and cultivars are two important factors which can affect the growth and yield of chickpea [11]. The most vital step towards enhancing yield of chickpea is to ensure that the phenology of the crop is well in line to resources and constraints of the production environment [6].

One method to enhance grain yield in chickpea would be to change the sowing time using existing cultivars which are resistant to biotic and abiotic disorders [12]. Winter sown chickpea produced higher seed yield than spring sown crop [13]. The maximum crop yield of winter crop has been recognized to the extended growing period and favorable climatic condition during winter and early spring [14]. Increased crop production, contributed by plant height, higher number of branches and pods per plant which was positively correlated with yield [15]. Optimum sowing time of chickpea vary from one cultivar to another and also from one region to another due to variation of agro-ecological conditions.

The present study was carried out to investigate the effect of different sowing dates on growth and grain yield of chickpea cultivars under agro-environment of Dokri, Sindh, Pakistan.

2. MATERIALS AND METHODS

2.1 Study Site

The present study was performed at the experimental field of Pulses Section, Rice Research Institute, Dokri (68° 07′ E, 27° 25′ N), Sindh, during Rabi season, 2012-13.

2.2 Experimental Design and Cultural Practices

The experiment was set up as a randomized complete block design in split plot arrangement with three replications. Plot size was 9 m^2 (1.8 m x 5 m). Treatments were: five sowing dates, i.e. S_l = 15th October 2012, S_2 = 30th October 2012, S_3 =15th November 2012, S_4 = 30th November 2012 and S_5 = 15th December 2012 and four cultivars, i.e. V_l = DG-92, V_2 = Chhola, V_3 = DG-89 and V_4 = Sanyasi. Sowing dates were randomized in the main plots while cultivars were randomized in subplots. After the harvesting of paddy crop the land was prepared for chickpea sowing as Dubari crop by giving two dry ploughings with disc plough followed by clod crushing to achieve fine seed bed. Monthly temperatures and rainfall of region is presented in Fig. 1. The soil of the experimental field was clay loam in texture under mono-cropping system (Table 1). The sowing of chickpea cultivars was done by means of single coulter hand drill in lines. The recommended seed rate of 100 kg ha^{-1} was used for chickpea cultivars. Nitrogen and phosphorus fertilizers were applied at 40 kg N ha^{-1} and 75 kg P_2O_5 ha^{-1}, respectively. Full doses of fertilizers were applied at the time of sowing. The pesticide Karate (Lambda-cyhalothri 2.5 EC) was applied at the rate of 625 mL ha^{-1}. Buctril Super 60 EC (Bromoxynil + MCPA) @ 700 ml ha^{-1} and Topik 15WP (Clodinafop propargyl) at 250 g ha^{-1} were applied to control both dicot and monocot weeds.

2.3 Development and Agronomic Parameters of Chickpea

Development and agronomic parameters of chickpea, including days to 50% flowering, days to 90% maturity, plant height (cm), branches per plant, pods per plant, 1000-seed weight (g) and seed yield per hectare (Mg) were recorded. The whole above ground harvested material was kept in bundles separately and to be dried up, it was threshed with sticks and weighed for recording grain yield. The growth and yield components were determined on twenty randomly selected plants from each plot.

2.4 Statistical Analysis

Data collected was analyzed statistically by applying Fisher's analysis of variance technique. Least significant difference (LSD) test was employed at 5% probability level to test the significance of the treatments' means [16].

Table 1. Soil analysis of experimental site during 2012-13

Soil analysis	Value
Mechanical analysis	
Sand (%)	38
Silt (%)	24
Clay (%)	39
Textural classes	Clay loam
Chemical analysis	
Soil pH	8.3
EC (dS m^{-1})	7.70
Organic matter (%)	0.30
Available N (g kg^{-1})	0.16
Available P (mg kg^{-1})	6.3
Available K (mg kg^{-1})	163

3. RESULTS AND DISCUSSION

Number of days taken to 50% flowering and 90% maturity show the genetic nature of a cultivar i.e. early, mid or late-season. Sowing dates and cultivars had significant effects on growth of chickpea (Table 2 & Fig. 2). Data revealed that interaction of days to 50% flowering was non-significant (Table 2) and days taken to 90 % maturity were significant (Fig. 2). Chickpea sown on 15th October took comparatively more time (87.9 days) to bear 50% flowering (Table 2). Days taken to 50% flowering at 30th October (85.2 days) were statistically similarwith15th November (84.7 days), while 30th November and 15th December had the lowest number of days (78.4 and 78.2 days) to 50% flowering. Significant differences of days to 50% flowering were recorded in varieties which may be due to their genetic characters; however, maximum number of days to 50% flowering was recorded by cultivar Sanyasi (84.5 days), followed by cultivar Chhola (83.5 days) while minimum number of days to 50% flowering was recorded by cultivar DG-92 (81.9 days). Chickpea sown at 15th October significantly took more time (165.2 days) to complete 90% maturity, while the lowest number of days (122.1 days) to 90% maturity was recorded at 15th December that was statistically similar with 30th November (129.3

days). Significant differences were recorded in cultivars; maximum number of days to 90% maturity was recorded by cultivar Sanyasi (145.5 days) followed by cultivar Chhola (143.2 days) while minimum number of days to 90% maturity was recorded by cultivar DG-89 (142 days) (Fig. 2). The interaction of sowing dates and cultivars (Fig. 2) envisaged that cultivar Sanyasi when sown on 15th October took maximum (169.0 days) number of days to maturity, followed by cultivars Chhola and DG-89 (167.7 and 163.3 days) and significantly lower in case of cultivar DG-92 under 15th December (120.7 days).

Sowing dates and cultivars had significant effects on yield parameters of chickpea (Table 2 & Fig. 3). Significantly higher plant height (53 cm) was recorded at 15th November and lower plant height (43.4 cm and 41.6 cm) at 30th October and 15th December sowing dates respectively. Number of branches per plant (4.8) at 15th October was significantly higher over other sowing dates. Significantly maximum (46.8) number of pods per plant was recorded at 15th October and minimum (40.3 and 40.7) was at 30th October and 15th December respectively. Sowing at 15th November produced significantly higher 1000-seed weight (224.8 g) while lower 1000-seed weight (209.7 g) was recorded at sowing date of 15th December (Table 2). Chickpea seed yield per hectare (Fig. 3) was significantly higher (2.3 Mg ha^{-1}) in 15th November while lower chickpea yield per hectare was recorded (1.4 g plot^{-1} and 1.1 Mg ha^{-1}) in 30th November and 15th December.

Cultivars had non-significant effect on plant height, number of branches per plant and pods per plant (Table 2). Significantly higher 1000-seed weight was recorded at DG-89 (263.9 g) that was 14, 21 and 34% greater as compared to in DG-92(226.7 g), Chhola (207.1 g) and Sanyasi (172 g) respectively. Chickpea seed yield per hectare was significantly higher in DG-92 (2.2 Mg ha^{-1}) and lower was recorded in Sanyasi (1.1 Mg ha^{-1}) (Fig. 3). In 2012-13 growing season, sowing dates and cultivars interaction significantly influenced the chickpea seed yield per hectare except plant height, number of branches and pods per plant and 1000-seed weight (Table-2). However, cultivar DG-89 at 15th November of sowing date and cultivar DG-92 at 15th October sowing data produced significantly higher growth and yield of chickpea as compared to other cultivar and sowing dates.

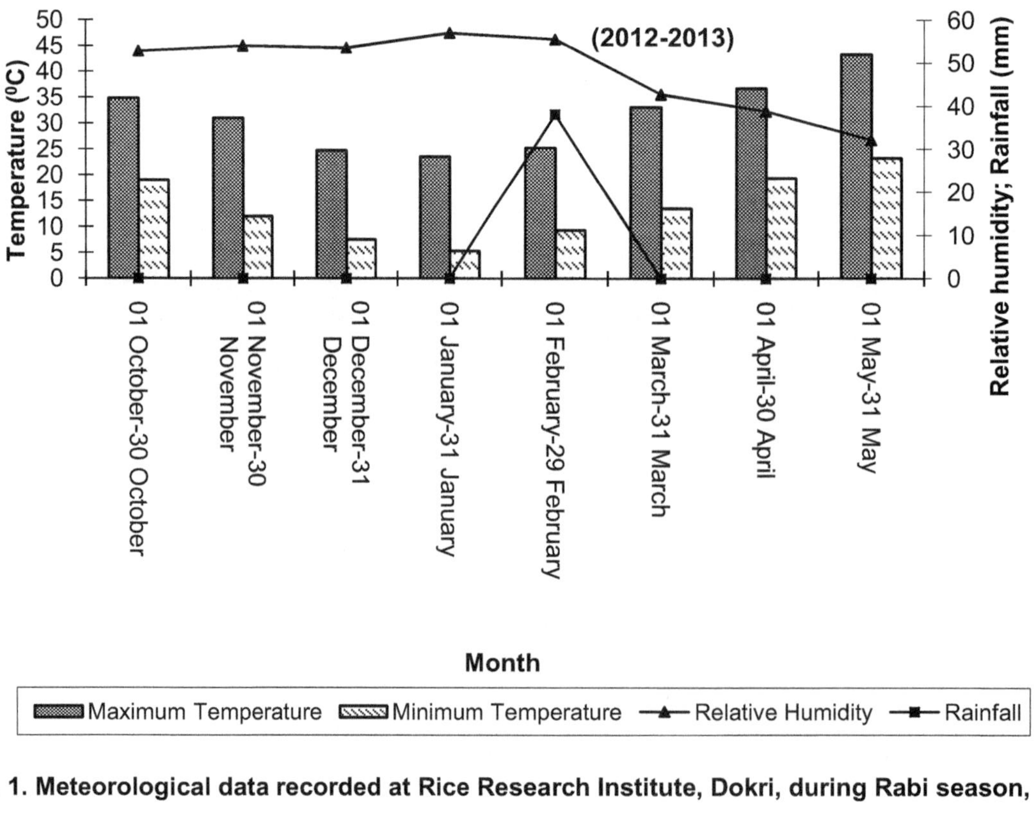

Fig. 1. Meteorological data recorded at Rice Research Institute, Dokri, during Rabi season, 2012-13

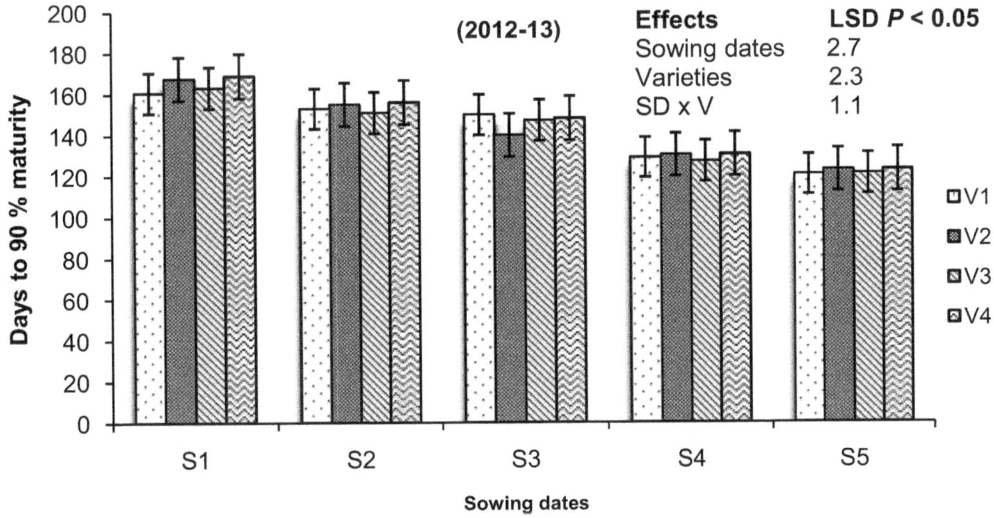

Fig. 2. Sowing dates and varieties effects on days to 90% maturity of chickpea (data over 2012-13)

(S_I= 15th October 2012, S_2 = 30th October 2012, S_3 =15th November 2012, S_4 = 30th November 2012 and S_5 =15th December 2012. V_I = DG-92, V_2 = Chhola, V_3 = DG-89 and V_4 = Sanyasi)

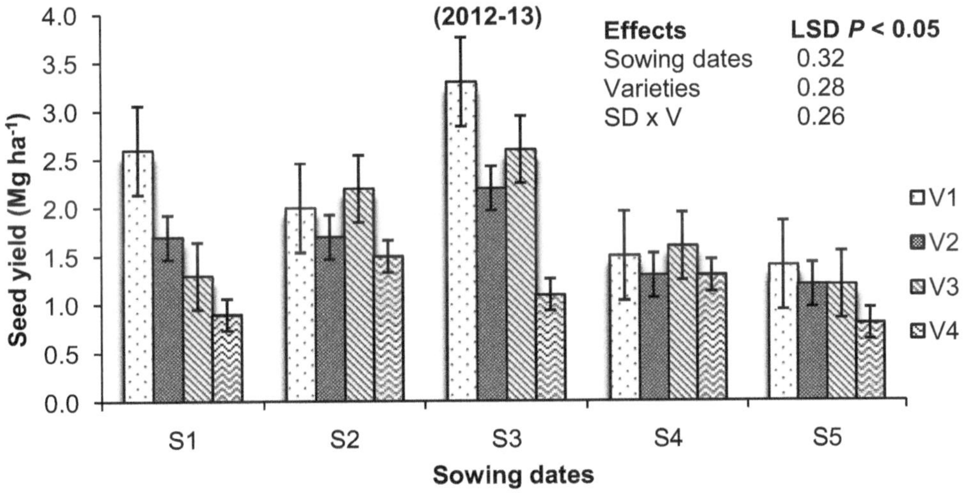

Fig. 3. Sowing dates and varieties effects seed yield (Mg ha^{-1}) of chickpea (data over 2012-13)

(S_I= 15th October 2012, S_2 = 30th October 2012, S_3 =15th November 2012, S_4 = 30th November 2012 and S_5 =15th December 2012. V_I = DG-92, V_2 = Chhola, V_3 = DG-89 and V_4 = Sanyasi)

The current study revealed that the interaction of different sowing dates and chickpea cultivars had significant effect on days to 90% maturity and seed yield per hectare while non-significant effect on days to 50% flowering, plant height, number of branches and pods per plant and 1000-seed weight. The variability in sowing dates and chickpea genotypes for yield and yield components also had been reported by other researchers [17,13,18]. Sowing dates had significant effect on days to flowering and maturity. Moreover, higher numbers of days taken to 50% flowering and 90% maturity in early autumn sown was due to the matching crop phenology with the availability of optimum temperature and moisture regimes and prolong growth period [19]. After a long delay at the seedling emergence stage due to low temperature at 15th December sowing while 15th November sown crops quickly started to develop

Table 2. Sowing dates and varieties effects on days to 50% flowering, plant height, branches per plant, pods per plant and 1000-seed weight of chick pea (data of 2012-13 growing season)

Treatments / Parameters	Days to 50% flowering	Plant height (cm)	Branches per plant	Pods per plant	1000-seed weight (g)
Sowing dates					
S_1 = 15th Oct. 2012	87.9a	49.6b	4.8a	46.8a	217.2c
S_2 = 30th Oct. 2012	85.2b	43.4c	3.1b	40.3b	222.0b
S_3 = 15th Nov. 2012	84.7b	53.0a	3.6b	46.1a	224.8a
S_4 = 30th Nov. 2012	78.4c	49.1b	3.1b	46.3a	213.6d
S_5 = 15th Dec. 2012	78.2c	41.6c	2.6b	40.7b	209.7e
LSD $P < 0.05$	2.51	2.10	1.12	2.72	2.41
Varieties					
V_1 = DG-92	81.9bc	49.6	3.6	44.3	226.7b
V_2 = Chhola	83.5ab	48.1	3.5	39.5	207.1c
V_3 = DG-89	82.0b	49.7	3.3	49.3	263.9a
V_4 = Sanyasi	84.5a	42.0	3.3	43.1	172.0d
LSD $P<0.05$	3.20	NS	NS	NS	2.11
Inter-action	NS	NS	NS	NS	NS

Any two means not sharing a letter in common differ significantly at 5 % probability level. NS = non-significant

a larger crop cover and grain yield in promising temperature and availability of soil moisture and longer growth duration [19]. However, delay in sowing resulted in adverse effect of climate which, resulted in poor crop stand and short time to complete their life cycle, specifically after 15th November [14]. The result showed that different sowing dates and chickpea cultivars had individually significant effect on yield and yield component [17].

Maximum yield in the mid-autumn sowing was as a result of favorable climatic condition which supported the vegetative and reproductive growth stages [20]. Mid-autumn sowing increased seeds weight and seed yield of chickpea compared than early and late sowing and this result was due to the moderate temperature regime during the seed filling stage. Chickpea sown on 15th November produced highest grain yield [19]. The lower number of 1000-seed weight and seed yield in early and late autumn sowing dates was due to encounter of flowering and fertilization stages with high and low temperatures respectively [18]. Overall the higher yields given by the 15th November sown chickpea, as compared with the 15th October and 15th December sown crop seem to be the result of better grain yield occurring as the result of a favorable growth period [21]. Under the low temperate, frost could be harmful for late-autumn sowing, because it usually occurs during the early stages of vegetative growth. Early spring sowing usually allows a crop to escape frost danger, unless there is a late frost. Autumn-sown chickpea may face water shortages in the later part of the crop's growth. We observed that the higher yield of 15th November related with favorable environmental conditions which can be lost due to unfavorable environmental conditions [22].

4. CONCLUSION

The current study confirms previous results showing that mid-autumn sowing improves the productivity of chickpea as compared with very early and late autumn sowing. Nonetheless, we observed that in very early autumn sowing date (October 15) resulted to plant lodging and flowering was encountered with low temperature which resulted low yield. In late autumn sowing date (December 15) resulted lower growth and yield of crop due to frost damage. After going through the results in detail, it was concluded that under climatic conditions of Taluka Dokri, District Larkana varieties DG-92 found best suitable when sown in the first fortnight of November.

ACKNOWLEDGEMENTS

Financial support from Higher Education Commission of Pakistan for this study is highly acknowledged.

COMPETING INTERESTS

Authors have declared that no competing interests exist.

REFERENCES

1. Gupta YP. Nutritive value of pulses. In: Pulse Crop. Baldev. S., Rawanujam and H. K. Jain (Eds.). Oxford IBH Publishing Co. Pvt. Ltd. New Dehli. 1988;561-601.
2. Galili S, Kitain S, Badani H. Developing new uses for chickpea emphasis on healthy food. Nir. Vatelem. 2008;8-10.
3. Govt. of Pakistan. Economic Survey of Pakistan, Ministry of Food, Agriculture and Livestock, Economic Wing, Islamabad. 2012;13.
4. Whish JPM, Castor P, Carberry PS. managing production constraints to the reliability of chickpea within marginal areas of the northern grains region of Australia. Aust J Agric Res. 2007;58(5):396-405.
5. Pacuacci G, Troccoli C, Leoni B. Effect of suplementary irrigation on yield of chickpea genotypes in a Mediterranean climate. Agricultural Engineering International: The CIGR Ejournal, Manuscrip LW 04 005. 2006;3.
6. Summerfield RJ, Virmani SM, Robertsand EH, Ellis RH. Adaptation of chickpea to agro-climatic constraints. In Chickpea in the Nineties: Proceedings of the 2nd International Workshop on Chickpea Improvement (Eds B.J. Walby& S.D. Hall), Patancheru, India: ICRISAT. 1990;61–72.
7. Knights EJ, Siddique KHM. Chickpea status and production constraints in Australia. In 'Integrated management of botrytis grey mould of chickpea in Bangladesh and Australia. Summary Proceedings of a Project Inception Workshop, Bangladesh Agricultural Research Institute, Joydebpur, Gazipur, Bangladesh'. (Eds MA Bakr, KHM Siddique, C Johansen) 2002a;33–41.
8. Clarke H, Siddique KHM. Growth and development in the Chickpea Book (eds. S. Loss, N. Brandon & K.H.M. Siddique), Agriculture Western Australia, Bulletin 1326; 1998.
9. Basu PS, Ali M, Chaturvedi SK. Terminal heat stress adversely affects chickpea productivity in Northern India-Strategies to improve thermos tolerance in the crop under climate change. ISPRS Archives XXXVIII-8/W3 Workshop Proceedings: Impact of Climate Change on Agriculture. New Delhi, India. 2009;189-193.
10. Matthews P, McCaffery D. Winter crop variety sowing guide. NSW DPI Management Guide; 2011.
11. Reddy BVS, Reddy PS, Bidinger F, Blummel M. Crop management factors influencing yield and quality of crop residues. Field Crops Res. 2003;84(1):57-77.
12. Janneli P, Bozzini A. Chickpea breeding for winter and spring sowing. ENED La Coltura Del Cece in Italia. 1987;96-106.
13. Iliadis C. Evaluation of six chickpea varieties for seed yield under autumn and spring sowing. J. Agric. Sci. Cam. 2001; 137(4):439-444.
14. O'Toole NFL, Stoddard O'Brien L. Screening of chickpea for adaptation to autumn sowing. J. Agron. Crop Sci. 2001; 186(3):193-207.
15. Singh KB, Bejiga G, Malhotra RS. Associations of some characters with seed yield in chickpea collections. Euphytica. 1990;93(1):83-88.
16. Steel RGD, Torrie JH, Dicky DA. Principles and procedures of statistics. A biometrical approach. 3rd Ed. McGraw Hill, Inc. Book Co. N.Y. 1997;352-358.
17. Yucel D, Anlarsal AE. Performance of some winter chickpea (*Cicer arietinum* L.) genotypes in Mediterranean conditions. Notulae Botanicae Horti Agrobotanici *Cluj*-Napoca. 2008;36(2):35-41.
18. Chaitanya SK, Chandrika V. Performance of chickpea varieties under varied dates of sowing in chittoor district of Andhra Pradesh. Legu. Res. 2006;29(2):137-139.
19. Ozdemir S, Karadavut U. Comparison of the performance of autumn and spring sowing of chickpeas in a temperate region. Turk J Agric For. 2003;27:345-352.
20. Valimohammadi F, Tajbakhsh M, Saeid A. Comparison winter and spring sowing dates and effect of plant density on yield, yield components and some quality, morphological traits of chickpea (*Cicer arietinum* L.) under environmental condition of Urmia. J. Agron. 2007;6(4): 571-575.

21. Saxena MC. Agronomic studies on winter chickpeas. In: Saxena, M.C., Singh, K.B. (Eds.), Proceedings of the Workshop on Ascochyta Blight and Winter Sowing of Chickpeas, ICARDA, 4-7 May 1981, Aleppo, Syria. 1984;123-137.

22. Oweis T, Hachum A, Pala M. Water use efficiency of winter-sown chickpea under supplemental irrigation in a Mediterranean environment. Agric Water Manag. 2004; 66(2):163-179.

Permissions

All chapters in this book were first published in AJEA, by SCIENCE DOMAIN International; hereby published with permission under the Creative Commons Attribution License or equivalent. Every chapter published in this book has been scrutinized by our experts. Their significance has been extensively debated. The topics covered herein carry significant findings which will fuel the growth of the discipline. They may even be implemented as practical applications or may be referred to as a beginning point for another development.

The contributors of this book come from diverse backgrounds, making this book a truly international effort. This book will bring forth new frontiers with its revolutionizing research information and detailed analysis of the nascent developments around the world.

We would like to thank all the contributing authors for lending their expertise to make the book truly unique. They have played a crucial role in the development of this book. Without their invaluable contributions this book wouldn't have been possible. They have made vital efforts to compile up to date information on the varied aspects of this subject to make this book a valuable addition to the collection of many professionals and students.

This book was conceptualized with the vision of imparting up-to-date information and advanced data in this field. To ensure the same, a matchless editorial board was set up. Every individual on the board went through rigorous rounds of assessment to prove their worth. After which they invested a large part of their time researching and compiling the most relevant data for our readers.

The editorial board has been involved in producing this book since its inception. They have spent rigorous hours researching and exploring the diverse topics which have resulted in the successful publishing of this book. They have passed on their knowledge of decades through this book. To expedite this challenging task, the publisher supported the team at every step. A small team of assistant editors was also appointed to further simplify the editing procedure and attain best results for the readers.

Apart from the editorial board, the designing team has also invested a significant amount of their time in understanding the subject and creating the most relevant covers. They scrutinized every image to scout for the most suitable representation of the subject and create an appropriate cover for the book.

The publishing team has been an ardent support to the editorial, designing and production team. Their endless efforts to recruit the best for this project, has resulted in the accomplishment of this book. They are a veteran in the field of academics and their pool of knowledge is as vast as their experience in printing. Their expertise and guidance has proved useful at every step. Their uncompromising quality standards have made this book an exceptional effort. Their encouragement from time to time has been an inspiration for everyone.

The publisher and the editorial board hope that this book will prove to be a valuable piece of knowledge for researchers, students, practitioners and scholars across the globe.

List of Contributors

K. Acheampong
Physiology/Biochemistry Division, Cocoa Research Institute of Ghana, P.O.Box 8, Akim Tafo, Ghana, West Africa

P. Hadley
Biodiversity, Crops and Agro-ecosystems Research Division, School of Agriculture, Policy and Development, The University of Reading, Whiteknights, Reading, RG66AR, UK

A. J. Daymond
Biodiversity, Crops and Agro-ecosystems Research Division, School of Agriculture, Policy and Development, The University of Reading, Whiteknights, Reading, RG66AR, UK

P. Adu-Yeboah
Physiology/Biochemistry Division, Cocoa Research Institute of Ghana, P.O.Box 8, Akim Tafo, Ghana, West Africa

A. Bangura
Agricultural Research Institute, Rokupr Agricultural Research Centre, Rokupr, PMB73, Freetown, Sierra Leone

Y. L. Oo
School of Agricultural and Wine Sciences, Charles Sturt University, Orange Campus, New South Wales, Australia

C. S. Kamara
Agricultural Research Institute, Rokupr Agricultural Research Centre, Rokupr, PMB73, Freetown, Sierra Leone

A. Raman
School of Agricultural and Wine Sciences, Charles Sturt University, Orange Campus, New South Wales, Australia

D. S. Hodgkins
School of Agricultural and Wine Sciences, Charles Sturt University, Orange Campus, New South Wales, Australia

H. I. Nicol
School of Agricultural and Wine Sciences, Charles Sturt University, Orange Campus, New South Wales, Australia

Barna Chakraborty
Department of Zoology, Vidyasagar University, Midnapore-721102, India

Asit Kumar Chanda
Central Sericultural Research and Training Institute, Berhampur, West Bengal, India

Susanta Kumar Chakraborty
Department of Zoology, Vidyasagar University, Midnapore-721102, India

A. A. Ansari
Department of Biology, University of Guyana, Georgetown, Guyana

M. Pereira
Department of Biology, University of Guyana, Georgetown, Guyan

S. Jaikishun
Department of Biology, University of Guyana, Georgetown, Guyana

A. Lado
Department of Agronomy, Bayero University, Kano, Nigeria

M. A. Hussaini
Department of Agronomy, Bayero University, Kano, Nigeria

S. G. Mohammed
Department of Agronomy, Bayero University, Kano, Nigeria

Daniele de Jesus Ferreira
Department of Animal Science, Federal University of Mato Grosso, MT, Brazil

Anderson de Moura Zanine
Department of Animal Science, Federal University of Mato Grosso, MT, Brazil

Edson Mauro Santos
Department of Animal Science, Federal University of Paraíba, PB, Brazil

Juliana Silva de Oliveira
Department of Animal Science, Federal University of Paraíba, PB, Brazil

Ricardo Martins Araujo Pinho
Department of Animal Science, Federal University of Paraíba, PB, Brazil

Antaryami Mishra
Department of Soil Science and Agricultural Chemistry, College of Agriculture, Orissa University of Agriculture and Technology (OUAT), Bhubaneswar-751003, India

Trupti Mayee Pattnaik
Departments of Chemistry, ITER, SOA University, Bhubaneswar, India

D. Das
Department of Soil Science and Agricultural Chemistry, College of Agriculture, Orissa University of Agriculture and Technology (OUAT), Bhubaneswar-751003, India

Mira Das
Departments of Chemistry, ITER, SOA University, Bhubaneswar, India

Rahim Darma
Department of Agricultural Socio-economics, University of Hasanuddin, Makassar, 90245, Indonesia

A. Majdah
Department of Agribusiness, University of Islam Makassar, Makassar, 90245, Indonesia

A. Zain
Department of Agribusiness, University of Islam Makassar, Makassar, 90245, Indonesia

A. Nixia Tenriawaru
Department of Agricultural Socio-economics, University of Hasanuddin, Makassar, 90245, Indonesia

Mogapi E. Madisa
Department of Crop Science and Production, Botswana College of Agriculture, Private Bag 0027, Gaborone, Botswana

Thembinkosi Mathowa
Department of Crop Science and Production, Botswana College of Agriculture, Private Bag 0027, Gaborone, Botswana

Christopher Mpofu
Department of Crop Science and Production, Botswana College of Agriculture, Private Bag 0027, Gaborone, Botswana

Thato A. Oganne
Department of Crop Science and Production, Botswana College of Agriculture, Private Bag 0027, Gaborone, Botswana

M. Haliru
Department of Soil Science and Agricultural Egineering, Usmanu Danfodiyo University, Sokoto, Nigeria

M. Audu
Department of Soil Science and Agricultural Egineering, Usmanu Danfodiyo University, Sokoto, Nigeria

S. A Lukman
Department of Soil Science and Agricultural Egineering, Usmanu Danfodiyo University, Sokoto, Nigeria

M. M Sauwa
Department of Soil Science and Agricultural Egineering, Usmanu Danfodiyo University, Sokoto, Nigeria

I. Aliyu
Department of Soil Science and Agricultural Egineering, Usmanu Danfodiyo University, Sokoto, Nigeria

N. G Hayatu
Department of Soil Science and Agricultural Egineering, Usmanu Danfodiyo University, Sokoto, Nigeria

Eliane T. Bodah
Division of Breeding, ProGene Plant Research LLC, 860 S. Crestline, Othello, WA 99344, USA

Kurt Braunwart
Division of Breeding, ProGene Plant Research LLC, 860 S. Crestline, Othello, WA 99344, USA

Brian W. Bodah
WSU Extension, Pierce County Extension, Washington State University, 3602 Pacific Ave, Tacoma, WA 98418, USA

Alcindo Neckel
Department of Geography, Faculdade Meridional University, Passo Fundo, RS 99100, Brazil

E. E. Ikenganyia
Department of Crop Science, Faculty of Agriculture, University of Nigeria, Nsukka, Nigeria

U. M. Ndubuaku
Department of Crop Science, Faculty of Agriculture, University of Nigeria, Nsukka, Nigeria

C. C. Onyeonagu
Department of Crop Science, Faculty of Agriculture, University of Nigeria, Nsukka, Nigeria

U. Ukonze
Department of Science Education, University of Nigeria, Nsukka, Nigeria

A. K. Mishra
Water Technology Center, Indian Agricultural Research Institute, New Delhi-110012, India

S. S. Parihar
Water Technology Center, Indian Agricultural Research Institute, New Delhi-110012, India

V. Visha Kumari
Indian Institute of Sugarcane Research, Lucknow, India

T. K. Srivastava
Indian Institute of Sugarcane Research, Lucknow, India

Chand Khan
Water Technology Center, Indian Agricultural Research Institute, New Delhi-110012, India

A. Matikiti
Crop Sciences Department, University of Zimbabwe, Zimbabwe

Patrick A. Ndakidemi
School of Life Sciences, Nelson Mandela African Institution of Science and Technology, P.O. Box 447, Arusha Tanzania

G. O. Adeyemo
Department of Animal Science, Faculty of Agriculture, University of Ibadan, Nigeria

F. A. Oluyede
Department of Animal Science, Faculty of Agriculture, University of Ibadan, Nigeria

C. C. Nnadi
Department of Crop Science, University of Nigeria Nsukka, Enugu State, Nigeria

C. C. Onyeonagu
Department of Crop Science, University of Nigeria Nsukka, Enugu State, Nigeria

S. C. Eze
Department of Crop Science, University of Nigeria Nsukka, Enugu State, Nigeria

John Peter Nyaaba Aetiba
Department of Crop and Soil Sciences, Kwame Nkrumah University of Science and Technology, Kumasi, Ghana

Enoch Adjei Osekre
Department of Crop and Soil Sciences, Kwame Nkrumah University of Science and Technology, Kumasi, Ghana

Amira M. S. Abdul Qados
Department of Biology, Faculty of Science, Princess Nora bint Abdul Rahman University, Saudi Arabia

M. R. Olojugba
Department of Soil Science and Management, Joseph Ayo Babalola University, Ikeji-Arakeji, Nigeria

D. O. Oke
Department of Forestry and Wood Technology, Federal University of technology, Akure, Nigeria

Edward Wilczewski
Department of Agrotechnology, Faculty of Agriculture and Biotechnology, UTP University of Science and Technology, 85-225 Bydgoszcz, Poland

Zbigniew Skinder
Department of Agrotechnology, Faculty of Agriculture and Biotechnology, UTP University of Science and Technology, 85-225 Bydgoszcz, Poland

Shahid Rasool
CSIR- Indian Institute of Integrative Medicine, Srinagar, Kashmir, India

R. H. Kanth
Division of Agronomy, Sher-e-Kashmir University of Agricultural Sciences and Technology of Kashmir, India

Shabana Hamid
Department of Agriculture, J&K Government, Kashmir, India

W. Raja
Division of Agronomy, Sher-e-Kashmir University of Agricultural Sciences and Technology of Kashmir, India

B. A. Alie
Division of Agronomy, Sher-e-Kashmir University of Agricultural Sciences and Technology of Kashmir, India

Z. A. Dar
Division of Plant Breeding and Genetics, Sher-e-Kashmir University of Agricultural Sciences and Technology of Kashmir, India

Atena Rahmani
Department of Agronomy and Plant Breeding, Young Researchers and Elite Club, Mashhad Branch, Islamic Azad University, Mashhad, Iran

Majid Nasrolah Alhossini
Department of Agronomy and Plant Breeding, Young Researchers and Elite Club, Mashhad Branch, Islamic Azad University, Mashhad, Iran

Seyed Mohsen Nabavi Kalat
Department of Agronomy and Plant Breeding, Mashhad Branch, Islamic Azad University, Mashhad, Iran

Hafeez-ur-Rehman
Agronomy Section, Agriculture Research Institute Tando Jam, Sindh, Pakistan

Rafi Qamar
Agronomy Section, Central Cotton Research Institute Sakrand, Sindh, Pakistan

Atique-ur-Rehman
Department of Agronomy, Bahauddin Zakariya University, Multan, Pakistan

Farhan Ahmad
Entomology Section, Central Cotton Research Institute Sakrand, Sindh, Pakistan

Jamshaid Qamar
Department of Agronomy, University College of Agriculture, University of Sargodha, Pakistan

Muhammad Saqib
Department of Agronomy, The Islamia University, Bahawalpur, Pakistan

Shah Nawaz
Entomology Section, Central Cotton Research Institute Sakrand, Sindh, Pakistan